The Froehlich / Kent

ENCYCLOPEDIA OF TELECOMMUNICATIONS

VOLUME 3

The Froehlich / Kent
ENCYCLOPEDIA OF TELECOMMUNICATIONS

Editor-in-Chief

Fritz E. Froehlich, Ph.D.

Professor of Telecommunications
University of Pittsburgh
Pittsburgh, Pennsylvania

Co-Editor

Allen Kent

Distinguished Service Professor of Information Science
University of Pittsburgh
Pittsburgh, Pennsylvania

Administrative Editor

Carolyn M. Hall

Pittsburgh, Pennsylvania

VOLUME 3

CODES FOR THE PREVENTION OF ERRORS to COMMUNICATIONS FREQUENCY STANDARDS

CRC Press
Taylor & Francis Group
Boca Raton London New York

CRC Press is an imprint of the
Taylor & Francis Group, an **informa** business

First published 1992 by Marcel Dekker, Inc.

Published 2021 by CRC Press
Taylor & Francis Group
6000 Broken Sound Parkway NW, Suite 300
Boca Raton, FL 33487-2742

© 1992 by Taylor & Francis Group, LLC
CRC Press is an imprint of Taylor & Francis Group, an Informa business

No claim to original U.S. Government works

ISBN 13: 978-0-8247-2902-8 (hbk)

**Visit the Taylor & Francis Web site at
http://www.taylorandfrancis.com**

**and the CRC Press Web site at
http://www.crcpress.com**

Library of Congress Cataloging-in-Publication Data

The Froehlich/Kent Encyclopedia of Telecommunications / editor-in-chief, Fritz E. Froehlich ; co-editor, Allen Kent.
　　p.　cm.
　　Includes bibliographical references and indexes.
　　ISBN 0-8247-2902-1 (v. 1 : alk. paper)
　　1. Telecommunication—Encyclopedias.　I. Froehlich, Fritz E.,
　　　　II. Kent, Allen.
　TK5102.E646　1990　　　　　　　　　　　　　90-3966
　384′.03—dc20　　　　　　　　　　　　　　　　CIP

LIBRARY OF CONGRESS CATALOG CARD NUMBER: 90-3966

CONTENTS OF VOLUME 3

CONTRIBUTORS TO VOLUME 3

Walter R. Beam, Ph.D. Independent Consultant, Alexandria, Virginia: *Command, Control, and Communications Systems*

Richard E. Blahut, Ph.D. IBM Fellow, IBM, Federal Sector Division, Owego, New York: *Codes for the Prevention of Errors*

Amalie J. Frank, Ph.D. Science Division, Widener University, Chester, Pennsylvania: *Coding of Facsimile Images*

Richard R. Goldberg District Manager, Bellcore, Red Bank, New Jersey: *Common Channel Signaling*

Daniel J. Hunt Director, Education Affairs, Northern Telecom, Research Triangle Park, North Carolina: *Communication in Education*

Monica Krueger, MST University of Pittsburgh, School of Library and Information Science, Pittsburgh, Pennsylvania: *Colpitts, Edwin H.*

Linda J. Laskowski Vice President and General Manager, Information Provider Market, U S West Communications, Denver, Colorado: *Communication Gateway Services for Videotex*

Nancy K. Metzler Manager, Videotex, Information Provider Market, U S West Communications, Denver, Colorado: *Communication Gateway Services for Videotex*

Bob Neveln, Ph.D. Associate Professor of Mathematics and Computer Science, Science Division, Widener University, Chester, Pennsylvania: *Comma-Free and Synchronizable Codes*

Victor L. Ransom Division Manager (Retired), Bell Communications Research, Inc., Red Bank, New Jersey: *Communication Aids for People with Special Needs*

Laura S. Redmann Member of Technical Staff, Bell Communications Research, Inc., Red Bank, New Jersey: *Communication Aids for People with Special Needs*

David C. Rife, Ph.D. Senior Principal Engineer, Hayes Microcomputer Products, Inc., Norcross, Georgia: *Communication with Intelligent Modems*

Carl V. Ripa Managing Director, Telesector Resources Group, White Plains, New York: *Communications and Information Network Service Assurance*

Samuel R. Stein, Ph.D. Ball Corp., Broomfield, Colorado: *Communications Frequency Standards*

Seymour Stein, Ph.D. SCPE, Inc., Newton, Massachusetts: *Communication over Fading Radio Channels*

Richard A. Thompson, Ph.D. Professor of Telecommunications, University of Pittsburgh, Pittsburgh, Pennsylvania: *Communication Terminals*

Heather S. Tooker Director, Videotex, Information Provider Market, U S West Communications, Denver, Colorado: *Communication Gateway Services for Videotex*

John R. Vig, Ph.D. U.S. Army Electronics Technology and Devices Laboratory, Fort Monmouth, New Jersey: *Communications Frequency Standards*

Keith D. Wallin   Manager, Videotex, Information Provider Market, U S West Communications, Denver, Colorado: *Communication Gateway Services for Videotex*

W. Bernard Wargotz, Ph.D.   Undersea Systems Laboratory, AT&T Bell Laboratories, Holmdel, New Jersey: *Communication Printed Wiring Interconnection Technology*

Taieb F. Znati, Ph.D.   Computer Science Department (Telecommunications), University of Pittsburgh, Pittsburgh, Pennyslvania: *Communication Protocols for Computer Networks: Fundamentals*

The Froehlich/Kent
ENCYCLOPEDIA OF TELECOMMUNICATIONS

VOLUME 3

Codes for the Prevention of Errors

Introduction

A profusion and variety of communication systems now exist to carry a massive amount of digital data between terminals. Alongside these communication systems are very large numbers of magnetic and optical tape and disk storage systems. The received signals in all communication and recording systems are always contaminated by thermal noise and, in practice, are also contaminated by various kinds of defects, non-Gaussian noise, burst noise, interference, and crosstalk. The communication system (or storage system) must transmit its data with very high reliability in the presence of these channel impairments. Bit error rates of as low as 1 bit error in 10^{10} bits (or even lower) are routinely specified.

Primitive communication and magnetic-storage systems seek to obtain low error rates by the simple expedient of transmitting high power in comparison to the noise. This simplistic approach may be adequate if the required bit error rate is not too stringent or if the data rate is low. Such systems, however, buy performance with the least expendable resources: power and spectral bandwidth.

In contrast, modern communication and storage systems obtain high performance by the use of elaborate message structures with complex cross checks built into the waveform. In some systems, it is not possible to determine where a particular bit resides in the channel waveform; the entire message is modulated into the waveform as a unit and an individual bit appears in a diffuse but recoverable way. The advantage of the modern communication waveforms is that high data rates can be transmitted reliably while keeping the transmitted power and spectral bandwidth small. This advantage is offset by the need for sophisticated computations in the receiver (and in the transmitter) to recover the message. However, such computations are now regarded as affordable using modern electronic technology. For example, some current telephone-line data modems use microprocessors in the receiver with approximately 500 machine cycles of computation per received data bit. Clearly, with this amount of computation in the receiver, the waveforms may have a very sophisticated structure and each individual bit can, indeed, be deeply buried in the waveform.

The codes described in this chapter are called *codes for the prevention of error*. This is a positive term that implies the role that such codes have in modern systems. The original term, *error-correcting codes*, is also used but suffers from the fact that it has a negative connotation. It implies that the code is used only to correct an unforeseen deficiency in the communication system whereas, in fact, the code is an integral part of the system design. Furthermore, the way some codes are usually used, it may be hard to find that point in the system where the errors that the code is correcting occur. It is a better description to say that the errors are prevented from ever happening.

Other, more neutral, terms for the codes in this article are *error-control codes* and *data-transmission codes*. The term *channel codes* may also be used, but this term includes other kinds of codes, such as the Morse code, that have other functions in the communication system.

Codes Based on Hamming Distance

Given two sequences of the same length of symbols from some fixed-symbol alphabet, perhaps the binary alphabet $\{0,1\}$, we discuss how different those two sequences are. The most suggestive way to measure the difference between the two sequences is to count the number of places in the sequence in which they differ. This is called the *Hamming distance* between the sequences. The reason for choosing the term "distance" is to appeal to geometric intuition when constructing codes.

For example, consider the two sequences of length 6 over the decimal alphabet given by 0,1,2,3,4,5 and 0,1,4,5,4,5. The Hamming distance between these sequences is 2, which we denote symbolically by

$$d(012345, 014545) = 2$$

or

$$d(\mathbf{v},\mathbf{v}') = 2$$

where the vectors $\mathbf{v}$ and $\mathbf{v}'$ denote the two sequences.

We may also have two infinitely long sequences over some symbol alphabet. Again, the Hamming distance is defined as the number of places in which the two sequences are different. This Hamming distance then will be infinite unless the sequences agree everywhere except on a finite segment.

Although user data may be treated on the bottom level as a sequence of bits, often it is treated as a sequence of bytes. Frequently, a byte consists of 8 bits, but sometimes a communication system may be designed around a byte of r bits for some value of r other than 8. This data structure within the communication system is transparent to the user because the datastream is reformatted at the input and output of the modem.

When sequences are described at the byte level, the Hamming distance between two sequences is the number of byte positions in which they differ. Two bytes are different if they are not the same. They may differ in any possible way involving one or more bit positions. Thus, we speak of byte errors rather than of bit errors.

A *datastream* is a sequence of user data symbols, either bits or bytes. A *codestream* is a sequence of channel symbols, either bits or bytes. The user perceives that the datastream is being sent through the channel, but the codestream is actually sent.

For constructing the code, additional structure is defined on the datastream by segmenting it into pieces called *datawords* or *data frames*. Likewise, the codestream is segmented into pieces called *codewords* or *code frames*.

The encoder maps the datastream into the codestream. Codes are of two types: block codes and tree codes; the distinction between them due to the memory in the encoders of the tree codes.

Elementary Block Codes

A block code breaks the datastream into datawords, each consisting of k data symbols. The encoder maps each dataword into a codeword consisting of n code symbols. In brief, this is called an (n,k) block code. The ratio k/n is called the *code rate* and is denoted by R. Each block is encoded independently without interaction with earlier or later blocks. The successive codewords are simply concatenated to form the codestream for passage through the channel. For example, the compact disk encodes a dataword consisting of 28 8-bit bytes into a codeword consisting of 32 8-bit bytes. This is a (32,28) code on the alphabet of 8-bit bytes and has a rate $R = 0.875$.

There are 2^k codewords in a binary block code and q^k codewords in a block code over a q-ary alphabet. For the alphabet of 8-bit bytes, $q = 256$, so there are 256^k codewords in such a code. For the (32,28) code used on the compact disk, there are 256^{28} codewords in the code. A block code is designed such that the codewords are very different from each other so as to make the code resistant to channel errors. This dissimilarity is measured by the smallest Hamming distance between the two most similar codewords. This leads to the definition of *minimum distance* d_{min} of a code as the smallest Hamming distance between any pair of codewords. If $d_{min} \geq 2t + 1$ and a channel makes t errors, then the received sequence differs from the transmitted codeword in t places and differs from every other codeword in $t + 1$ or more places. Therefore, the codeword closest to the received sequence is the correct codeword.

A simple block code, known as the (7,4) Hamming code is shown in Fig. 1. There are 16 binary codewords in this code; each codeword has a blocklength 7 and the minimum distance of the code is 3. Each of the 16 codewords can be

```
0   0   0   0   0   0   0
0   0   0   1   0   1   1
0   0   1   0   1   1   0
0   0   1   1   1   0   1
0   1   0   0   1   1   1
0   1   0   1   1   0   0
0   1   1   0   0   0   1
0   1   1   1   0   1   0
1   0   0   0   1   0   1
1   0   0   1   1   1   0
1   0   1   0   0   1   1
1   0   1   1   0   0   0
1   1   0   0   0   1   0
1   1   0   1   0   0   1
1   1   1   0   1   0   0
1   1   1   1   1   1   1
```

FIG. 1 Hamming (7,4) code.

assigned to represent 1 of the 16 4-bit binary datawords. Therefore, the code represents 4-bit datawords ($k = 4$) by 7-bit codewords ($n = 7$). Any one-to-one assignment of datawords to codewords can be used. A natural choice is to assign codewords so that the first four bits of the codeword are identical to the four bits of the dataword. Examination of Fig. 1 shows that this choice is consistent with the structure of the (7,4) Hamming code by choosing the first four code bits as data bits. This kind of encoder is called a *systematic encoder*. The three bits other than the four data bits are then called *parity-check* bits.

Notice that there is a careful distinction between the notion of a code and the notion of the encoder. The code shown in Fig. 1 is simply a set of codewords and is independent of the method of encoding. The term *encoder* refers to the assignment of datawords to codewords. The term also applies to the mechanism for implementing this assignment.

The code is used as follows. Four bits of data are encoded into a 7-bit codeword. The channel makes at most 1 bit error. Because the minimum distance of the code is 3 (which can be verified by inspection of Fig. 1), the received word differs from the true codeword in at most 1 bit position and differs from every other codeword in at least 2 bit positions. The decoder can easily deduce the correct codeword and then recover the 4 data bits.

Convolutional Codes

An encoder for a tree code encodes a stream of data symbols into a stream of codeword symbols. The duration of the datastream is so long that, effectively, it is infinite and does not enter into the design of the encoder and decoder. Beginning at time zero and continuing indefinitely into the future, a data sequence is shifted into the encoder and a code sequence is shifted out. A tree code breaks the datastream into segments called data frames each consisting of k data symbols, k being a small integer. The encoder is a finite-state machine that retains some memory of earlier data frames; in the simplest case, it simply stores the m most recent data frames unchanged. A single code frame consisting of n symbols is computed from the mk data symbols of the m frames stored in memory and the k symbols in the incoming data frame; these n symbols are shifted out to the channel as the k new data symbols are shifted into the encoder. The ratio k/n is called the code rate of the tree code and is denoted by R. The code frames are concatenated to form the codestream for passage through the channel. Tree codes with a special memory and linearity structure, which is defined below, are called *convolutional codes*. These are the most common types of tree codes used in practice.

The *constraint length* v of a convolutional code is defined as the number of memory cells in a minimum encoder. The minimum distance $d_{\min}$ of a convolutional code is defined as the minimum Hamming distance between any two codewords.

One way to describe a convolutional code is by a polynomial representation. We describe a simple binary convolutional code with $k = 1$, $n = 2$, and $v = 2$. Let

$$a(x) = a_0 + a_1 x + a_2 x^2 + a_3 x^3 + \ldots$$

denote the infinite sequence of binary data; each coefficient is either a 0 or a 1. Let

$$g_0(x) = x^2 + x + 1$$
$$g_1(x) = x^2 + 1.$$

These two polynomials are called *generator polynomials*. The encoder is described by the polynomial product (with modulo-2 arithmetic in the coefficients)

$$c_0(x) = g_0(x)a(x)$$
$$c_1(x) = g_1(x)a(x).$$

These polynomial products produce the two codeword polynomials

$$c_0(x) = c_{00} + c_{01}x + c_{02}x^2 + c_{03}x^3 + \ldots$$
$$c_1(x) = c_{10} + c_{11}x + c_{12}x^2 + c_{13}x^3 + \ldots$$

The binary coefficients of $c_0(x)$ and $c_1(x)$ are interleaved to form the binary codestream and sent to the channel. Thus, there are two code bits for every data bit and the code rate is 0.5.

The minimum distance of this code is 5. Two code sequences at Hamming distance 5 are defined by the data polynomial

$$a(x) = 0$$

corresponding to code polynomials

$$c_0(x) = 0$$
$$c_1(x) = 0$$

and the data polynomial

$$a(x) = 1$$

corresponding to code polynomials

$$c_0(x) = x^2 + x + 1$$
$$c_1(x) = x^2 + 1.$$

The Hamming distance between these two codestreams is clearly 5. It is easy to check that no two codestreams are closer and, consequently, the minimum distance of this code is 5. Because $2 < d_{min}/2$, this convolutional code can correct two errors. Furthermore, this convolutional code can correct multiple error events, each having one or two bit errors, provided the error events are spaced far enough apart that they do not interact in the structure of the code.

In general, a convolutional code is described by a vector of data polynomials over some field $GF(q)$

$$\mathbf{a}(x) = [a_1(x), a_2(x), \ldots, a_k(x)].$$

The k data symbols in the ℓth frame provide the coefficients of x^ℓ. The nature of the set of generator polynomials defining the convolutional code is described by the matrix

$$\mathbf{G}(x) = \begin{bmatrix} g_{11}(x) & \cdots & g_{1n}(x) \\ g_{21}(x) & \cdots & g_{2n}(x) \\ & \vdots & \\ & \vdots & \\ g_{k1}(x) & \cdots & g_{kn}(x) \end{bmatrix}$$

The matrix-vector product

$$\mathbf{c}(x) = \mathbf{a}(x)\mathbf{G}(x)$$

then gives the vector of codeword polynomials

$$\mathbf{c}(x) = [c_1(x), c_2(x), \ldots, c_n(x)].$$

The n symbols forming the coefficients of x^ℓ provide the n code symbols in the ℓth frame. They are interleaved to form the codestream.

Good convolutional codes—that is, good matrices of generator polynomials $\mathbf{G}(x)$—are found by computer search. Strong algebraic constructions analogous to those described below for Reed-Solomon codes have never been discovered.

Arithmetic in a Galois Field

The most successful rules for constructing good block codes for error prevention make use of a special kind of arithmetic known as the arithmetic of Galois fields. A Galois field with q elements, denoted $GF(q)$, is an unconventional arithmetic system containing the operations of addition, subtraction, multiplication, and division defined in such a way that most familiar arithmetic and algebraic procedures remain valid. The arithmetic operations of a Galois field have the enormous advantage that there is no overflow or round-off error. Because the error-control code is used to protect bit packages but not to do real computations, it does not matter that the arithmetic rules are unconventional.

Figure 2 shows the addition and multiplication tables for several simple Galois fields. Notice that $GF(2)$ and $GF(3)$ are modulo-2 and modulo-3 arithmetic, respectively, but $GF(4)$ is *not* modulo-4 arithmetic. [Modulo-4 arithmetic cannot form a field because $2 \cdot 1 = 2 \cdot 3 \pmod 4$, so division by 2 would not behave properly in modulo-4 arithmetic.] The larger field $GF(256)$ is very important in practice because it provides a closed arithmetic structure for the

GF(2)

+	0	1
0	0	1
1	1	0

·	0	1
0	0	0
1	0	1

GF(3)

+	0	1	2
0	0	1	2
1	1	2	0
2	2	0	1

·	0	1	2
0	0	0	0
1	0	1	2
2	0	2	1

GF(4)

+	0	1	2	3
0	0	1	2	3
1	1	0	3	2
2	2	3	0	1
3	3	2	1	0

·	0	1	2	3
0	0	0	0	0
1	0	1	2	3
2	0	2	3	1
3	0	3	1	2

FIG. 2   Arithmetic tables for several simple Galois fields.

set of 8-bit bytes, but the addition and multiplication tables are too large to show here. Addition and multiplication tables in large fields usually are described by rules rather than tables.

We discuss only the binary power Galois fields $GF(2^m)$ in this article. The elements of the field $GF(2^m)$ can be defined as the set of m-bit bytes. Thus, $GF(2^8)$ is the set of 8-bit bytes and $GF(16)$ is the set of 4-bit bytes (or hexadecimal characters). Addition is defined simply as componentwise modulo-2 addition (bit-by-bit exclusive-or). For example, two of the elements of $GF(16)$ are 1101 and 1110. Their addition is

$$(1101) + (1110) = (0011).$$

Multiplication is more complicated to define. The multiplication in $GF(2^m)$ must be consistent with addition in the sense that the distributive law

$$(a + b)c = ac + bc$$

holds for any a, b, and c in $GF(2^m)$. Because addition is a bit-by-bit exclusive-or, this suggests that multiplication should have the structure of a shift and exclusive-or rather than the conventional structure of shift and add.

We may try to define multiplication in accordance with the following product:

```
        1 1 1 0
        1 1 0 1
        1 1 1 0
      0 0 0 0
    1 1 1 0
  1 1 1 0
  1 0 0 0 1 1 0
```

However, the arithmetic system must be closed under multiplication. The product of two m-bit numbers must produce an m-bit number; the wordlength does not increase. If a shift and exclusive-or structure is the right definition for the multiplier, then we also need a rule for folding back the overflow bits into the m-bits of the product. The trick is to define the overflow rule so that $b = c$ whenever $ab = ac$ and a is nonzero. Otherwise, division could not be defined and the arithmetic system would not be satisfactory.

The overflow rule is constructed in terms of polynomial division. Let $p(x)$ be an irreducible polynomial over $GF(2)$ of degree m. This means that $p(x)$ can have only coefficients equal to 0 or 1, and that $p(x)$ cannot be factored into the product of two such polynomials over $GF(2)$. Factoring $p(x)$ means writing

$$p(x) = p^{(1)}(x)p^{(2)}(x)$$

where polynomial multiplication uses modulo-2 arithmetic on the coefficients.

Multiplication of two elements a and b in $GF(2^m)$ to produce the element $c = ab$ can be described as a polynomial multiplication modulo the irreducible polynomial $p(x)$. Let a and b be numbers in $GF(2^m)$. These are m-bit binary numbers with the binary representations

$$a = (a_0, \ldots, a_{m-1})$$
$$b = (b_0, \ldots, b_{m-1}).$$

They also have the polynomial representations

$$a(x) = \sum_{i=0}^{m-1} a_i x^i$$

$$b(x) = \sum_{i=0}^{m-1} b_i x^i$$

where a_i and b_i are the ith bits of a and b, respectively. Then the product is defined as the sequence of coefficients of the polynomial

$$c(x) = a(x)b(x) \qquad \text{mod } p(x).$$

Because the coefficients are added and multiplied by the bit operations of $GF(2)$, the polynomial product is equivalent to the shift and exclusive-or operations mentioned above. The modulo-$p(x)$ operation specifies the rule for folding overflow bits back into the m-bits of the field element. Simply divide the product polynomial by the polynomial $p(x)$ and keep the remainder. With this definition of multiplication and the definition of addition noted above, the description of the Galois field $GF(2^m)$ is complete.

As an example of a Galois field, we construct $GF(2^4)$. The 16 elements are the set of 4-bit bytes

$$GF(2^4) = \{0000, 0001, 0010, \ldots, 1111\}$$

and addition of two elements is bit-by-bit modulo-2 addition. To define multiplication, we use the polynomial

$$p(x) = x^4 + x + 1.$$

To verify that this polynomial is irreducible, we can check that x, $x + 1$, $x^2 + 1$, $x^2 + x + 1$ are not factors, whereas $p(x)$ must have either a first-degree factor or a second-degree factor if it is reducible.

Then, to multiply 0101 by 1011, for instance, we represent these by $a(x) = x^2 + 1$ and $b(x) = x^3 + x + 1$ and write

$$c(x) = (x^2 + 1)(x^3 + x + 1) \qquad \mathrm{mod}\, p(x)$$
$$= x^5 + x^2 + x + 1 \qquad \mathrm{mod}\, x^4 + x + 1$$

The modulo-$p(x)$ operation consists of division by $p(x)$ and keeping the remainder polynomial. Carrying out the division for the sample calculation gives

$$c(x) = 1$$

which is the polynomial representation for binary 0001, giving the product in $GF(16)$

$$(0101)(1011) = (0001).$$

Because the product happens to be equal to 1, this example also tells us how to divide in $GF(16)$. We see that

$$(0101)^{-1} = (1011)$$

because $(0101)^{-1}$ is defined to be the field element for which $(0101)^{-1}(0101) = 1$. Likewise

$$(1011)^{-1} = (0101).$$

To divide by 0101 we multiply by 1011, while to divide by 1011 we multiply by 0101. For example

$$\frac{(0110)}{(1011)} = (0110)(0101)$$
$$= (1101)$$

where the product is calculated as described above.

Division will be possible for every nonzero element a if a^{-1} exists. The inverse a^{-1} is the value of b that solves

$$ab = 1.$$

This equation always has a solution (except when $a = 0$) if $p(x)$ is chosen to be an irreducible polynomial, so the use of an irreducible polynomial ensures that division is defined. The reason why this is so is quite involved and is not explained further.

Although the multiplication and division rules of a Galois field may appear unfamiliar to us, logic circuits or computer subroutines to implement them are straightforward. One could even build a programmable computer with Galois-field arithmetic as primitive instructions.

Algebraic manipulations in the Galois field $GF(2^m)$ behave very much like algebraic manipulations in the fields more usually encountered in engineering problems such as the real field or the complex field. The conventional algebraic properties of associativity, commutativity, and distributivity all hold. Methods of solving linear systems of equations are valid, including matrix algebra, determinants, and so forth. There is even a discrete Fourier transform in $GF(2^m)$ and it has all the familiar properties of the discrete Fourier transform. The Fourier transform is particularly important to our purposes because it is the basis of the definition of the Reed-Solomon code.

Hamming Codes on Bytes

A simple application of Galois fields is to construct those single-byte-error-correcting codes known as Hamming codes. In the alphabet of m-bit bytes, the codes are constructed with the arithmetic of $GF(2^m)$. For each value of r, it is possible to construct a Hamming code for $GF(2^m)$ with r parity bytes and block-length $n = (2^{mr} - 1)/(2^m - 1)$. Thus, each codeword can encode $k = [(2^{mr} - 1)/(2^m - 1)] - r$ data bytes. For example, for $m = 8$ and $r = 2$, the code symbols are 8-bit bytes; there are $r = 2$ parity-check symbols, the blocklength n equals 257, and there are $k = 255$ data symbols. This Hamming code can be shortened by one symbol (by permanently setting one data symbol to 0 mathematically, and thereafter deleting it from the codeword) to produce a (256,254) Hamming code over 8-bit bytes. Such a code could be used to find and correct 1 byte error in a block of 256 8-bit bytes.

The values n and k for some Hamming codes are given in Fig. 3. Any of these codes can be shortened to a convenient blocklength by deleting data symbols.

The construction is simple to describe when $r = 2$. For example, we describe

GF(2)	GF(4)	GF(16)	GF(256)
(7,4)	(5,3)	(17,15)	(257,255)
(15,11)	(21,18)	(273,270)	(65793,65790)
(31,26)	(85,81)		
(63,57)	(341,336)		
(127,120)			

FIG. 3   Parameters (n,k) for some Hamming codes.

the code over $GF(16)$ with (hexadecimal) data symbols $(a_1, a_2, \ldots, a_{15})$. Write the parity-check symbols as

$$p_1 = a_1 + a_2 + a_3 + \ldots + a_{14} + a_{15}$$
$$p_2 = a_1 + 2a_2 + 3a_3 + \ldots + Ea_{14} + Fa_{15}.$$

The 15 coefficients in p_2 are the 15 nonzero elements of $GF(16)$. The equations can be abbreviated as

$$p_1 = \Sigma_i a_i$$

$$p_2 = \Sigma_i g_i a_i$$

where g_i is the field element pointing at the ith component. After every block of 15 (hexadecimal) data symbols, these 2 (hexadecimal) parity-check symbols are inserted to form the codeword $(c_1, c_2, \ldots, c_{15}, c_{16}, c_{17}) = (a_1, a_2, \ldots, a_{15}, p_1, p_2)$.

The decoder can correct a single symbol error in the block of 17 symbols as follows. The received word is $(v_1, v_2, \ldots, v_{15}, v_{16}, v_{17})$, the first 15 symbols correspond to data symbols, the last 2 symbols correspond to parity-check symbols, and at most 1 symbol is in error. Let e_i denote the error in component i. Then

$$v_i = c_i + e_i.$$

Compute the following terms, known as *syndromes*

$$s_1 = v_1 + v_2 + v_3 + \ldots + v_{14} + v_{15} - v_{16}$$
$$s_2 = v_1 + 2v_2 + 3v_3 + \ldots + Ev_{14} + Fv_{15} - v_{17}$$

If both s_1 and s_2 equal 0, there is no error. If either s_1 or s_2 equals 0, the single error occurs in a parity-check symbol. If both s_1 and s_2 are nonzero, the single error e_i occurred in a data symbol, and

$$s_1 = e_i$$
$$s_2 = g_i e_i.$$

Therefore, s_1 gives the byte value of the error and $s_2/s_1 = g_i$ points to the ith component, the component that is in error.

Such a procedure with $r = 2$ applies to any field $GF(2^m)$ to give a code with $2^m - 1$ data bytes and 2 parity-check bytes. When the number of data bytes is larger than $2^m - 1$, there are not enough field elements for one field element to point to a unique component. That is why more than 2 parity-check bytes are then necessary.

Reed-Solomon Codes and BCH (Bose-Chaudhuri-Hocquenghem) Codes

The most important block codes in applications are those codes known as Reed-Solomon codes and the closely related codes known as BCH (Bose-Chaudhuri-Hocquenghem) codes. The Reed-Solomon codes are codes on the byte level, while the BCH codes are (usually) codes on the bit level.

The Reed-Solomon codes are defined using the mathematics of the discrete Fourier transform in a Galois field. This Fourier transform is defined in exactly the same way as the Fourier transform in the complex field. Specifically, if $\mathbf{v}$ is a vector of blocklength n of symbols from the field $GF(q)$, then the Fourier transform of $\mathbf{v}$ is another vector, denoted $\mathbf{V}$, also of blocklength n, given by

$$V_j = \sum_{i=0}^{n-1} \omega^{ij} v_i \quad j = 0, \ldots, n - 1$$

where ω is an element of $GF(q)$ of order n. That is, $\omega^n = 1$ and $\omega^k \neq 1$ if $k < n$. Notice that this is a close analog of the discrete Fourier transform in the complex field.

$$V_k = \sum_{i=0}^{n-1} (e^{-j2\pi/n})^{ik} v_i \quad k = 0, \ldots, n - 1$$

The only qualification that needs to be made is that a Fourier transform of blocklength n exists in the Galois field $GF(q)$ only if $GF(q)$ contains an element ω of order n. Only if n divides $q - 1$ does such an ω exist. In particular, if q is a power of 2, n is always odd. For example, in $GF(256)$, n can only take the values 255, 85, 51, 17, 15, 5, and 3.

The Fourier transform in a Galois field has all of the properties that one might expect and they can be proved in the natural way. Specifically, there is a modulation/delay theorem that says that if $v_i' = v_i \omega^r$, then $V_j' = V_{j-r}$, a convolution theorem that says that if $v_i = u_i w_i$, then

$$V_j = \sum_{k=0}^{n-1} U_k W_{((j-k))},$$

an inverse Fourier transform, and so on. The inverse Fourier transform has the usual form

$$v_i = \frac{1}{n} \sum_{j=0}^{n-1} \omega^{-ij} V_j$$

where n in the denominator multiplying the sum is interpreted as an integer of $GF(q)$. That is, n is the sum of 1 with itself n times. Because $1 + 1 = 0$ in $GF(2^m)$ for any m, and n is odd, the integer n in the Galois field $GF(2^m)$ is equal to 1 and can be suppressed in the equation for the inverse Fourier transform.

The t-error-correcting Reed-Solomon code of blocklength n is the set of all vectors $\mathbf{c}$ of blocklength n whose spectrum satisfies $C_j = 0$ for $j = n - 2t$, ..., $n - 1$. This code is described briefly as an $(n, n - 2t)$ Reed-Solomon code.

One way to find the Reed-Solomon codewords is to encode in the frequency domain as is suggested by the definition. This means setting $C_j = 0$ for $j = n - 2t, \ldots, n - 1$, and setting the remaining $n - 2t$ components of the transform equal to the $n - 2t$ data symbols given by $a_0, \ldots, a_{n-2t-1}$. That is

$$C_j = \begin{cases} a_j & j = 0, \ldots, n - 2t - 1 \\ 0 & j = n - 2t, \ldots, n - 1 \end{cases}$$

An inverse Fourier transform then produces the codeword $\mathbf{c}$. The number of data symbols encoded equals $n - 2t$ and there are $2t$ extra symbols in the codeword to correct t errors. A Reed-Solomon code always uses two overhead symbols for every error to be corrected.

Using the inverse Fourier transform is not the only way to encode the $n - 2t$ data symbols into the codewords — others may yield a simpler implementation — but the frequency-domain encoder is the most instructive because it exhibits very explicitly the notion that the codewords are those vectors with the same set of $2t$ zeros in the transform domain.

An alternative encoder in the time domain works as follows. The $n - 2t$ data symbols are expressed as a polynomial

$$a(x) = a_{n-2t-1}x^{n-2t-1} + a_{n-2t-2}x^{n-2t-2} + \ldots + a_1 x + a_0$$

where $a_0, a_1, \ldots, a_{n-2t-1}$ are the $n - 2t$ data symbols. Then the n codeword symbols are given by the coefficients of the polynomial product

$$c(x) = g(x)a(x)$$

where $g(x)$ is a fixed polynomial called the *generator polynomial*. The generator polynomial is the unique monic (leading coefficient equals 1) polynomial of degree $2t$ that has zeros at $\omega^{n-2t}, \omega^{n-2t+1}, \ldots, \omega^{n-1}$. It can be obtained by multiplying out the expression

$$g(x) = (x - \omega^{n-2t})(x - \omega^{n-2t+1}) \ldots (x - \omega^{n-1}).$$

We can verify that this time-domain encoder does indeed produce a Reed-Solomon code. The Fourier transform of the codeword is formally the same as evaluating the polynomial $c(x)$ at ω^j. That is,

$$C_j = \sum_{i=0}^{n-1} c_i \omega^{ij} = c(\omega^j) = g(\omega^j)a(\omega^j).$$

By the definition of $g(x)$, $g(\omega^j)$ equals 0 for $j = n - 2t, \ldots, n - 1$. Consequently, $C_j = 0$ for $j = n - 2t, \ldots, n - 1$. Therefore, the encoding in the time domain does produce legitimate Reed-Solomon codewords. The set of codewords produced by the time-domain encoder is the same as the set of codewords produced by the frequency-domain encoder, but the mapping between datawords and codewords is different.

 Thus far, both methods of encoding discussed have the property that the symbols of the dataword do not appear explicitly in the codeword. Another method of encoding, known as *systematic encoding*, leaves the data symbols unchanged and contained in the first $n - 2t$ components of the codeword. Multiplication of $a(x)$ by x^{2t} will move the components of $a(x)$ left $2t$ places. Thus, we can write an encoding rule as

$$c(x) = x^{2t}a(x) + r(x)$$

where $r(x)$ is a polynomial of degree less than $2t$ appended to make the spectrum be a legitimate codeword spectrum. The spectrum will be correct if $c(x)$ is a multiple of $g(x)$ and this will be so if $r(x)$ is chosen as the negative of the remainder when $x^{2t}a(x)$ is divided by $g(x)$. Thus, because it gives a multiple of $g(x)$, the equation

$$c(x) = x^{2t}a(x) - R_{g(x)}[x^{2t}a(x)]$$

defines a systematic encoder where the operator $R_{g(x)}$ takes the remainder under division by $g(x)$. This definition of $c(x)$ is indeed a codeword because it has zero remainder when divided by $g(x)$. Thus

$$\begin{aligned} R_{g(x)}[c(x)] &= R_{g(x)}[x^{2t}a(x)] - R_{g(x)}\{R_{g(x)}[x^{2t}a(x)]\} \\ &= 0 \end{aligned}$$

because remaindering can be distributed across addition.

 A decoder for a Reed-Solomon code does not depend on how the codewords are used to store information except for the final step of reading the data symbols out of the corrected codeword.

 To prove that an $(n, n - 2t)$ Reed-Solomon code can correct t symbol errors, it is enough to prove that every two codewords in the code differ from each other in at least $2t + 1$ places. If this is true, then changing any t components of a codeword will produce a word that is different from the correct codeword in t components and is different from every other codeword in at least $t + 1$ components. If, at most, t errors occur, then choosing the codeword that differs from the noisy received word in the fewest number of components will recover the correct codeword. If each symbol is more likely to be correct than to be in error, then choosing the codeword that differs from the noisy received word in the fewest places will recover the most likely codeword and will minimize the probability of decoding error.

By definition of the code,

$$C_j = 0 \quad j = n - 2t, n - 2t + 1, \ldots, n - 1.$$

By linearity of the Fourier transform, the difference in two codewords (computed componentwise) then must also have a spectrum that is zero for $j = n - 2t, \ldots, n - 1$ and so itself is a codeword. We only need to prove that no codeword has fewer than $2t + 1$ nonzero components unless it is zero in every component. Let

$$C(y) = \sum_{j=0}^{n-2t-1} C_j y^j.$$

This is a polynomial of degree at most $n - 2t - 1$, so, by the fundamental theorem of algebra, it has at most $n - 2t - 1$ zeros. Therefore,

$$\begin{aligned} c_i &= \frac{1}{n} \sum_{j=0}^{n-1} \omega^{-ij} C_j \\ &= \frac{1}{n} C(\omega^{-i}) \end{aligned}$$

can be zero in at most $n - 2t - 1$ places, and so it is nonzero in at least $2t + 1$ places. Therefore, we can conclude that an $(n, n - 2t)$ Reed-Solomon code can correct t symbol errors.

Future Developments

A code is judged by its rate R, minimum distance $d_{\min}$, and blocklength n. For any fixed rate R, which determines the percentage of coding overhead, the best code is the one with the largest value of $d_{\min}$, which determines the performance of the code. The blocklength n is also important. A code of blocklength $10n$ that protects against $10t$ channel errors is much better than 10 uses of a code of blocklength n that protects against t errors per block, because the latter code requires the errors to be distributed with at most t errors to a block.

The block codes and tree codes that are known and in use have shown their worth in many applications. One may naturally ask whether better codes are possible. Remarkably, we know that, when the blocklength is large, the known codes are not nearly as good as the best possible codes. Indeed, if messages longer than a few thousand bits are to be transmitted, the known codes can be surprisingly weak in comparison to the best codes. However, despite more than 40 years of intense effort, there is little known about how to find optimum codes, nor is much known about good encoding and decoding algorithms for such codes.

Let $\delta = d_{\min}/n$ denote the fractional minimum distance of the code. The code can prevent decoded errors even if a fraction $\frac{1}{2}\delta$ of the channel symbols are in error. It is known that if the blocklength is large enough, codes exist that have R and δ satisfying (within ϵ) the equation

$$R = 1 + \delta \log_2 \delta + (1 - \delta) \log_2 (1 - \delta)$$

which is known as the *Gilbert bound* and is shown in Fig. 4. For example, with $R = 0.5$, $\delta = 0.11$. Thus, if one allows that 50% of channel bits are overhead, a code exists that will decode correctly even if any 5.5% of the channel bits are in error. With such a code, were it known, a message codeword of 1 million code bits would be received correctly if any 55,000 channel bits were in error, for instance the first 55,000. Similarly, with $R = 0.9$, $\delta = 0.013$. If one allows that 10% of channel bits are overhead, a code exists that will decode correctly even if any 0.65% of channel bits are in error. With such a code, were it known, a message codeword of 1 million code bits would be received correctly even if any 6500 bits were in error.

Unfortunately, though we know that it is possible in principle to do at least as well as the Gilbert bound, we have no idea how to construct such codes. Furthermore, we do not even know that the Gilbert bound is a tight bound; it may be possible to construct binary codes that are even better than the Gilbert bound.

Attempts to find the codes of exceptional performance promised by the Gilbert bound are continuing and are likely to succeed eventually. The major effort employs the tools of algebraic geometry and was launched by the discovery of the *Goppa codes*. There has also been slight progress in the development of nonlinear block codes launched by the discovery of the *Preparata codes*.

Algorithms for Decoding

A practical decoder for an error-control code cannot compare exhaustively the received word to every codeword to see which codeword most closely agrees

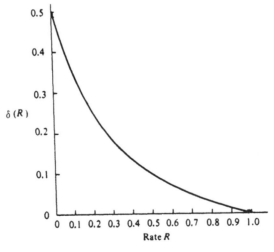

FIG. 4 Gilbert bound.

with the received word. This is because the number of codewords in many codes is astronomical. For example, the (32,28) code over $GF(256)$ used by the compact disk has 256^{28} (about 2.69×10^{67}) codewords. A decoder for this code is practical only if it uses some algorithmic procedure that computes the correct codeword from the received word.

Strong algorithms, such as the *Berlekamp-Massey algorithm*, for decoding Reed-Solomon codes are known and are in use. This algorithm can be described using the language of spectral estimation, but applied in a Galois field.

The best algorithm for decoding convolutional codes, known as the *Viterbi algorithm*, is essentially an efficient, computational algorithm for searching all possible codewords exhaustively.

Algorithms such as the Berlekamp-Massey algorithm for decoding Reed-Solomon codes have an essentially quadratic computational cost in the performance of the code. Algorithms such as the Viterbi algorithm for decoding convolutional codes have an essentially exponential computational cost in the performance of the code.

Spectral Estimation in a Finite Field

A practical decoder for a Reed-Solomon code requires a computational procedure to find the transmitted codeword. The codeword $\mathbf{c}$ is transmitted and the channel makes an error described by the vector $\mathbf{e}$, which is nonzero in not more than t places. The received word $\mathbf{v}$ is written componentwise as

$$v_i = c_i + e_i \qquad i = 0, \ldots, n - 1.$$

The decoder must process the received word $\mathbf{v}$ to remove the error word $\mathbf{e}$; the dataword is then recovered from $\mathbf{c}$.

The received noisy codeword $\mathbf{v}$ has a Fourier transform

$$V_j = \sum_{i=0}^{n-1} \omega^{ij} v_i \qquad j = 0, \ldots, n - 1$$

with components $V_j = C_j + E_j$ for $j = 0, \ldots, n - 1$. But, by construction of a Reed-Solomon code,

$$C_j = 0 \qquad j = n - 2t, \ldots, n - 1.$$

Hence

$$V_j = E_j \qquad j = n - 2t, \ldots, n - 1.$$

This block of $2t$ components of $\mathbf{V}$ gives us a window through which we can look at $2t$ of the n components of the error-pattern transform. The decoder must find all n components of $\mathbf{E}$ given a segment consisting of $2t$ components of **E**, and the additional information that, at most, t components of the time-domain error pattern **e** are nonzero. Once **E** is known, the computation is trivial because $C_j = V_j - E_j$. From **C** one can compute **c**. The data symbols are recovered easily in various ways, depending on the method of encoding.

To find a procedure to compute **E**, we use properties of the Fourier transform. Suppose, for the moment, that there exists a polynomial $\Lambda(x)$ of degree at most t and with $\Lambda_0 = 1$ such that

$$\Lambda(x)E(x) = 0 \qquad (\mathrm{mod}\ x^n - 1)$$

where

$$E(x) = \sum_{j=0}^{n-1} E_j x^j.$$

The polynomial product is equivalent to a cyclic convolution. Using the assumed properties of $\Lambda(x)$ we can rewrite the convolution as

$$\Lambda_0 E_j + \sum_{k=1}^{t} \Lambda_k E_{j-k} = 0$$

or, because $\Lambda_0 = 1$,

$$E_j = - \sum_{k=1}^{t} \Lambda_k E_{j-k} \qquad j = 0, \ldots, n-1.$$

This equation may be recognized as a description of a kind of filter know as an *autoregressive filter* with t taps. It defines component E_j in terms of the preceding t components $E_{j-1}, \ldots, E_{j-t}$. Since we know $2t$ components of **E**, by setting $j = n - t, \ldots, n - 1$, we can write down the following set of $n - 1$ equations:

$$
\begin{aligned}
E_{n-t} &= -\Lambda_1 E_{n-t-1} - \Lambda_2 E_{n-t-2} - \ldots - \Lambda_t E_{n-2t} \\
E_{n-t+1} &= -\Lambda_1 E_{n-t} - \Lambda_2 E_{n-t-1} - \ldots - \Lambda_t E_{n-2t+1} \\
&\ \ \vdots \\
E_{n-1} &= -\Lambda_1 E_{n-2} - \Lambda_2 E_{n-3} - \ldots - \Lambda_t E_{n-t-1}
\end{aligned}
$$

Here, there are t equations, linear in the unknown components Λ_k and involving only known components of **E**. Hence, provided there is a solution, we can find **Λ** by solving this system of linear equations. The solution will give the polynomial $\Lambda(x)$ that was assumed earlier. Once $\Lambda(x)$ is known, all other values of **E** can be obtained by recursive computation using the equation

$$E_j = - \sum_{k=1}^{t} \Lambda_k E_{j-k} \qquad j = 0, \ldots, n - 2t - 1$$

recalling that the indices are modulo-n.

To verify that the system of equations has a solution, suppose that there are $\nu \leq t$ nonzero errors at locations with indices i_ℓ for $\ell = 1, \ldots, \nu$. Define the polynomial $\Lambda(x)$, known as the *error-locator polynomial*, by

$$\Lambda(x) = \prod_{\ell=1}^{\nu} (1 - x\omega^{i_\ell})$$

which has at most degree t and $\Lambda_0 = 1$. The vector $\mathbf{\Lambda}$ of length n whose components Λ_j are coefficients of the polynomial $\Lambda(x)$ has an inverse transform

$$\lambda_i = \frac{1}{n} \sum_{j=0}^{n-1} \Lambda_j \omega^{-ij} = \frac{1}{n} \Lambda(\omega^{-i})$$

$$= \frac{1}{n} \prod_{\ell=1}^{\nu} (1 - \omega^{-i}\omega^{i_\ell})$$

which is zero if $i = i_\ell$. Therefore, $\lambda_i e_i = 0$ for all i. Because a product in the time domain corresponds to a cyclic convolution in the frequency domain, we see that the cyclic convolution in the frequency domain is equal to zero.

$$\mathbf{\Lambda} * \mathbf{E} = \mathbf{0}.$$

Hence, a polynomial $\Lambda(x)$ solving $\Lambda(x)E(x) = 0 \pmod{x^n - 1}$ does exist.

The Berlekamp-Massey Algorithm

The procedure for decoding Reed-Solomon codes requires solution of the matrix-vector equation

$$\begin{bmatrix} E_{n-t-1} & E_{n-t-2} & \cdots & E_{n-2t} \\ E_{n-t} & E_{n-t-1} & \cdots & E_{n-2t+1} \\ \cdot & & & \\ \cdot & & & \\ \cdot & & & \\ E_{n-2} & E_{n-3} & \cdots & E_{n-t-1} \end{bmatrix} \begin{bmatrix} \Lambda_1 \\ \Lambda_2 \\ \cdot \\ \cdot \\ \cdot \\ \Lambda_t \end{bmatrix} = - \begin{bmatrix} E_{n-t} \\ E_{n-t+1} \\ \cdot \\ \cdot \\ \cdot \\ E_{n-1} \end{bmatrix}$$

The matrix is a special kind of matrix known as a *Toeplitz matrix* because the elements in any subdiagonal are equal. The computational problem is one of inverting a Toeplitz system of equations. This kind of computational problem arises frequently in digital-signal processing not only in decoding Reed-Solomon codes, but also in the design of autoregressive filters and in spectral analysis. To invert a Toeplitz system of equations, special computational algorithms usually are used because they are computationally more efficient than computing the matrix inverse by a general method.

The Berlekamp-Massey algorithm is the preferred method for solving this system of equations in a Galois field. The algorithm is easy to execute, though it is difficult to derive. At the rth of $2t$ iterations, the algorithm computes a shortest-length autoregressive filter with length L_r and taps $\Lambda_i^{(r)}$ that satisfies

$$E_j = -\sum_{i=1}^{L_r} \Lambda_i^{(r)} E_{j-i}$$

for $j = 1, \ldots, r$. The shift register for the $(r + 1)$th iteration is then computed from the shift register for the rth iteration. There are $2t$ iterations and each has computational complexity t. Thus, the algorithm has computational complexity proportional to t^2. The Berlekamp-Massey algorithm can be implemented easily either in hardware or in software.

The Viterbi Algorithm

The most important method of decoding convolutional codes is the Viterbi algorithm. The computational complexity of this algorithm is exponential in the constraint length so it can be used only for convolutional codes of small constraint length. The algorithm is best described as an efficient method for searching a trellis.

A *trellis* is a graphical representation of the memory that exists in many kinds of data sequences, in particular the codestream of a convolutional code. An example of a very small trellis is shown in Fig. 5. Each node in one column denotes one possible state of a finite-state machine. Each column denotes a different time frame. In a practical problem of decoding a convolutional code of constraint length ν, there are 2^ν nodes in each column and the trellis extends indefinitely to the right.

A convolutional code generates a sequence of numbers by starting at the left-most node of the trellis and moving to the right along any path, one node per unit of time. A time unit might be as small as 100 nanoseconds, so the source might move through 10 million nodes per second. As the encoder runs through the trellis, it encounters a series of labels that then become the

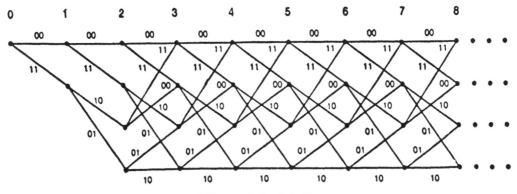

FIG. 5 A simple trellis.

codeword symbols. By observing the channel output, the decoder must deduce which path the encoder took through the trellis. The Viterbi algorithm finds the path whose symbol sequence best agrees with the received sequence in the sense of minimum Hamming distance.

In principle, it is easy to find the best path. Simply lay the received sequence along every possible path through the trellis, compute the Hamming distance to that codeword, and choose the path whose Hamming distance is smallest. The Viterbi algorithm is a recursive procedure to do this computation based on the general principle that if the optimal path from point A to point C passes through point B, then that section of the path from point B to point C must correspond to the optimal path from point B to point C.

The Viterbi algorithm operates frame-by-frame tracing through the trellis. An iteration begins with the assumption that the minimum-distance path to every node in that frame is known. Then, in the next frame, the algorithm determines the most likely path to each of the new nodes. To get to any one of the nodes in the new frame, the path must pass through one of the nodes in the previous frame. One can get the candidate paths to a new node by extending to this new node each of the old paths that can be so extended. The minimum-distance path is found simply by updating a table of possible paths and choosing the best path to the new node. This process is repeated for each of the nodes in the new frame. The algorithm continues in this way, computing frames indefinitely.

Performance on Noisy Channels

Codes for the prevention of errors are used to communicate over noisy channels. When evaluated for a specific channel, a code is judged by its rate and by its probability of decoding failure. By the nature of noise, errors are randomly distributed. There is always some possibility that more errors will occur than the code can correct. Then the decoder fails on that block or on that segment of the codestream. There are two kinds of decoding failure: a decoding default and a decoding error. When the decoder defaults, it announces to the user that a portion of the message is bad and cannot be corrected. It does not give the user bad data as if it were correct. This is the less-serious failure mode.

The decoder may also fail by "correcting" the wrong symbols in a bad message, thereby making the message even worse and passing the message to the user as a good message. This is a decoding error. In most well-designed systems, the probability of decoding error is many orders of magnitude smaller than the probability of decoding default.

Coding Gain

A systems designer may elect to use a code not to reduce bit error rate, but rather to enable the system to operate at a lower signal-to-noise ratio at the same bit error rate. The reduction in the required signal-to-noise ratio obtained by coding is called *coding gain*. For example, a simple binary phase-shift-keyed (BPSK) communication system operates at a bit error rate of 10^{-5} at a signal-to-noise ratio (E_b/N_0) of 9.6 dB on an additive Gaussian-noise channel. By adding a sufficiently strong error-control code to the communication system, the ratio E_b/N_0 could be reduced. If the code requires only 6.6 dB for a bit error rate of 10^{-5}, then we say that the code has a coding gain of 3 dB at a bit error rate of 10^{-5}. It is important to emphasize that this coding gain is just as real as doubling the size of an antenna or increasing the transmitted power by 3 dB. Because it is usually far easier to obtain coding gain than to increase received power, the use of a code is preferred.

In most applications, the average transmitted power is fixed. Parity-check symbols can be inserted into the channel codestream only by taking power from the data symbols. Therefore, coding gain is defined in terms of required E_b/N_0. The energy-per-bit E_b is defined as the total energy in the message divided by the number of data bits. This is equal to the energy received per channel bit divided by the code rate. The channel spectral-noise density is denoted N_0.

Figure 6 illustrates the probability of error versus E_b/N_0 for a simple binary code known as a Golay code. The coding gain at a bit error rate of 10^{-5} is 2.1 dB. For a larger code, the coding gain will be larger. The discussion, however, will be more complicated if one makes a careful distinction between decoding errors and decoding defaults.

Arbitrarily large coding gains are not possible. The illustration in Fig. 7 indicates the region where a coding gain must lie. There is one region for a hard-decision demodulator given by $E_b/N_0 > 0.4$ dB and another for a soft-decision demodulator given by $E_b/N_0 > -1.6$ dB.

Error Detection and Retransmission

A code with minimum distance d_{min} can detect every pattern of $d_{min} - 1$ or fewer errors, but can correct only $\frac{1}{2}(d_{min} - 1)$ or fewer errors. Furthermore, the complexity of a decoder that only detects errors is much less than a decoder that corrects errors. To detect errors when the code is systematic, one simply recomputes all the parity-check symbols from the data symbols. If any disagree with the received parity-check symbols, an error is detected.

Accordingly, many systems early on were designed only to detect errors. The errors are not corrected; rather, the transmitter is asked to retransmit the message. There are several reasons, however, why error detection and retransmission is an outdated approach. Powerful error-correction decoders are now inexpensive; on the other hand, retransmission has an adverse effect on system protocol and, accordingly, can actually be more complex than forward error correction.

Retransmission protocols require message buffering at the transmitter, a feedback link for the repeat request, and a scheme for inserting the retransmitted message into the received datastream. Normal data flow must be interrupted for the retransmission. The flow control can become extremely complex for a multi-access system, especially for one that uses relays and packetized messages.

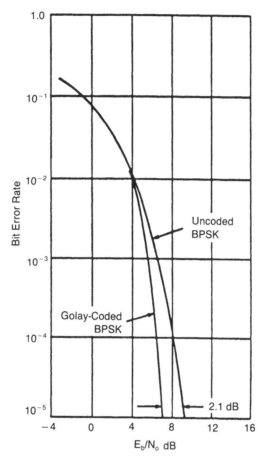

FIG. 6 Coding gain.

Another consideration is that retransmission techniques can only be used successfully if the majority of code blocks are received error-free. This is contrary to the modern trend of using the full power of a code to reduce the required signal-to-noise ratio at the receiver, or to increase spectral density of the waveform. Low-power messages will contain many errors that can be removed by error correction. Retransmission, however, only would lead to another noisy message.

Channel Capacity

Information theory tells us that every communication channel can be described concisely by a number, called the channel's *capacity* (*C*), that tells the data rate (in bits per second) that can be transmitted reliably over that channel. Thus, for any rate *R* smaller than *C*, one can construct a coding scheme for that channel that will prevent errors with any reliability desired; the probability of bit error can be made as small as desired.

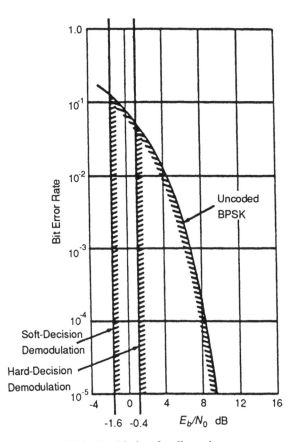

FIG. 7   Limits of coding gain.

Suppose that a binary channel makes random bit errors independently and with bit error probability ϵ. The capacity of this channel is

$$C = 1 + \epsilon \log_2 \epsilon + (1 - \epsilon) \log_2(1 - \epsilon)$$

bits per channel use. The theory says that a code exists to transmit data reliably (at an arbitrarily small probability of error) at any rate R smaller than the capacity.

For example, suppose $\epsilon = 0.11$. This means that each channel bit is wrong with a probability of 11%. Then, $C = 0.5$ bits per channel symbol. Codes of rate 0.5 exist that have any decoding-failure probability desired. However, such a code may be impractically complex if the probability is chosen very small, say 10^{-10}.

The basic theorems of information theory tell us that good codes exist. These theorems do not tell us, however, that good codes for reliable communication are simple to implement, nor do those theorems tell us how to find the good codes.

Many binary channels actually are made from an additive Gaussian-noise channel by a BPSK modulator/demodulator (modem). We may wish to know whether the choice of modem has appreciably decreased the channel's capacity. The capacity of the original additive Gaussian-noise channel is illustrated in Fig. 8, together with the capacity of the derived binary channel. The difference in capacity between these cases is significant and was the motivation for the development of codes based on Euclidean distance, discussed in the following section.

Codes Based on Euclidean Distance

Some codes are designed to maximize the minimum Euclidean distance between any pair of codewords instead of maximizing the minimum Hamming distance. This is appropriate when the channel input and output is a real or complex number and the channel noise is additive Gaussian noise.

Given two sequences of the same length of symbols from the real or complex number system, the squared Euclidean distance between the sequences is the sum of the (magnitude) squares of the componentwise differences. The Euclidean distance between the sequences, then, is the square root of the squared Euclidean distance. For example, consider the two sequences of length 6 of real numbers given by $\mathbf{v} = (0,1,2,3,4,5)$ and $\mathbf{v}' = (0,1,4,5,4,5)$. The Euclidean distance between these sequences is

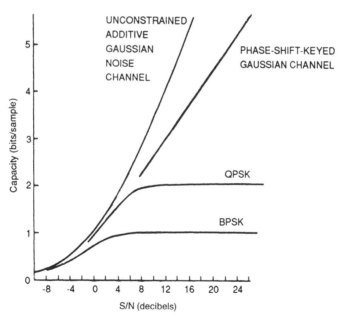

FIG. 8 Channel capacity.

$$d_E(\mathbf{v},\mathbf{v}') = \sqrt{(0 - 0)^2 + (1 - 1)^2 + (2 - 4)^2 + (3 - 5)^2 + (4 - 4)^2 + (5 - 5)^2}$$
$$= 2\sqrt{2}$$

We may also have two infinitely long sequences of real or complex numbers. Their Euclidean distance is defined in a similar way as the square root of an infinite sum of squared differences.

Soft-Decision Decoders

Every binary code (with symbols in $\{0,1\}$) can be used on the real bipolar channel (with symbols in $\{-1,1\}$). Simply represent a 0 by the real number -1 and a 1 by the real number 1. Two binary codewords are then mapped into two bipolar codewords. The squared Euclidian distance between two sequences is four times the Hamming distance

$$d_E^2(\mathbf{c},\mathbf{c}') = 4d_H(\mathbf{c},\mathbf{c}').$$

Consequently, the minimum squared Euclidean distance of the code is four times the minimum Hamming distance.

The output of the additive Gaussian-noise channel is the sequence of real random variables

$$v_i = c_i + n_i$$

where n_i is a Gaussian-noise random variable. Whereas a hard-decision demodulator estimates the binary symbols 0 or 1 from v_i, a soft-decision demodulator provides v_i as a real number directly to the decoder. The decoder is then known as a soft-decision decoder.

A soft-decision decoder calculates codeword $\mathbf{c}$ such that the Euclidean distance $d_E(\mathbf{v},\mathbf{c})$ is smallest. It is quite straightforward to modify the Viterbi algorithm to use Euclidean distance in place of Hamming distance, therefore convolutional codes are quite suitable for Gaussian channels.

Reed-Solomon codes are in a nonbinary alphabet; there is more than one way to modulate nonbinary symbols into the real (or complex) number systems, resulting in more than one way to introduce a Euclidean-distance structure. These possibilities are not discussed in this article other than to mention that the Berlekamp-Massey algorithm can be augmented to serve as a soft-decision decoder, though with some increase in complexity.

The probability of error of a soft-decision decoder can be approximated by

$$P_e \sim Q\!\left(\frac{d_{\min}}{2\sigma}\right)$$

where σ is the noise variance and $d_{\min}$ is the minimum Euclidean distance of the code. Clearly, increasing $d_{\min}$ by choosing a better code is as good as reducing the channel-noise variance σ. This argument shows the role of an error-prevention code in many systems.

Trellis Codes

Rather than use a convolutional code designed for maximum Hamming distance on a Euclidean-distance channel, one can design the code directly for maximum Euclidean distance. The Ungerboeck trellis codes are designed in this way for either real or complex Euclidean-distance channels. Figure 9 portrays a trellis for an Ungerboeck trellis code in contrast to uncoded QPSK, which is so popular for satellite communications; the code has the same bandwidth but requires 4.1 dB less signal-to-noise ratio. The trellis code replaces the signal constellation in the QPSK waveform with an 8-ary PSK signal constellation so as to introduce eight points in the complex plane that represent amplitude and phase modulation (or, in this example, phase modulation only). Instead of using the eight points to increase the data rate by modulating three bits at a time into a channel symbol, the trellis code modulates only two bits of data at a time into a three-bit channel symbol. The three code bits define one of eight possibilities that are labeled in the phase diagram on the right-hand side of Fig. 9. Every pair of incoming data bits is mapped into three code bits by the trellis encoder in order to create a waveform. This action is performed by shift-register circuits that are encoding data based on the memory of the two previous pairs of bits into the encoder, rather than on the current two bits alone.

The transmitted trellis-coded waveform has the same data rate as a QPSK waveform and the code is transparent to the user. The waveform has the same

1) UNCODED QPSK

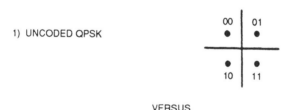

VERSUS

2) UNGERBOECK CODED 8-ARY PSK
   -PARITY IN SIGNALING CONSTELLATION NOT IN BANDWIDTH

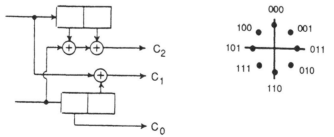

- SAME BANDWIDTH
- SAME DATA RATE
- 4.1 dB LESS POWER

FIG. 9 Ungerboeck codes for reduction of signal-to-noise ratio.

bandwidth and the same data rate, but the power requirement is reduced by 4.1 dB. We are free to spend that 4.1 dB of now unneeded power in any desired fashion.

Ungerboeck codes can be used as a means of increasing the transmitted data rate at a fixed signal-to-noise ratio. Much more complicated signal constellations are required for this purpose. Signal constellations with as many as 256 points are now considered practical.

Matched Spectral-Null Codes

So far, we have only considered memoryless channels. Except for additive errors or noise, the channel output is the same as the channel input. In this section, we comment briefly on methods developed recently to construct codes that exploit the spectral characteristics of the channel.

Matched spectral-null codes combine the methods of trellis codes with the methods of partial-response signaling. The codewords, which are the input to the channel, are designed to obtain large minimum distance at the output of the channel. These codes exploit the memory in a channel to improve performance. Matched spectral-null codes are designed to exploit the observations that the minimum distance of a real-valued trellis code can be bounded by the order of the spectral null, and spectral nulls can be created in a sequence by constraining running digital sums.

The first-order running digital sum of a sequence $\mathbf{c}$ is defined as

$$s_\ell = \sum_{i=0}^{\ell} c_i$$

and the Kth-order running digital sum is defined recursively in terms of the $(K - 1)$th-order running digital sum as

$$s_\ell^{(K)} = \sum_{i=0}^{\ell} s_i^{(K-1)}.$$

The Kth-order running digital sum can be understood as the output of the cascade of K accumulators.

For example, the sequence of coefficients

$$\mathbf{c} = (1,1,1,-7,2,2)$$

has a running digital sum given by

$$\mathbf{s}^{(1)} = (1,2,3,-4,-2,0).$$

The running digital sum of the running digital sum is

$$\mathbf{s}^{(2)} = (1,3,6,2,0,0)$$

The reason that the running digital sum is important is that the minimum Euclidean distance satisfies

$$d_{\min}^2 \geq 2K$$

if the Kth-order running digital sum is constrained to remain in a bounded interval. Moreover, if the channel itself has a spectral null of order L at zero frequency, then

$$d_{\min}^2 \geq 2(K + L).$$

By designing code sequences to satisfy the Kth-order running-digital-sum condition, coding gain can be realized.

This brief discussion suggests that unlimited coding gain is available simply by making K large. However, this is not so. Not only does the encoder become too complex, but other neglected effects become important and ultimately limit the achievable gain.

Synchronization Codes

A communication system that uses a block code (or a tree code) for data transmission will concatenate codewords (or code frames) into an indefinitely long string of codeword symbols. To decode an individual codeword, the decoder must be able to identify the boundaries between the codeword blocks (or frames). This requires a method to acquire block synchronization initially. Moreover, a channel may insert one or more false symbols into the datastream or delete one or more symbols from the datastream causing the decoder to misframe subsequent blocks. This requires a method to maintain or recover block synchronization.

A simple synchronization procedure is to insert a fixed sequence of symbols, known as a *marker*, into the codestream periodically, perhaps after every codeword. For example, the sequence 10111000 is an 8-bit binary marker that could be inserted after every codeword as an additional overhead. The codestream can be broken into codewords by recognizing the periodic occurrence of this pattern. However, the use of a marker in this way has two weaknesses. The marker is used outside of the error correction; in very noisy applications that use powerful error-prevention codes, there will be an unacceptable probability that all markers in a long interval will be in error. A second weakness is that nothing prevents a sequence of symbols of the codeword from forming a marker by chance. Indeed, this occasionally will happen if the data is random. More rarely, it can even happen in each of a long sequence of successive blocks. In a system transmitting millions of bits per second, even if the probability of a false marker is very low, it can still cause occasional unacceptable system interrupts.

Block codes can be constructed that are capable of recovering synchronization even in the presence of channel errors. If the system is initially synchronized and thereafter no synchronization slippage of more than a few symbols will be encountered, then a self-synchronizing code may be employed to recover synchronization. Such codes can be designed to combine error correction and synchronization.

In the absence of errors, the preferred class of self-synchronizing codes are those for which the synchronization correction is not data dependent. This means that no concatenation of two codewords can form a codeword at the boundary. A *comma-free code* is a set of q-ary n-tuples such that for every pair of codewords $\mathbf{c} = (c_0, c_1, \ldots, c_{n-1})$ and $\mathbf{c}' = (c_0', c_1', \ldots, c_n')$, the n-tuple $(c_i', \ldots, c_{n-1}', c_0, \ldots, c_{i-1})$ is not a codeword for any nonzero i. For example, the set

$$= \{10000, 10001, 10011, 10111\}$$

is a binary $(5,2)$ comma-free code. A concatenated stream of such codewords can always be synchronized, even if the start of the sequence is lost. Thus

$$\ldots 0110011100011011110000010 \ldots$$

can only be punctuated as

$$\ldots 01, 10011, 10001, 10111, 10000, 10 \ldots$$

Long, comma-free codes are more powerful than synchronization markers because there is no possibility of false synchronization, and more efficient than markers because comma-free codes with high rates exist. However, powerful methods for constructing such codes are not yet known nor has error correction been incorporated into such codes. Good constructive encoding and decoding algorithms for long, comma-free codes have not yet been developed. Encoding of a $(100,92)$ comma-free code by a look-up table would not be practical.

A Reed-Solomon code is not a comma-free code because a comma-free code cannot have a codeword that is a cyclic shift of another codeword. However, a Reed-Solomon code can be modified to correct random errors and some synchronization errors simultaneously.

Suppose that a concatenated stream of codewords from a t-error-correcting Reed-Solomon code has a q-ary symbol deleted from the codestream. Then, all codewords that follow the codeword containing the deletion will be misframed. The first symbol of a received word will be the second symbol of a transmitted word, and so forth. The last symbol of the received word will be the first symbol of the next transmitted codeword. The last symbol will appear to be an error and, if there are at most $t-1$ other symbol errors, will be changed by the Reed-Solomon decoder to the missing first symbol of that codeword. Thus, the concatenated stream of corrected output words from the decoder will be correct, but each will appear cyclically shifted by one symbol position within its timeslot. Subsequent logic will eventually recognize that the last symbol of almost every codeword requires correction and will interpret this as a synchronization offset that then can be corrected in both future data and past data. In this way, a t-error-correcting Reed-Solomon code usually can correct ν random errors and a synchronization slippage of $t - v$ symbols simultaneously. If the synchronization slippage is more than t symbols, the Reed-Solomon code cannot recover synchronization.

The Reed-Solomon code can be modified so that synchronization is not data dependent. A *coset code* of a code is the set $\{c + h\}$ where c is a codeword and h is a fixed vector. To encode into a coset code, first encode into the underlying Reed-Solomon code, then add the components of h to the components of the codeword. The sum forms the channel word. To decode, first subtract h, then decode to a codeword. Adding h and then subtracting it has no effect on the ability to correct random errors. It does, however, ensure that synchronization slippages are correctable.

Specifically, choose h given by $h_0 = 1$ and $h_i = 0$ for $i \neq 0$. After a synchronization error of r symbols, the received word is partly one codeword and partly the following codeword and the 1 appears in the rth component. After the template is subtracted prior to the Reed-Solomon decoder, there will be two extraneous 1s, one in component 0 and one in the rth component, that are added to the Reed-Solomon codeword. These particular error patterns (two 1s separated by r symbols) occurring in every block are an indication of a synchronization error of r symbols.

Bibliography

Blahut, R. E., *Theory and Practice of Error Control Codes*, Addison-Wesley, Reading, MA, 1983.

Blahut, R. E., *Digital Transmission of Information*, Addison-Wesley, Reading, MA, 1990.

Clark, G. C., Jr., and Cain, J. B., *Error-Correction Coding for Digital Communications*, Plenum, New York, 1981.

Lin, S., and Costello, D. J., Jr., *Error Control Coding: Fundamentals and Applications*, Prentice-Hall, Englewood Cliffs, NJ, 1983.

RICHARD E. BLAHUT

Coding of Facsimile Images

Introduction and Perspective

In digital facsimile systems we scan an image, convert the resulting analog signal to binary digital form, and then encode it to reduce the number of bits that represent the image. Advantages of coding include reduced transmission time and bandwidth requirements, with concomitant smaller transmission costs and speedier user service. Numerous facsimile coding studies have been made and various coding methods developed, each with particular advantages under specified conditions. In this article, we review the major approaches taken for coding of bilevel facsimile images and discuss issues relating to their performance, use, and future development.

We start with the initial capture of the source image, progressing through the processes of sensing, sampling, and quantization. Next we consider mathematical modeling of image generation, which provides an important tool in the analysis and design of coding methods. The models are based on information theory and the underlying theory of random or stochastic processes. The next section discusses the major approaches to *exact* or *lossless* coding. In this type of coding, the decoded image is exactly the same as the original digitized image. Included in this section is an overview of the recommendations for facsimile coding adopted by the International Telegraph and Telephone Consultative Committee (CCITT) and also a discussion of figures of merit for comparing coding methods. A coding process that is not exact is called an *approximate* or *lossy* one. The last section gives the major approaches to approximate coding and also discusses associated distortion issues.

Compression performance has often played the primary, if not sole, role in determining the relative worth of different coding methods. A host of other considerations are also of importance. Among these are design, production, and marketing concerns, including the implementation cost of both logic and storage, and the extent of design and testing effort. Of operational importance are speed of encoding and decoding, sensitivity to channel errors, and code synchronizability. Yet other figures of merit are the processability of the encoded data and the robustness and compatibility of the coding method. *Processability* is the ability to perform image operations directly on the encoded data. *Robustness* is the ability to handle image data variations without compression penalty. *Compatibility* is the ability of a method to convert into or evolve into methods that permit pattern-recognition functions, approximate codings, or other image operations, and the ability to work smoothly with other modules addressing wider functionality. For lossy encodings, an important figure of merit is perceived image quality.

The initial development of digital facsimile systems was based on several assumptions, including that the system should operate in a standalone mode, the system should handle only bilevel (black and white) images, and the received output should be displayed on hardcopy, usually paper. The underlying assumption was that typewritten pages and drawings comprise the major use of facsimile. Based on these assumptions, the CCITT adopted a set of recommended facsimile coding methods. Most facsimile systems today meet one or more of these standards, and the resulting compatibility has encouraged substantial growth of the facsimile industry. However, system designers currently are addressing a number of issues not envisioned in the initial period of development.

Because of technological advances and new consumer demands, significant changes are underway in the uses and designs of facsimile systems. These changes are likely to have an impact on our view of coding processes. A particularly important trend is the migration of the facsimile function into computer systems. This currently takes on a variety of architectural forms, affecting where and how the facsimile coding methods are implemented. Along with the increasing role of computers, there is an emerging view of an integrated capability that embraces disparate communication modes, including electronic mail, voice mail, and facsimile, and requires the conversion from the format of any one mode to the format of any other (1). An allied trend is the use of soft facsimile, which is the display of received transmissions on erasable media, such as cathode-ray tubes (CRTs) and liquid-crystal display (LCD) panels, instead of paper. An important application of soft facsimile is browsing of remote large image databases. For this purpose, it is necessary to address coding needs for both storage and transmission, which may differ, as in the use of individual versus ensemble code books, for example (2). Also, browsing is likely to occur on low-capacity channels, where progressive transmission and display are necessary to make both informational and entertainment services attractive. In progressive modes, a crude representation of an image is transmitted first and finer detail is added in stages.

Still other facsimile-system trends include increased demand for higher resolutions, for color images, and also for increased multilevel capability. Some facsimile systems today offer digital halftone capability, using a bilevel representation of a multilevel image that gives the appearance of a gray-level image similar to pictures printed in newspapers. Current devices simulate from about 16 to 64 gray levels. In a soft facsimile environment, demand for 256 and more gray levels can be expected. Several color facsimile machines are now available. Additionally, there is a growing demand for composite documents, consisting of separately encoded scanned material, ASCII (U.S.A. Standard Code for Information Interchange) or other coded characters, graphics commands, and computer-generated images (3). Future scenarios may also include the extension of facsimile to provide such services as combined voice/picture mail and, more generally, multimedia messaging systems and other services that mix facsimile, still images, text, voice, and motion video. Such services are in step with advances in scanning and display technology and increased available bandwidth, such as that with broadband Integrated Services Digital Networks (ISDNs).

In response to increased capabilities and demands, many new facsimile features and architectures are already emerging. The traditional standalone facsimile unit is now supplemented with a variety of other ways to handle the facsimile function. These include numerous facsimile modem boards, alone or bundled with software packages that convert computer files into facsimile format and provide numerous other features. Some of these boards are equipped with a small computer system interface (SCSI) for linking to scanners, printers, and mass storage units, and some boards have both data and facsimile modem capability. In another approach, the facsimile modems are external portable units that work from any personal computer (PC) serial port. Another configuration consists of a full facsimile machine that interfaces to a PC, local-area network (LAN) station, or mainframe host. It can operate in a standalone mode or as a scanner and backup printer for the computer system. Also of note are software packages for facsimile network servers equipped with facsimile modem boards that connect to standalone facsimile machines for scanning purposes and to computer printing devices for output. Offered more recently are products, referred to as *hydra* systems, that combine such multiple functions as scanning, facsimile, laser printing, digital voice recording, data modem communications, graphics display, and copying. In some cases, a hydra system also links to a computer. Also available are special chip products for the original equipment manufacturer (OEM) market, including a single chip that combines functions of modem signal processing, image compression, image enhancement, protocol processing, and control of mechanical parts.

In addition to the offerings above, there are also special software packages for image compression, some of which include the CCITT facsimile coding methods. These packages often are coupled with general graphics processors and with such peripherals as scanners, random-access storage devices, and streaming-tape drives. Raw data consists of both non-image and image types, and the image type includes bilevel, multilevel, and color images. Of particular interest, the coding methods generally used are more advanced than the standard CCITT facsimile methods, and they yield significantly higher compressions than do the CCITT methods. This further indicates that newer coding methods may prove desirable in an integrated systems approach that addresses both storage and transmission.

In addition to the basic facsimile functions, the innovations above include a rich offering of other features. Presentation abilities include monitor previewing of outgoing documents, monitor viewing of incoming documents with a choice of printing and/or computer storage, the ability to rotate images to correct for upside-down scanning, and to zoom at one-third, one-half, and full size. Other user conveniences are preformatted cover sheets (including letterhead, logo, and signature), phone directory services, background facsimile reception, and automatic notification of incoming messages. Conversion features include converting ASCII text to facsimile images and converting facsimile images to ASCII text, which enables editing of received messages. Among networking abilities are sending composite facsimile, text, and binary files in one transmission, sending in a broadcast mode to multiple destinations, delayed sending to take advantage of off-hours, unattended sending and receiving, automatic dialing and automatic redialing in case of a busy destination or an interruption, automatic time and date stamping, automatic logging of facsimile documents sent and received, store-and-forward capability for facsimile traffic from other terminals, and automatic extraction of routing instructions from incoming documents and transfer to indicated destinations.

In connection with this diversity of options, there is a need for continued systems analysis to integrate different solutions and to provide for graceful growth of this industry, including reassessment of facsimile coding needs and the need for expanded standards that incorporate new functional parameters and appropriate coding methods. This article reviews some of the options available for bilevel facsimile coding; additional sources of information are listed in the Bibliography.

Capturing the Source

Introduction

The facsimile coding process consists of separate encoding and decoding phases. These are part of a sequence of functions performed in a digital facsimile system, which are shown in a generic form in the diagram in Fig. 1. In traditional standalone facsimile systems, the functions shown generally are incorporated in a single unit. As discussed above, in newer architectures, these functions are distributed in various ways among computers, their peripherals, and other units, including self-contained facsimile systems.

In Fig. 1, the first facsimile function is the scanning of a source image. Scanning consists of three phases. First, the image is sensed, obtaining an electrical analog signal that varies in accordance with the optical intensity variations of the image. Then, the analog signal is sampled, producing a discrete set of real-valued sample values of the analog signal. Finally, these samples are quantized, resulting in a discrete sequence of digital values consisting of equal-length binary words. These three processes are discussed in separate sections below, particularly as they refer to bilevel (two-tone, usually black and white) and multilevel (monochrome or gray-level) images.

Sensing the Source Image

To sense an image, it is first illuminated by light. For paper images, this light is reflected from the medium; for film images, it passes through the medium. The reflected or transmitted light varies in intensity according to the image content. In either case, a photoelectric device converts the light variations into an analog electrical signal. There are a variety of devices that perform illumination. They vary in the portion of the image that is illuminated by the source light at any given time, in the way the source light is delivered to the image, in the type of source light, and in the way of sensing the reflected or transmitted light.

In a common type of device, the image is illuminated and sensed in strips or lines across the width of the image, one strip at a time, from the top to the

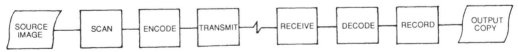

FIG. 1   Digital facsimile system functions.

bottom of the image (see Fig. 2-*a*). In one arrangement, the light source is a fluorescent tube extending across the strip being sensed (4). The reflected light is focused through a lens system onto a smaller field containing a linear array of photodiodes, charge-coupled devices (CCDs), or charge-injection devices (CIDs), which convert light into electrical form. In many of these devices, the image lies on a flat bed and remains stationary while the light source and sensors are moved to access successive image strips. In another arrangement, light from an incandescent source travels to the image through a set of optical fibers fitted closely to the image strip being sensed (5). A second set of optical fibers, also fitted closely to the image strip being sensed, pick up the reflected light and transmit it to a CCD device, which again provides the conversion to electrical form. In this arrangement, the imaging components are stationary and the image steps through a rollfeed mechanism to access successive image strips.

Lasers or light-emitting diodes (LEDs) also can supply the source light. Or, the scanning device may be a video camera that focuses an entire optical image onto a plate of photosensitive material, which acquires an electrical pattern varying in accordance with the optical image. An electron beam then scans the plate in a line-by-line fashion, causing the pattern to be converted into signal form. Also available are two-dimensional arrays of CCDs, onto which an entire image is focused. Yet another approach illuminates and senses a single small

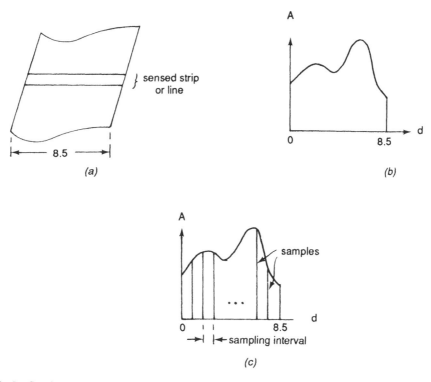

FIG. 2 Sensing and sampling the source image (A = amplitude; d = distance from left edge or time of sensing): *a*, document being sensed; *b*, sensed signal for one image line; *c*, sampled signal for one image line.

spot at a time. A version of this is the rotating-drum scanner, often used in graphic-arts applications. The image is wrapped around the drum, and is illuminated by a spotlight. The light source and the sensor mechanism are contained in a housing mounted on a carriage that permits linear travel parallel to the drum axis. Scanning is provided in the X direction by the drum rotation and in the Y direction by the linear movement of the sensor. Flying-spot scanners are another version of illuminating and sensing a small spot at a time. Other arrangements are also possible.

In most scanning devices, the output signal is nonuniform over the image area. This is called *shading*, and can be due to uneven lighting, lens curvature, electron-beam positioning error, or variations in photosensor sensitivity. Correction circuitry compensates for these aberrations. Some newer devices also provide for calibration of each particular device the first time it is used; direct current (DC) offset voltage and individual photosite responses are stored in separate tables and used dynamically to correct data values as they are acquired (6).

Sampling the Analog Signal

The result of the sensing operation is an analog electrical signal (see Fig. 2-*b*). To convert this signal to digital form, it is first sampled, and the resulting samples are then quantized. *Sampling* is the process of obtaining signal amplitude values taken at successive and, usually, evenly spaced points along the distance or time coordinate of the analog signal (Fig. 2-*c*). The *sampling interval* is the distance between two successive points on the coordinate, and its inverse is the *sampling rate*.

The question arises as to what sampling rate should be used. In theory, if the highest frequency contained in the signal is W (cycles per unit of distance or time), then the sampling rate should be at least $2W$. This follows from the classical sampling theorem proposed by Nyquist (7), which states that it is possible to reconstruct completely a signal from its samples if the sampling rate is equal to twice the signal bandwidth, $2W$. This is called the *Nyquist rate*. The theorem also holds if the sampling rate is greater than $2W$. If a sampling rate less than $2W$ is used, a condition called *aliasing* occurs, wherein the spectrum of the sampled signal contains overlaps of adjacent spectral segments in the periodic spectrum. Figure 3 illustrates the effect of sampling rate on an ideally sampled one-dimensional signal. The spectrum of the signal to be sampled is shown in Fig. 3-*a*. The spectrum of the ideally sampled signal using a sampling rate greater than $2W$ (Fig. 3-*b*) is reconstructed exactly by an ideal low-pass filter of range W. This also applies where the sampling rate exactly equals $2W$ (Fig. 3-*c*). However, when the sampling rate is less than $2W$, adjacent spectral segments overlap, resulting in the spectrum indicated by the solid line at the top of Fig. 3-*d*. Applying an ideal low-pass filter of range W to this spectrum yields the spectrum illustrated in Fig. 3-*e*. This clearly is not the spectrum of the original signal depicted by Fig. 3-*a*; the altered spectrum of Fig. 3-*d* indicates the aliasing error. Figure 3 illustrates aliasing for a one-dimensional signal. A

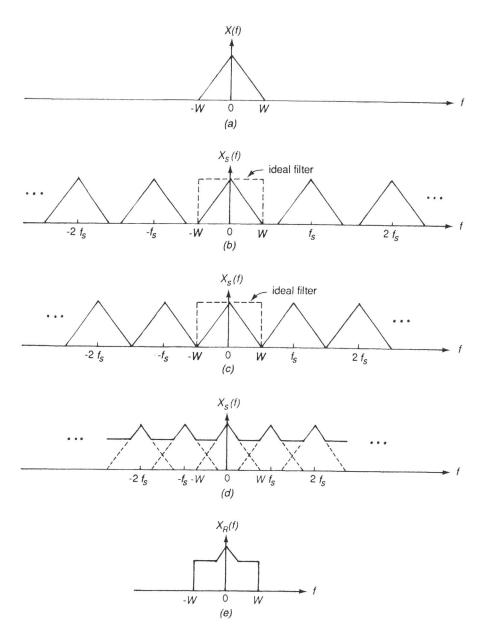

FIG. 3 Effect of sampling rate: *a*, spectrum of signal; *b*, spectrum of ideally sampled signal at sampling rate of $f_s > 2W$; *c*, spectrum of ideally sampled signal at sampling rate of $f_s = 2W$; *d*, spectrum of ideally sampled signal at sampling rate of $f_s < 2W$; *e*, spectrum of ideally reconstructed signal at sampling rate of $f_s < 2W$, showing aliasing.

similar analysis applies for an image signal defined in two dimensions, where aliasing results in moire patterns in the reconstructed image (8).

Aliasing can be reduced by increasing the sampling rate, which also increases the number of bits needed to represent an image. An alternative is to low-pass filter the analog image signal before sampling. Filtering is mathematically equivalent to finding the convolution of the image signal with a filter function. A common filter function has a value of 1 over a unit square, a value of 0 elsewhere, and effectively performs local neighborhood averaging throughout an image. Area averaging is effective along polygon edges, but is less effective on lines that are thinner than the sampling interval; Gaussian filters give better total performance (9). Averaging also can be accomplished in optical spot scanners by designing the filtering function into the lens system or by slightly defocusing the spot beam.

In practice, sampling rates suitable for scanning images have been determined by subjective criteria. The needs differ depending on the type of image and its intended use. The results of several subjective studies on bilevel material follow. For typewritten pages, 100 samples, or dots per inch (dpi), achieve legibility, but 150 dpi are required for acceptability (10). This increases to 200 dpi for Chinese characters (11). For weather maps, the requirement is 100–200 dpi, and for fingerprints it is 200–400 dpi (12). For engineering drawings, 200–250 dpi is used (13,14). Text in newspaper pages requires 400 dpi for satisfactory results (15). Material of graphic-arts quality, such as that found in the White Pages and Yellow Pages of telephone directories, requires from about 400 dpi to 2000 dpi (16,17).

Quantizing the Samples

The result of the sampling operation is a set of real-valued samples of the analog signal such as those illustrated by Fig. 2-c. The next facsimile function to be performed is quantization. In this discussion, we address *scalar quantization*, which maps each real-valued sample separately into one of a finite set of quantized values. An extension of this is *vector quantization*, which maps a vector of real-valued samples (corresponding to a block of image picture elements) into one of a finite set of vectors of quantized values. *Binary-vector quantization* is a form that maps a vector of quantized values into one of a smaller set of vectors of quantized values; this is discussed below in the section concerning coding.

For scalar quantization, the possible range of sample values is divided into a discrete set of ranges called *quantization intervals*. Each interval is assigned a unique symbol from a symbol alphabet that usually consists of a set of equal-length binary words. Quantization is the process of mapping each sample value into the symbol of the interval containing the value. For a bilevel quantization, there are only two alphabet symbols, 0 and 1. An example of bilevel quantization intervals and symbols is shown in Fig. 4-a.

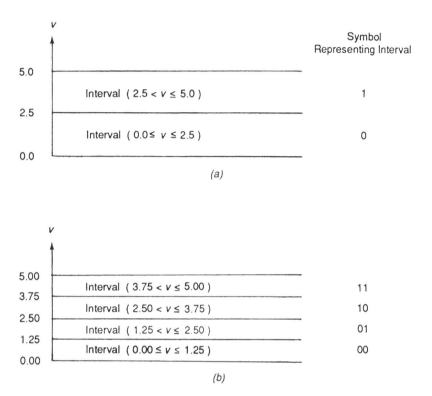

FIG. 4 Quantization intervals and symbols (v = signal amplitude or sample value): *a*, bilevel quantization; *b*, multilevel quantization.

The boundaries of the intervals are called threshold values or decision levels because they are used to decide the interval to which a sample belongs. It is not necessary or even optimal for the two intervals to be of the same size. For typewritten characters, the position of the decision level between the two intervals is critical. Setting it too far in one direction causes the reproduced type to appear bloated, and setting it too far in the other direction causes it to appear spindly. The latter case results in fewer bits required to code the image. In making comparisons of image coding methods, it is important that the same digitized material be used for each method, otherwise an apparent advantage of one method may be due to a favorable decision level.

When more than two quantization intervals are used, the process is called *multilevel* or *gray-level quantization*. Figure 4-*b* offers an example of multilevel quantization that defines four intervals, represented by the symbols 00, 01, 10, and 11.

The set of symbols or digital values resulting from the quantization of an image are collectively referred to as the image *bit map*. Each symbol corresponds to a subarea of the image termed a *picture element*, contracted to *pel* or *pixel*. Figure 5 presents excerpts of bit maps arranged in matrix form corresponding to the pels of the original image; Fig. 5-*a* is an example of a bilevel image and Fig. 5-*b* illustrates a multilevel image digitized to eight levels.

0	1	1	0	0	0
0	1	1	1	0	1
0	0	1	1	1	1
1	0	0	1	1	0

001	000	010	001	000	011
010	011	010	011	100	011
010	011	101	100	101	010
110	111	111	110	101	011

(a)   (b)

FIG. 5   Image bit maps: *a*, bilevel image; *b*, multilevel image with eight levels.

Upon reconstructing the image, each symbol is replaced by a reconstruction value, also called a representation level, which is a signal value assigned to be representative of all values in the interval corresponding to the symbol. In the case of bilevel images, the reconstruction values generally are chosen to be the extremes of the two quantization intervals so that the reconstructed picture provides maximum contrast between black and white. For multilevel images, the reconstruction values usually are set at the midpoints of the intervals. The image is reconstructed by filling or painting each pel with the gray-level intensity indicated by the reconstruction value for the pel. Assuming each pel has a square or rectangular shape, this process results in a blocky appearance of the image with discontinuities at pel boundaries, and introduces high-frequency components not in the original image. To overcome this effect, some systems subject the reconstructed values to a digital low-pass filter before outputting them onto a display device. An alternative remedy is to increase the sampling rate and, because this is also one of the approaches to reduce aliasing, the blockiness of reconstructed images is sometimes also referred to as aliasing (18).

The difference between an original sample value and the associated reconstruction value is called the *quantization error*. The quantization error may be modeled as a random variable; then the random quantization errors are referred to as *quantization noise*. If it is assumed that the signal range is finite, that it is divided into intervals of equal size d, and that all values of the random error are equally likely, then it may be shown that the mean-squared value of the quantization noise is $d^2/12$ (19). It follows that the root-mean-squared (rms) value of the quantization noise increases linearly with the size of the quantization interval.

For gray-level quantization, if the number of gray levels is too few, a condition called contouring may occur in the reconstructed image. This is the appearance of clearly noticeable contours or boundaries between adjacent areas with differently quantized gray levels; it is particularly apparent in areas corresponding to large regions of slowly changing intensity in the original image. In the reconstructed image, these areas appear patchy. In subjective tests, 64 gray levels are sufficient to avoid this condition and 32 levels are sometimes adequate; below this amount, contouring becomes evident (20,21). Contouring can be mitigated by adding pseudorandom noise or a dither signal prior to quantization to break up the indicated contours (22,23). Other methods include use of filters on the reconstructed image (such as a Butterworth low-pass filter) (24), use of high-frequency preemphasis (25), use of feedback in the quantizing process (26), and use of nonuniform quantization intervals. In this last method, the aim is to use smaller intervals in areas of slowly changing intensity, and larger intervals elsewhere.

One way of selecting nonuniform quantization intervals is to size them in accordance with estimates of the frequencies of occurrence of the signal sample values. Thus, smaller intervals are used for ranges with frequently occurring sample values. This is called *tapered quantization*. There are also several other ways of selecting nonuniform quantization intervals (27,28).

Nonuniform quantization may also be achieved on an adaptive basis, using different quantization intervals for different parts of an image. This may be used to enhance images, such as a bilevel image originally containing black text (or drawings) on a white background that has aged with time, resulting in some background areas becoming darker and some type becoming lighter. Use of a constant quantization may result in erroneously treating darker background area as black type and lighter type as white background. Other uses of nonuniform quantization include removing spurious paper noise and gaps on the edges of characters or pinholes within them. The removal of noise may also be done after quantization by preprocessing techniques (28). Currently, many scanners change the quantization decision level as the image is being scanned.

Nonuniform quantization may be achieved by companding (a contraction of compression/expansion). In this process, the sample is fed to a compressor that maps it in a nonlinear fashion to an output value, which is then quantized uniformly. Upon reconstruction, the uniformly quantized value is fed to an expander, which does the inverse function of the compressor.

Together, the sampling rate (or spatial resolution) and the number of quantization levels (or intensity resolution) determine the number of bits in the bit map. Subjective studies show the existence of various tradeoffs between the two resolutions in order to achieve the best subjective quality. These tradeoffs depend strongly on image content. In one study that judged aesthetic quality, a combination of higher intensity resolution and lower spatial resolution was preferred for images with relatively little detail, while the reverse held for images with much detail (20). Preferences may also vary depending upon the intended use of an image. In another study involving medical diagnosis using ultrasound images, a combination of higher intensity resolution and lower spatial resolution was also preferred (21). Interestingly, in both studies the preferred combinations required a smaller number of bits than that of several other combinations.

Current Scanning Capabilities

Commercial scanners currently available perform at various levels of scanning ability to meet the needs indicated above. Examples of monochrome flatbed scanners with scanning ability up to 300 dpi and 256 levels of gray are the Datacopy GS Plus of Xerox Imaging Systems (Sunnyvale, CA) (29), the Scanjet Plus of Hewlett-Packard (Cupertino, CA), and the AVR 3000/GS Grayscale

Image Scanner of Advanced Vision Research, Inc. (San Jose, CA) (30). Portable scanners are also available, including the Scanman Model 32 from Logitech Inc. (Fremont, CA), which scans at 400 dpi with 32 levels of gray (31), and the A4Scan AS-8000P model of ECA C&C Products, Inc. (Lodi, NJ), with 100, 200, 300, or 400 dpi resolution (32).

Various technologies are used to achieve the resolutions required for graphic-arts applications. One of these uses a rotating drum scanner capable of spot sizes as small as 12 microns, and yields scanning rates over 2000 dpi (33). Some manufacturers of graphic-arts scanners are Skantek (Warren, NJ), Optigraphics (San Diego, CA), Metagraphics (Woburn, MA), and SysScan (a joint venture of Messerschmitt-Bolkow-Blohm, Germany, and Kongsberg Vaapenfabrik, Norway) (5).

In the arena of color scanning, the Clearscan Color of NCL America (Sunnyvale, CA) is a portable scanner digitizing at 400 dpi with 256 colors that can be converted to 256 gray levels (31). The JX-600 Commercial Color Scanner of Sharp Electronics, Inc. (Mahwah, NJ) scans transparencies as large as 11 in by 16.5 in at 600 dpi, and also handles 35 mm film (34). The LS-3500 scanner of Nikon Inc. Electronic Imaging Products (Garden City, NY) produces a 6144 pel by 4096 pel digitization of 35 mm film (35).

Various scanners on the market today also provide a number of functions related to the scanning process (36). These may be embodied in firmware or in software within the scanning equipment. Included are such image-enhancement capabilities as blackening gray areas, adding hatchlines, completing lines, and modifying the gray scale or color shades. Another feature is the ability to recognize areas of text in the image, to convert the characters to ASCII code, and to convert the remaining portion into bit-map form. Additionally, many image-manipulation functions are being provided by software packages for personal computers. In some cases, a software package and a scanner interface board are now bundled with a scanner unit. Also, independent offerings of software packages that can operate on the output of scanners of several different manufacturers are being made. Some of the software operations include image enhancement, character recognition, page composition, conversion of multilevel image data to bilevel (halftone) form, and conversion of ASCII data to bit-map form.

Mathematical Models and Information Theory

Introduction

Significant autocorrelations are present in most images, leading to redundancy in the associated digital bit-map data. An efficient coding process capitalizes on this redundancy to develop an alternative representation with a smaller number of bits. The number of bits prior to coding are called *uncoded bits* and the number of bits after coding are called *coded bits*. In almost all cases, a coding method gives a number of coded bits that is less than the number of uncoded bits, and for this reason efficient coding is also referred to as *compression*. The ratio of uncoded bits to coded bits is called the *compression ratio* or *compression factor*; its inverse is called the *compression coefficient*. For bilevel images, the number of uncoded bits is equal to the number of pels.

A main objective of coding studies is to identify good coding methods. As indicated above, the goodness of a method is measured along several dimensions, one of which is compression performance. With respect to compression performance in particular, there are several ways of searching for good coding methods. In one approach, a new coding method is invented and then implemented or simulated; performance on a chosen ensemble of images is measured relative to other methods. Of course, the particular ensemble used is only a sample and the performance may differ for a wider and more representative image population. In another approach, mathematical models of image generation are formulated and coding methods based on the models are designed. The models used are primarily those defined in information theory and are instances of random or stochastic processes. Other theoretical approaches that do not rely on stochastic models also exist (37–39).

The uses and importance of modeling are discussed in the section below. Information-theory concepts covering the definition of image models and their major properties are also summarized in that section. The remaining two sections of this article describe typical coding methods that have emerged from the models, several important coding methods that are less amenable to modeling, and approaches that do not use stochastic models.

Importance of Image Modeling

The importance of image modeling centers on the existence for each model of a measure called entropy, the definition of which is discussed in the next section. Its importance to coding is that, for any given model, it serves as a lower bound on the average number of coded bits achievable by any method that embodies the same assumptions of the model. This knowledge can be used in several ways to aid the search for good coding methods (40,41).

One direct use is that the structure of the model often is suggestive of the construction of one or more coding methods that embody the same assumptions as the model. A related useful function is the comparison of different models based on their bounds. A model with a low bound would more likely engender methods with better compressions than models with higher bounds. Knowing this, design effort can be directed to formulating methods associated with the most promising model. Furthermore, the bound can become a benchmark in the design process. A large gap between the coded bits of a method and the bound of the model would point toward tweaking of the method or looking for another approach based on the same model.

Models can also be used to aid in the comparison of methods devised independently of the modeling process. To do this, a corresponding model based on the assumptions inherent in the method is constructed. This is sometimes difficult to do, but where it is feasible it is possible to compare methods by comparing the bounds of their respective models. This gives a more general result than the comparison of performance on a sample ensemble of images.

Generally, models vary in their bounds on coded bits. This variance occurs because the models account for more or less correlation present in images. In the range of models, one that accounts for less correlation (i.e., has a higher bound) may be preferred for reasons other than coding efficiency, (e.g., methods based on it may be amenable to low-cost implementation). However, it is important to realize that these methods still encode the images exactly.

Information-Theory Concepts

The purpose of this section is to present a framework for understanding the information-theory concepts relevant to image modeling and coding. Additional coverage of this material can be found in several other sources (42–52).

Definition of Source Models

The commonly used image models based on information theory are referred to as *source models*. A source model is a *random process* or a stochastic process, as defined in classical probability theory (53–57). One definition of a random process is that it is a collection of random variables, one for each instance of time, along with the joint-probability distributions of all orders of the random variables. In image modeling, a random process is usually a discrete-time process, in which the indicated instances of time are discrete values, denoted as t_i with $i = \ldots -2, -1, 0, 1, 2, \ldots$. Also, for image modeling, the random process is usually a discrete-amplitude process, in which the random variable values are finite in number. With these specifications, an experiment concerning the random process consists of a separate experiment for each of the distinct random variables, yielding a sequence of values, one corresponding to each random variable. Such a sequence of values is a realization or a sample function of the process. (For a continuous-time process, the sample function is a continuous waveform instead of a sequence of values.) The set of all possible sample functions is called an *ensemble*.

An equivalent way of describing a random process is to define the ensemble (i.e., the entire set) by sample functions, along with a probability measure on the set. The set of sample functions may be defined by an analytical expression or by statistical rules. The ensemble and the associated probability measure, in turn, determine the collection of random variables and their statistics at each instance of time. Figure 6 is a representation of a source model as a discrete-time, discrete-amplitude random process.

For simplicity, the random processes used in image modeling generally are assumed to be *ergodic*, by which the time average of the values of any sample function equals the expected value of any of the random variables almost everywhere. Ergodicity assumes stationarity, in that the joint-probability distributions of the random variables do not vary with time.

In information theory, a source model is viewed as a probabilistic mechanism that selects and outputs sequences of source symbols (or characters or letters) from a source alphabet. A source symbol corresponds to a simple event

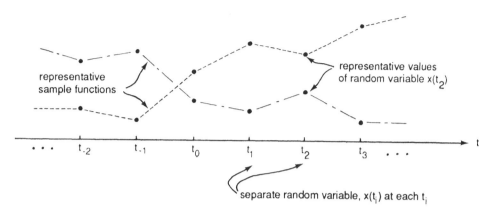

FIG. 6 Representation of a source model as a discrete-time, discrete-amplitude random process.

(i.e., a sample point) and maps into a random-variable value. A sequence of source symbols corresponds to a sample function of the random process. The selections are made based on probabilistic rules specified for each particular model. In one kind of model, each symbol in the sequence is selected independently of every other symbol; in other words, the random variables of the random process for this particular model are independent of each other. In this case, the source is said to be memoryless. To specify a particular memoryless model, the probability of each symbol in the alphabet is given and, since stationarity is assumed, each random variable of the random process has the same probability distribution. In another kind of model, the selection of successive symbols is not independent, and the source is said to have memory. To specify a particular model with memory, either joint- or conditional-probability distributions are given.

Entropy

An important concept of information theory, as indicated above, is that of the entropy of a source model. The material that follows introduces the concept of entropy for a single random variable. This is then extended to a collection of random variables for which a joint entropy and a conditional entropy are defined. Several relations between the different entropies are then presented. Finally, the entropy for a source model is given and an important theorem linking entropies and coding is discussed.

Entropy of a Random Variable.   Consider a single random variable X with a finite number of values J. Denote x_j as the jth value of the random variable and $P(x_j)$ as the probability of the random variable. Then a measure called the *self-information* of x_j is defined as

$$I(x_j) = -\log_2 P(x_j) \quad \text{(bits)}. \tag{1}$$

This mathematical definition of information shares some of the properties of the common usage of the word "information." For example, if the random variable can assume only one value, then its probability $p(x_1)$ is 1, and $I(x_1)$ is 0. This agrees with the common usage in that the same value is always expected and its occurrence gives no new information. Also, as the probability of a value decreases, $I(x_j)$ increases. Again, this is in line with common usage, for a less frequently occurring value is less expected and, when it does occur, it comes as a surprise and is indicative of more information than that imparted by a frequently occurring value.

The average value of self-information, taken over all the possible values of a random variable X is defined as the entropy $H(X)$ of the random variable, giving

$$H(X) = -\sum_{j=1}^{J} P(x_j) \log_2 P(X_j) \quad \text{(bits/symbol)}. \tag{2}$$

As a simple example, consider a random variable having three values with the probabilities $P(x_1) = 0.7$, $P(x_2) = 0.2$, $P(x_3) = 0.1$. The entropy is

$$H(X) = -0.7 \log_2 0.7 - 0.2 \log_2 0.2 - 0.1 \log_2 0.1$$

$$= 1.157 \text{ (bits/symbol)}.$$

(Entropies are shown rounded to three decimals.)

For another example, a random variable having two values with the probabilities $P(x_1) = 0.9$ and $P(x_2) = 0.1$ has an entropy of

$$H(X) = -0.9 \log_2 0.9 - 0.1 \log_2 0.1$$

$$= 0.469 \text{ (bits/symbol)}.$$

Entropy is thus defined for a sample space consisting of a finite number of mutually exclusive events with probabililties summing to 1. This is utilized in determining the entropies involving more than one random variable, as discussed below.

Joint Entropy.   Consider first a joint-probability system with two random variables X and Y with corresponding sample spaces S_X and S_Y. Denote the events defined on S_X by $A_1, A_2, \ldots, A_J$, and the events defined on S_Y by $B_1, B_2, \ldots, B_K$. An event of the joint or product sample space consists of the combination of one event defined on S_X and one event defined on S_Y. There are a total of JK events in the joint sample space, which are shown in the matrix that follows.

$$
\begin{array}{cccc}
A_1\ B_1 & A_1\ B_2 & \cdots & A_1\ B_K \\
A_2\ B_1 & A_2\ B_2 & \cdots & A_2\ B_K \\
\cdot & \cdot & & \cdot \\
\cdot & \cdot & & \cdot \\
\cdot & \cdot & & \cdot \\
A_J\ B_1 & A_J\ B_2 & \cdots & A_J\ B_K
\end{array}
$$

Since each event is a different combination, they are mutually exclusive. A joint event $A_j\ B_k$ has an associated probability $P(A_j, B_k)$ and, since all the joint events in the matrix cover an entire sample space, their probabilities sum to 1. Note that the joint probability $P(A_j, B_k)$ is equal to the product $P(A_j)\ P(B_k)$ only if the random variables X and Y are independent. In defining a model, the joint probabilities are estimated from data chosen as typical for a given application. As indicated above, since the joint sample space consists of a finite number of mutually exclusive events with probabilities summing to 1, an entropy may be defined. This is called the *joint entropy*, and takes the following form:

$$
H(X,Y) = -\sum_{j=1}^{J} \sum_{k=1}^{K} P(x_j, y_k)\ \log_2\ P(x_j, y_k) \tag{3}
$$

As an example, consider the random variable X with values x_1 and x_2 and the random variable Y with values y_1, and y_2. Assume the joint probabilities shown below in matrix form.

$$
\begin{array}{ll}
P(x_1, y_1) = 0.855 & P(x_1, y_2) = 0.045 \\
P(x_2, y_1) = 0.045 & P(x_2, y_2) = 0.055
\end{array}
$$

The joint entropy is

$$
H(X,Y) = -0.855\ \log_2\ 0.855 - 2\ (0.045\ \log_2\ 0.045) - 0.055\ \log_2\ 0.055
$$
$$
= 0.826\ (\text{bits/2-symbol pair}).
$$

The first-order probabilities of each variable X and Y can be derived from the joint probabilities by summing separately on each value for each of the variables as follows:

$$
P(x_n) = \sum_{k=1}^{K} P(x_n, y_k), \qquad P(y_n) = \sum_{j=1}^{J} P(x_j, y_n) \tag{4}
$$

These are also called the *marginal* probabilities because they are the row and column sums, which are often shown in the margins of the joint probability matrix, as shown below for the example above.

$$P(x_1,y_1) = 0.855 \quad P(x_1,y_2) = 0.045 \quad \bigg| \quad P(x_1) = 0.9$$
$$P(x_2,y_1) = 0.045 \quad P(x_2,y_2) = 0.055 \quad \bigg| \quad P(x_2) = 0.1$$

$$P(y_1) = 0.9 \qquad P(y_2) = 0.1$$

In general, the first-order probability distributions of X and Y need not be identical. However, if X and Y are random variables from a stationary random process, their first-order distributions are the same (as shown in the example above).

As shown above, the entropy for the first-order probability distribution with probabilities of 0.9 and 0.1 is 0.469 bits per value (or symbol) of the first-order distribution. The joint-probability distribution assumed above has an entropy of 0.826 bits per value but, since this value comprises two symbols of the first-order distribution, the entropy of 0.826 must be divided by 2 to get an effective per-symbol joint entropy. Doing this, we obtain 0.413, which is less than 0.469, the entropy of the first-order distribution, an expected result since dependencies are indicated in the second-order statistics. However, if the two random variables are independent, then the per-symbol joint entropy is exactly equal to the first-order entropy.

The development above for two random variables may be extended to give an Nth-order joint entropy for N random variables $X_1, X_2, \ldots, X_N$ as

$$H(X_1, X_2, \ldots, X_N)$$

$$= - \sum_{x_1, x_2, \ldots, x_N} P(x_1, x_2, \ldots, x_N) \log_2 (x_1, x_2, \ldots, x_N) \quad (5)$$

In this expression, the summation is made over all combinations of values of the random variables $X_1, X_2, \ldots, X_N$. Also, if all of the random variables are independent, the Nth-order joint entropy divided by N is exactly equal to the first-order entropy.

Conditional Entropy.   Consider again two random variables X and Y with corresponding sample spaces S_X and S_Y and events $A_1, A_2, \ldots, A_J$ defined on S_X, and events $B_1, B_2, \ldots, B_K$ defined on S_Y. We next consider the set of events that may occur under the assumption that a particular event, say B_1, has occurred. The set, called $(A \mid B_1)$, includes a separate event for each value of A_j, as follows:

$$\{A \mid B_1\} = \{A_1 \mid B_1, A_2 \mid B_1, \ldots, A_J \mid B_1\} \quad (6)$$

For each of these events, a conditional probability $P(X = A_j \mid Y = B_1)$, or simply $P(A_j, B_1)$, is defined by

$$P(A_j \mid B_1) = \frac{P(A_j, B_1)}{P(B_1)}. \quad (7)$$

Accordingly, the probabilities of the set $\{A \mid B_1\}$ are

$$P(A_1 \mid B_1), \quad P(A_2 \mid B_1), \quad \ldots, \quad P(A_J \mid B_1)$$

or,

$$\frac{P(A_1,B_1)}{P(B_1)}, \quad \frac{P(A_2,B_1)}{P(B_1)}, \quad \ldots, \quad \frac{P(A_J,B_1)}{P(B_1)}.$$

Using the identity for marginal probabilities, we obtain the sum of these probabilities:

$$\frac{P(A_1,B_1)}{P(B_1)} + \frac{P(A_2,B_1)}{P(B_1)} + \ldots + \frac{P(A_J,B_1)}{P(B_1)} = \frac{P(B_1)}{P(B_1)} = 1 \tag{8}$$

The set $\{A \mid B_1\}$ is thus a sample space consisting of a finite number of mutually exclusive events with probabilities summing to 1. Therefore, as indicated above, an entropy may be defined and takes the form

$$H(X \mid y_1) = -\sum_{j=1}^{J} P(x_j \mid y_1) \log_2 P(x_j \mid y_1). \tag{9}$$

A similar sample space and associated entropy may be defined for each value of Y. In general, the entropy of the sample space for a value y_k is

$$H(X \mid y_k) = -\sum_{j=1}^{J} P(x_j \mid y_k) \log_2 P(x_j \mid y_k). \tag{10}$$

The average of all the entropies $H(X \mid y_k)$, with $k = 1, 2, \ldots, K$, is formed as shown below. This is called the *average conditional entropy* or, simply, *conditional entropy* of the entire probability system defined by the events $A \mid B$.

$$H(X \mid Y) = \sum_{k=1}^{K} P(y_k) H(X \mid y_k) \tag{11}$$

$$= -\sum_{k=1}^{K} \sum_{j=1}^{J} P(y_k) P(x_j \mid y_k) \log_2 P(x_j \mid y_k)$$

or, equivalently,

$$H(X \mid Y) = -\sum_{k=1}^{K} \sum_{j=1}^{J} P(x_J,y_k) \log_2 P(x_J \mid y_k). \tag{12}$$

In a similar manner, $H(Y \mid X)$ is obtained as

$$H(Y \mid X) = -\sum_{j=1}^{J} \sum_{k=1}^{K} P(x_j, y_k) \log_2 P(y_k \mid x_j). \qquad (13)$$

As an example, consider again the random variable X, with values x_1 and x_2, and the random variable Y, with values y_1 and y_2. The entropy $H(X \mid Y)$ is given by

$$
\begin{aligned}
H(X \mid Y) = \ & -P(y_1) \ P(x_1 \mid y_1) \log_2 P(x_1 \mid y_1) \\
& -P(y_1) \ P(x_2 \mid y_1) \log_2 P(x_2 \mid y_1) \qquad (14) \\
& -P(y_2) \ P(x_1 \mid y_2) \log_2 P(x_1 \mid y_2) \\
& -P(y_2) \ P(x_2 \mid y_2) \log_2 P(x_2 \mid y_2).
\end{aligned}
$$

Assume the following values (estimated from real data) are given: $P(x_1 \mid y_1) = 0.95$, $P(x_1 \mid y_2) = 0.45$, $P(y_1) = 0.9$. The remaining components in the entropy equation are obtained from the given values as follows:

$$
\begin{aligned}
P(x_2 \mid y_1) &= 1 - P(x_1 \mid y_1) = 0.05 \\
P(x_2 \mid y_2) &= 1 - P(x_1 \mid y_2) = 0.55 \\
P(y_2) &= 1 - P(y_1) \quad\ \ = 0.10
\end{aligned}
$$

The entropy is

$$
\begin{aligned}
H(X \mid Y) = \ & -(0.9)(0.95) \log_2 (0.95) - (0.9)(0.05) \log_2 (0.05) \\
& -(0.1)(0.45) \log_2 (0.45) - (0.1)(0.55) \log_2 (0.55) \\
= \ & -0.357 \ \text{(bits/symbol)}.
\end{aligned}
$$

Alternatively, if the given values are of three conditional entropies, such that the conditioning values are all different, then $P(y_1)$ need not be explicitly given since it can be derived from the given values. For example, assume the given values are $P(x_1 \mid y_1) = 0.95$, $P(x_1 \mid y_2) = 0.45$, $P(y_2 \mid x_1) = 0.05$. To derive $P(y_1)$, form the identities

$$
\begin{aligned}
P(y_1 \mid x_1) \ P(x_1) &= P(x_1 \mid y_1) \ P(y_1) \qquad (15) \\
P(y_2 \mid x_1) \ P(x_1) &= P(x_1 \mid y_2) \ P(y_2).
\end{aligned}
$$

Upon substituting $1 - P(y_1)$ for $P(y_2)$, the result is two linear equations in the two unknowns $P(x_1)$ and $P(y_1)$. Solving these equations for $P(x_1)$ yields

$$P(x_1) = \frac{1}{[P(y_1 \mid x_1)/P(x_1 \mid y_1)] + [P(y_2 \mid x_1)/P(x_1 \mid y_2)]}. \qquad (16)$$

Substituting the given values and $1 - P(y_2 \mid x_1)$ for $P(y_1 \mid x_1)$ results in a value of 0.9 for $P(y_1)$.

In a simpler case, where it is known that X and Y have the same probability distribution, that is, $P(x_1) = P(y_1)$ and $P(x_2) = P(y_2)$, then $P(x_1)$ may be obtained by solving Eqs. 17 and 18. Equation 17 may be interpreted as a state equation, where the probability of being in state x_1 is equal to the probability of being in state x_1 and remaining there, plus the probability of being in state x_2 and transitioning to state x_1.

$$P(x_1) = P(x_1)\,P(x_1 \mid x_1) + P(x_2)\,P(x_1 \mid x_2) \tag{17}$$

$$P(x_1) + P(x_2) = 1 \tag{18}$$

The solution is

$$P(x_1) = \frac{1}{1 + [P(x_2 \mid x_1)/P(x_1 \mid x_2)]} = \frac{P(x_1 \mid x_2)}{P(x_1 \mid x_2) + P(x_2 \mid x_1)}. \tag{19}$$

The definition of conditional entropy for two random variables extends to N random variables $X_1, X_2, \ldots, X_N$, giving an Nth-order conditional entropy

$$H(X_N \mid X_1, X_2, \ldots, X_{N-1})$$

$$= -\sum_{x_1, x_2 \ldots, x_N} P(x_1, x_2, \ldots, x_N) \log_2 P(x_N \mid x_1, x_2, \ldots, x_{N-1}). \tag{20}$$

In the equation above, $P(x_N \mid x_1, x_2, \ldots, x_{N-1})$ is the conditional probability of x_N, given $x_1, x_2, \ldots, x_{N-1}$, and the summation is made over all combinations of values of the random variables $X_1, X_2, \ldots, X_N$.

Relations between Entropies. The relationship between the joint, marginal (or first-order), and conditional entropies is given by a number of identities and inequalities. Several of these are described in this section. These are important because they define a hierarchy of entropies, thus identifying models with the lowest entropies that can lead to coding methods with the best compressions.

First, the following inequalities are derived directly from the entropy definitions as originally presented by Shannon (42).

$$H(X) \geq H(X \mid Y) \tag{21}$$

$$H(Y) \geq H(Y \mid X) \tag{22}$$

The equality condition holds if, and only if, X and Y are independent random variables. Assuming inequality, these relations indicate that a coding method based on a conditional model can achieve better average compression than a coding method based simply on first-order statistics.

Next are two identities relating the three kinds of entropies, again for the two random variables X and Y (51):

$$H(X, Y) = H(X \mid Y) + H(Y) \tag{23}$$

$$H(X,Y) = H(Y \mid X) + H(X) \tag{24}$$

The first identity is derived using the definition of conditional probability.

$$P(x_j,y_k) = P(x_j \mid y_k) P(y_k) \tag{25}$$

$$\log_2 P(x_j \mid y_k) = \log_2 P(x_j \mid y_k) + \log_2 P(y_k)$$

Substitution of the last identity in the definition of $H(X \mid Y)$ produces

$$H(X \mid Y) = -\sum_{j=1}^{J} \sum_{k=1}^{K} P(x_j,y_k) \log_2 P(x_j,y_k)$$

$$= -\sum_{j=1}^{J} \sum_{k=1}^{K} P(x_j,y_k) [\log_2 P(x_j \mid y_k) + \log_2 P(y_k)]$$

$$= -\sum_{j=1}^{J} \sum_{k=1}^{K} P(x_j,y_k) \log_2 P(x_j,y_k)$$

$$-\sum_{j=1}^{J} \sum_{k=1}^{K} P(x_j,y_k) \log_2 P(y_k)$$

$$= H(X \mid Y) + H(Y). \tag{26}$$

In justification of this last step, the first double-summation term in the prior identity is $H(X \mid Y)$ by definition. The second double-summation term can be shown to be equal to $H(Y)$ as follows. First, the term is written as K single sums:

$$-\sum_{j=1}^{J} \sum_{k=1}^{K} P(x_j,y_k) \log_2 P(y_k) = -\log_2 P(y_1) \sum_{j=1}^{J} P(x_j,y_1)$$

$$-\log_2 P(y_2) \sum_{j=1}^{J} P(x_j,y_2)$$

$$- \tag{27}$$

$$\cdot$$

$$\cdot$$

$$\cdot$$

$$-\log_2 P(y_k) \sum_{j=1}^{J} P(x_j,y_k)$$

Using the identity

$$P(y_n) = \sum_{j=1}^{J} P(x_j,y_n),$$

this becomes

$$-\sum_{j=1}^{J} \sum_{k=1}^{K} P(x_j,y_k) \log_2 P(y_k) = -P(y_1) \log_2 P(y_1)$$

$$-P(y_2) \log_2 P(y_k)$$

$$-$$

$$.$$

$$.$$ (28)

$$.$$

$$-P(y_k) \log_2 P(y_k)$$

$$= -\sum_{k=1}^{K} P(y_k) \log_2 P(y_k)$$

$$= H(Y).$$

As an example, assume that the set of consistent statistics shown below is given.

$$P(x_1) = 0.9 \qquad P(x_2) = 0.1$$

$$P(y_1) = 0.9 \qquad P(y_2) = 0.1$$

$$P(x_1,y_1) = 0.855 \qquad P(x_1 \mid y_1) = 0.95$$

$$P(x_1,y_2) = 0.045 \qquad P(x_1 \mid y_2) = 0.45$$

$$P(x_2,y_1) = 0.045 \qquad P(x_2 \mid y_1) = 0.05$$

$$P(x_2,y_2) = 0.055 \qquad P(x_2 \mid y_2) = 0.55$$

These data are consistent in that they obey all of the defining relations

$$\sum_{j=1}^{J} P(x_j) = 1, \qquad \sum_{k=1}^{K} P(y_k) = 1, \qquad \sum_{j=1}^{J} P(x_j \mid y_n) = 1$$

$$\sum_{j=1}^{J} P(x_j,y_n) = P(y_n), \qquad P(x_j,y_k) = P(x_j \mid y_k) P(y_k).$$ (29)

The various entropies based on these data, as computed in the examples above, are $H(X,Y) = 0.826$, $H(X \mid Y) = 0.357$, $H(Y) = 0.469$. From these entropies, it is seen that the stated relation holds:

$$H(X,Y) = H(X \mid Y) + H(Y)$$ (30)

For coding purposes, it is advantageous to know how the conditional entropy compares with the per-symbol joint entropy. For the example given above, the conditional entropy $H(X \mid Y)$, with a value of 0.357, is less than the per-symbol joint entropy $H(X \mid Y)/2$, with a value of 0.413. More generally, the conditional entropy is always less than or equal to the per-symbol joint entropy whenever the random variables have the same probability statistics, as they do for source models that assume stationarity. For the two-variable case, this is shown by using the equality of $H(X)$ and $H(Y)$, along with the Shannon inequality $H(X) \geq H(X \mid Y)$ noted above, as follows:

$$H(X,Y) = H(X \mid Y) + H(Y)$$
$$= H(X \mid Y) + H(X)$$
$$\geq H(X \mid Y) + H(X \mid Y)$$
$$\geq 2 H(X \mid Y)$$

or,

$$H(X,Y)/2 \geq H(X \mid Y) \tag{31}$$

A similar result holds for any number of variables, assuming the process is stationary. The derivation is based on the chain rule of probability:

$$P(x_1, x_2, \ldots, x_N)$$
$$= P(x_1) P(X_2 \mid x_1) P(x_3 \mid x_1, x_2) \ldots P(x_N \mid x_1, x_2, \ldots, x_{N-1}) \tag{32}$$

Applying this identity to the corresponding definition of information and taking the average yields the following:

$$H(X_1, X_2, \ldots, X_N) = H(X_1) + H(X_2 \mid X_1) + H(X_3 \mid X_1, X_2)$$
$$+ \ldots + H(X_N \mid X_1, X_2, \ldots, X_{N-1}) \tag{33}$$

The following inequalities result under the assumption of stationarity:

$$H(X_1) \geq H(X_2 \mid X_1) \geq H(X_3 \mid X_1, X_2) \geq \ldots$$
$$\geq H(X_N \mid X_1, X_2, \ldots, X_{N-1}) \tag{34}$$

Applying these inequalities to the preceding identity gives

$$H(X_1, X_2, \ldots, X_N) \geq N H(X_N \mid X_1, X_2, \ldots, X_{N-1}) \tag{35}$$

or,

$$H(X_1, X_2, \ldots, X_N) / N \geq H(X_N \mid X_1, X_2, \ldots, X_{N-1}). \tag{36}$$

An important implication is that, under the assumption of stationarity, coding methods based on a conditional model can yield better compressions than those based on a joint model.

Entropy of a Source.   The entropy definitions given above assume a finite number of random variables. However, a source is defined for an infinite sequence of symbols, implying an infinite number of random variables. An entropy $H(S)$ for a stationary source may be defined in two ways. In one of these, the source entropy is the limiting value of the per-symbol joint entropy as the number of random variables becomes infinite:

$$H(S) = \lim_{N \to \infty} \frac{1}{N} H(X_1, X_2, \ldots, X_N) \tag{37}$$

In the other definition, the source entropy is the limiting value of the conditional entropy, again as the number of random variables becomes infinite:

$$H(S) = \lim_{N \to \infty} H(X_N \mid X_1, X_2, \ldots, X_{N-1}) \tag{38}$$

Both of the limits above have been proven to exist and to be equal to each other (45).

The sequence of per-symbol joint entropies and of conditional entropies, considered as N increases, are monotonically nonincreasing and, thus, each member of these sequences is at least as small as the preceding member of its respective sequence and at least as large as the source entropy. In general, as N increases, the amount of the source dependencies that are accounted for by the per-symbol joint entropy and by the conditional entropy also increases. In the limit, the source entropy accounts for all possible source dependencies.

In one kind of image modeling, $X_1, X_2, \ldots, X_N$ is associated with a block of pels that becomes increasingly larger in size in the limiting process. The assumption under a source model is that the image is of infinite size. Frequently, however, the image model is defined with the assumption that the memory is limited and that dependencies over widely spaced image areas do not exist. In such a case, the value of $H(S)$ is assumed for a finite N. The assumption of limited memory may be invalid, but generally leads to simpler analyses and implementations. In the extreme case of a memoryless source that is also stationary so that all the random variables have the same probability distributions, the source entropy is equal to the entropy of a representative random variable.

Noiseless Coding Theorem.   Coding of a source sequence consists of substituting a sequence of binary digits for the source-symbol sequence. Assuming that the coding is exact or *noiseless*, then decoding the coded bits reconstructs the original source sequence exactly. The relative performance of various exact coding methods may be obtained by comparing their average bits per source symbol. It would also be very useful to have a measure of absolute performance. This in fact is given by Shannon's noiseless coding theorem, which states the following about exact coding of a discrete ergodic source:

(a)   The source entropy is a lower limit on the average bits per source symbol of any exact encoding.

(b) Exact encodings exist with average bits per source symbol arbitrarily close
to the source entropy.

Coding methods that asymptotically achieve this goal are discussed in the
next section.

Entropy Coding

Entropy coding refers to methods that, in the limit of some parameter, achieve
the goal of having the average bits per source symbol equal to the source entropy. Two of these methods, Huffman coding and arithmetic coding, are discussed in this section. Both of these methods apply directly to a stationary
memoryless source, for which the source entropy reduces simply to the entropy
of a representative variable. In this case, the coding problem becomes one of
encoding a sequence of source symbols drawn from an alphabet with given
probabilities. The application of these methods for entropy coding of sources
with memory is more complex than entropy coding for memoryless sources,
and some examples are given in the section on exact coding under methods for
encoding of images. Another entropy coding approach is permutation coding,
which is described by Berger (58).

Variable-Length Coding.   Given a source-symbol alphabet and a probability
distribution for the alphabet, one way of encoding sequences of symbols is to
use variable-length codes. Compression is achieved by assigning short codewords to symbols with high probabilities and long codewords to symbols with
low probabilities. There are many variable-length codes that may be constructed. These have varying properties and are discussed below.
    A necessary property for general coding purposes is that the codewords be
distinct and uniquely decodable. Codewords are distinct if only one source
symbol maps into any given codeword. Codewords are uniquely decodable if
any code bit stream can be decoded into only one unique source-symbol stream.
Shown below is an example in which the codewords are not uniquely decodable.

Source Symbol	Codeword
A	1
B	01
C	10

These codewords are not uniquely decodable, as exemplified by the bit stream
1101, which can be generated by either the source-symbol stream AAB or the
stream ACA.
    For ease of decoding, a desirable property is that a code be instantaneous.
An *instantaneous code* is a uniquely decodable code in which each codeword in
a coded bit stream can be decoded without referring to succeeding bits in the
stream. An example of a uniquely decodable code that is not instantaneous is
shown here.

Source Symbol	Codeword
A	1
B	10

This code is uniquely decodable because, upon getting a 1, if it is followed by a 1, it maps into the symbol A; if it is followed by a 0, it maps into the symbol B. It is not instantaneous because, upon getting a 1, reference must be made to the next bit to determine if it maps into A or B. It has been shown that a code is instantaneous if no codeword comprises the initial part, or prefix, of any other codeword. This is called the *prefix-free* property.

A *compact* or *most-efficient* code is one that yields the least average number of coded bits per source symbol. If the source symbols are equally probable, a most-efficient code has constant-length codewords. If the source symbols are not equally probable, a most-efficient code is provided by a Huffman coding (59), which is also uniquely decodable and instantaneous.

Huffman Coding.   Given a symbol alphabet and a probability distribution on the alphabet, a Huffman code may be constructed in two phases, as indicated below. Alternative construction methods also exist.

Encoding Phase I.

1. List the probabilities in descending order.
2. Add the two smallest probabilities at the bottom of the list and form a new list with the two smallest items deleted and with their sum inserted in the list, maintaining the descending order. Mark the summed item (e.g., with an * if done manually, as in the example below). Tied items may be listed in any order.
3. Continue Steps 1 and 2, forming successively smaller lists, until a list of only two items remains.

The example below shows the process for an alphabet of four symbols.

List 1 (Symbol Probabilities)	List 2	List 3
0.50	0.50	0.50
0.20	0.30*	0.50*
0.15	0.20	
0.15		

Encoding Phase II.   In this phase, the process obtains the set of codewords, which is called a codebook. The codewords may be obtained directly or the codeword sizes may be obtained and the codewords derived from the sizes. For one set of codeword sizes, there may be several realizations of the codewords, some of which are more structured or orderly than the others. Deriving the codewords from the sizes permits the construction of an orderly codebook. If it is desired to store or transmit the codebook information, an encoding of the codeword sizes comprises a considerably more compact form of this information than the codewords themselves. The steps to obtain the codeword sizes follow.

1. Starting with the smallest list, which is the first current list, mark a codeword size of 1 for each of the two items.
2. Take the codeword size of the summed item (marked with an *) from the current list, add 1 to it, and use this for the codeword size of each of the two items at the bottom of the next larger list. For the other items, simply copy the codeword sizes from the current list to the next larger list.
3. Rename the larger list formed in Step 2 as the current list and repeat Steps 2 and 3 until the initial list has codeword sizes marked.

For the example given above, Phase II yields

List 1	List 2	List 3
1	1	1
2	2*	1*
3	2	
3		

The codewords are derived from the codeword size List 1 as follows:

1. Starting with the smallest codeword size, mark down a number of 0s equal to the codeword size.
2. If the next codeword size on the list is the same as the one before it, then simply add a 1 to the codeword using binary addition.
3. If the next codeword size is different from the one before it, then first add a 1 using binary addition and then attach as many 0s to the right as the difference between the two codeword sizes.
4. Repeat Steps 2 and 3 for the remaining codeword sizes. The final codeword should always be all 1s.

For the example above, the codewords using this method are 0, 10, 110, 111.

Decoding can be done by a tree-tracing technique, whereby one bit at a time is processed, each of which causes a path to one of two branches of the tree to be taken until an end node is reached, signifying that a codeword has been decoded and giving its value. This can be cumbersome in that the maximum number of branches that must be traced is equal to the largest codeword size. Also, the decoding time varies for different size codewords. Another method that is faster and has more uniform decoding time per codeword is reported in Ref. 60.

Numerous improvements and variations on Huffman coding have been developed. An important technique is the dynamic capture of the symbol probabilities, or even the symbol set itself, as the symbol stream is encoded (61–66). Another valuable technique is the construction of optimal Huffman-like codes under the restriction of a maximum codeword length (67). Implementations using very-large-scale integration (VLSI) have also been considered (68). For information about other work on Huffman coding, see Refs. 69 and 70.

Huffman Coding with Extensions. Consider the example given in the section on entropy above of an alphabet of three symbols with probabilities 0.7, 0.2, and 0.1. A Huffman encoding gives an average bits per symbol of 1.3, as indicated below.

Source Symbol	Source-Symbol Probability	Code Length (In Bits)	Probability × Code Length
A	0.7	1	0.7
B	0.2	2	0.4
C	0.1	2	0.2

Average bits-per-symbol = 1.3

In comparison, the entropy for this source is 1.157. In general, equality is reached only if each probability is a negative integral power of 2. When that is not the case, it is necessary to encode extensions of the source in order to achieve an average bits per symbol that approaches the entropy. An Nth-order extension is the set of all groups of $N + 1$ source outputs. The average bits-per-extension symbol is divided by $N + 1$ to obtain the average bits per symbol (with no extension). For the example above, the first-order extension decreases the average bits per symbol to 1.165, as shown below. Note that the pairing of the source outputs does not imply dependency and should not be confused with the joint sample spaces described above.

1st-order Extension	Probability		Code Length (In Bits)	Probability × Code Length
AA	(0.7) (0.7)	= 0.49	1	0.49
AB	(0.7) (0.2)	= 0.14	3	0.42
BA	(0.2) (0.7)	= 0.14	3	0.42
AC	(0.7) (0.1)	= 0.07	4	0.28
CA	(0.1) (0.7)	= 0.07	4	0.28
BB	(0.2) (0.2)	= 0.04	4	0.16
BC	(0.2) (0.1)	= 0.02	5	0.10
CB	(0.1) (0.2)	= 0.02	6	0.12
CC	(0.1) (0.1)	= 0.01	6	0.06

Average bits-per-extension symbol = 2.33
Average bits per symbol = 1.165

Huffman coding of a binary source without extensions yields codewords of 0 and 1, resulting in no compression. If the two symbols are not equally probable, the entropy is less than 1. In this case, extensions are always required to achieve an average bits per symbol of less than 1. For example, a binary source with symbol probabilities of 0.9 and 0.1 has an entropy of 0.469. The average bits per symbol for first-order, second-order, and third-order extensions are 0.645, 0.533, and 0.493, respectively.

Huffman Codebooks. For an Nth-order extension of a K symbol source, the number of Huffman codebook values is K^{N+1}. Even with no extensions, Huffman codebooks can be quite large, resulting in unwieldy encoder implementations.

For image coding, another codebook consideration relates to storage of images. There is the possibility of reducing storage requirements by using individual image codebooks instead of, or in addition to, ensemble codebooks (2,16). For some images, the sum of the bits for the individual codebook and the encoded image using that codebook is less than the encoded image bits using an ensemble codebook. Storage of the individual codebook and the associated encoded image bits for these cases results in smaller total required storage. An efficient storage of the codebook information consists of an encoding of the codeword lengths. This may be a run-length encoding of the codeword lengths, assuming they are ordered as in the Huffman coding process described above.

Another way to reduce large codebooks is by using a method called *Huffman modulo coding* (16). This method starts with the ensemble statistics, consisting of the frequencies of different values. Each such value is divided by a base B, obtaining a multiple of the base and a non-negative remainder less than the base. Next, separate frequency counts of the multiples and of the remainders are formed. A probability distribution for the multiples is estimated based on the frequency counts of the multiples, and a separate probability distribution for the remainders is estimated based on the frequency counts of the remainders. Huffman codes are derived separately for the multiple distribution and for the remainder distribution. If the original codebook has C different values, then the multiple codebook has $\text{int}(C/B) + 1$ values, and the remainder codebook has B values. Their sum is $\text{int}(C/B) + 1 + B$, with a minimum at $B = C^{1/2}$. This gives a sum of $2\,C^{1/2} + 1$ for the number of multiple and remainder codebook values. Under the assumption that the bits to store a codebook are proportional to the number of values in it, the ratio of the codebook bits for the Huffman modulo code to the codebook bits for the Huffman code becomes $2/C^{1/2}$ for large C. For example, for C equal to 625, the codebook for Huffman modulo coding requires only 8% of the bits needed for the codebook for Huffman coding.

Alternative approaches that do not require codebooks encode each source value by forming a function of the source value. The functions generally are simple, but the codes are non-instantaneous and suboptimal, although they may be nearly optimal for particular distributions (71). Included are a code proposed by Elias (72), codes well suited to geometric distributions (73), and codes pertinent to negative-power distributions (74).

Arithmetic Coding.   As described above, in Huffman coding, or in Huffman coding using extensions, the source-symbol stream is partitioned into separate, equally sized blocks of source symbols and a codeword is substituted for each block. The entropy is approached by taking increasingly higher source extensions. In arithmetic coding, the entire source-symbol stream is not partitioned into blocks. Instead, the source symbols are processed one at a time and, for each source symbol, none, one, or several coded bits may be produced, depending on the value of the source symbol and the value of all the source symbols that preceded it in the source stream. The entropy is approached with increasing source-symbol-stream size, and source extensions do not apply.

The basic idea of arithmetic coding is summarized below, followed by an example showing an earlier process of how the coded bits are developed on a dynamic basis. It is assumed that the source symbols are binary, although the method extends to multilevel symbols as well.

Assume that the probability of the two values of a binary source symbol are given, and that a sequence of N source symbols is to be encoded. For a given N, there are 2^N different sequences. Each of these sequences has a defined probability of occurrence equal to the product of the probabilities of the N individual values in the sequence. The totality of the probabilities of all the sequences of size N is 1. The totality of these probabilities may be represented by the real-line segment from 0 to 1. Consider a division of this line segment into 2^N nonoverlapping, half-open subintervals $[j,k)$, where each subinterval corresponds to the probability of one of the 2^N source-symbol sequences. This achieves a mapping of each source-symbol sequence of size N onto a distinct half-open subinterval of the real line between 0 and 1. There is a family of such mappings, one for each value of N. In the limit, as N becomes infinite, each subinterval reduces to a point on the real line from 0 to 1.

The order of the subintervals corresponding to the symbol sequences on the real-line segment is not important, except that it must be the same for both the encoder and the decoder. A particular order is usually assumed; Fig. 7 shows examples of the mappings of the source-symbol sequences for values of N equal to 1, 2, and 3. In this example, the binary source-symbol values are 0 and 1, with the probabilities of 0.8 and 0.2, respectively. Figure 7-a illustrates the mapping for a sequence consisting of a single symbol, $N = 1$. The subinterval for the symbol 0 is $[0,0.8)$, corresponding to $P(0) = 0.8$, and the subinterval for the symbol 1 is $[0.8,1)$, corresponding to $P(1) = 0.2$. To obtain the mapping for $N = 2$ (Fig. 7-b), each subinterval shown in Fig. 7-a is divided into two parts in the proportion of the symbol probabilities 0.8 and 0.2. Thus, the segment $[0,0.8)$ divides into $[0,0.64)$ and $[0.64,0.8)$. The first segment, with width 0.64, corresponds to the source-symbol sequence 00, which has a probability of $(0.8)(0.8)$, or 0.64. The second segment, with width 0.16, corresponds to the

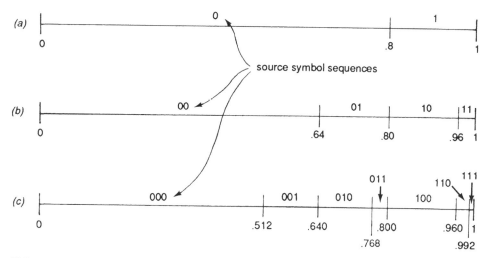

FIG. 7 Mapping source-symbol sequences into subintervals of the real line between 0 and 1: $a, N = 1; b, N = 2; c, N = 3$.

source-symbol sequence 01, which has a probability of (0.8)(0.2), or 0.16. The remaining subintervals shown in Figs. 7-*b* and 7-*c* are obtained in a similar manner.

A finite source-symbol sequence may be represented by any of the real numbers in its associated half-open subinterval. (Of course, an infinite source-symbol sequence is represented by only one real number in the line segment from 0 to 1.) An encoding then may consist of replacing a source-symbol sequence with a real number in its associated subinterval. Upon decoding, each real number in the line segment from 0 to 1 points to a unique source-symbol sequence.

In practice, an equivalent binary encoding is used. In Fig. 7, the endpoints of the subintervals are shown as decimal fractions. To arrive at the coded bits for a particular source-symbol sequence, we compare the endpoints of the associated subinterval, which are expressed as binary fractions. (For completeness, the value 1 may be taken to be .111. . . .) The leading bits that are the same for both endpoints are the same for all real numbers in the subinterval and are unique to that subinterval. They are taken to be the coded bits for the indicated source-symbol sequence. In the decoding process, these coded bits identify a unique subinterval, which in turn maps back into the original source-symbol sequence.

Figure 8 shows an example of arithmetic coding for a binary source-symbol stream of 000010, with $P(0) = 0.8$ and $P(1) = 0.2$. The encoder starts by delineating the two subintervals shown in Fig. 8-*a*. The first source symbol of 0 causes the encoder to choose the subinterval [0,0.8), specified in decimal, or [.000,.110) in binary. Three bits of accuracy suffice for this example. Since the endpoints do not have any leading digits in common, no coded bits are generated at this point. Whatever subsequent source symbols are received, their corresponding subintervals will all lie within this first subinterval [0,0.8). Therefore, the second subinterval [0.8,1) is discarded, and the encoder divides the subinterval [0,0.8) into two pieces in the proportions 0.8 and 0.2, yielding the subintervals [0,0.64) and [0.64,0.8) (Fig. 8-*b*). The second source symbol 0 causes the encoder to choose the subinterval [0,0.64), which corresponds to the source stream 00 so far. The subinterval [0,0.64) is [.000,.101) in binary. Again, the endpoints do not have any leading digits in common, so no coded bits are generated. This condition occurs also for the third source symbol 0 (Fig. 8-*c*). However, as shown in Fig. 8-*d*, the fourth source symbol 0 causes the encoder to chose the subinterval [0,0.4096). In binary, this is [.000,.011), with a single leading digit of 0 that is common to both endpoints. At this point, the encoder outputs a corresponding 0 as the first coded bit. The fifth source code symbol 1 causes the encoder to choose the subinterval [0.32768,0.4096) (Fig. 8-*e*). In binary, this is [.010,.011), with the two leading digits 01 in common to both endpoints. The leading 0 has already been put out, as indicated by the enclosing parentheses. The encoder outputs the 1 as the second coded bit. The chosen subinterval [0.32768,0.4096) is further subdivided at Fig. 8-*f*. The sixth source symbol does not provide any coded bits.

Some insight into the compression dynamics can be obtained by noting that it takes a string of four 0s to drive the dividing point from 0.8 down to below 0.5. At this point, the endpoints of the subinterval have a leading digit in

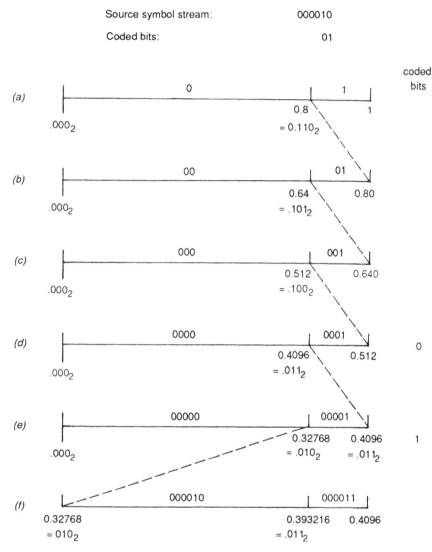

FIG. 8 Example of arithmetic coding with $P(0) = 0.8$ and $P(1) = 0.2$.

common, causing the first coded bit to be generated. For $P(0) = 0.9$, it takes a string of seven 0s to arrive at this point.

The basic ideas indicated above are attributed to Elias (44,75) and Shannon (43). As can be observed from the example above, the method requires an additional decimal digit to specify the endpoints with each new source symbol processed. This poses serious implementation difficulties. Subsequent developments include enumerative coding (76–78) and more recent forms of arithmetic coding based on enumerative coding (79–82). This later work, particularly that of Rissanen, removed the earlier difficulties and made the method practical. An important addition is the ability to change the probability distributions instantaneously. This allows for adaptation to changing source statistics, as well as for specifying probability distributions of a source symbol conditioned on preceding symbols (83,84), a feat difficult to achieve with Huffman coding with extensions. With this adaptation, the method gives a 21% improvement in the compression factor for the total CCITT standard images over that obtained using the CCITT Group 4 method. For a set of digital halftone images, this increases to 207% (84). For other details on arithmetic coding and related coding efforts, see Refs. 85–87.

Exact Coding

Introduction

In this section, major approaches to exact bilevel image coding applicable for facsimile systems are described. The intent of this material is to illustrate basic principles of image coding, including the use of the mathematical models and associated entropy codings covered above, as well as the use of other important approaches. There are many coding methods that are not mentioned specifically here. Additional material providing comprehensive reviews of image coding methods, including those for bilevel, multilevel, halftone, and color images, and for special presentations, such as progressive transmission and display, are included in the Bibliography.

Major coding thrusts have emerged from the source models defined in information theory. The mathematical models themselves do not indicate the interpretations or meanings of the source symbols with respect to images. These are specified by the designer and a particular interpretation of the source symbols, along with assumptions regarding dependencies of successive source-symbol outputs, defines a particular image model. Once an interpretation is specified, it links particular image parameters and their probability statistics to the mathematical model. The statistics may be obtained by making assumptions and evaluating analytical expressions or by estimating them with frequency counts of particular properties in a representative ensemble of images. The sections below describe various image models and associated coding methods first, and then discuss figures of merit and aspects of compression performance.

Coding Based on Independent Probability Models

The simplest image models assume independence of successive source-symbol outputs. In this section, two such models are discussed. The first model, though theoretically correct, leads to infeasible coding methods and illustrates the need for attaching an appropriate interpretation to a mathematical model. The second image model is based on the same mathematical model as the first image model, but gives a different interpretation to the source symbols.

Coding a Global Image Model

Consider an image model in which the successive source-symbol outputs are independent and each source symbol of the source alphabet corresponds to a distinct, entire bilevel image. This is called a global image model (37). Assume that the image is 8.5 in by 11 in, and is scanned at horizontal and vertical resolutions of 200 dpi, giving a total of (8.5)(11)(200)(200), or 3,740,000 pels in the image. Each source symbol then consists of 3,740,000 bits, and the number of different source symbols is $2^{3,740,000}$.

Possible coding methods suggested by the model consist of the entropy codings. For Huffman coding without extensions, a distinct codeword must be assigned to each source symbol (i.e., to each possible image). If the images are equally probable, then the entropy is $\log_2 3^{,740,000}$ or simply 3,740,000 bits per symbol. This is achieved if the codeword assigned to an image is simply the digitized image itself, so no compression results. If the images are not equally probable, then the entropy is less than 3,740,000 bits per symbol, and compression is theoretically possible. But, in reality, the method is not feasible because of the prohibitive size of the codebook requiring $2^{3,740,000}$ codewords and the prohibitive task of estimating the probabilities of the images and deriving the associated Huffman codebook.

Similarly, arithmetic coding is not realistically defined for this model. Normally, arithmetic coding maps a succession of source symbols onto successively smaller intervals of the 0–1 real-line segment. For the model as stated, this implies a mapping of a sequence of complete images, a formidable computational task. Of course, the process could stop after mapping only one symbol. Even in this case, the encoder must be able to partition the 0–1 real-line segment into $2^{3,740,000}$ pieces, each with length in proportion to the probability of occurrence, and must be able to identify the subinterval associated with each particular input source symbol. Clearly, the storage and processing is prohibitive, as well as the initial estimation of the probabilities.

Coding a Local Image Model

Next, we consider the other extreme to a global image model, namely, a model in which there are only two source symbols. One source symbol is 0 and corresponds to a single white pel; the other source symbol is 1 and corresponds to a single black pel. It is assumed that successive source-symbol outputs are independent of each other.

Coding methods suggested by this model are again the entropy codings. As discussed above, Huffman coding without extensions yields no compression for a binary source such as that assumed for this local model. To achieve compression, several extensions may be required, but for this case several extensions pose neither a storage nor a processing burden. Also, it is usually a reasonable task to obtain a probability distribution suitable for a given application. However, the method is limited in performance to the entropy of the binary independent source. To use an example given above, assume that the probability of a white pel is 0.9 and that of a black pel is 0.1. Then the entropy is 0.469. A third-order extension, requiring a codebook of 16 codewords, yields an average of 0.493 bits per source symbol. As noted above, we can achieve an average bits per source symbol as close to the entropy as desired by taking appropriately higher-order extensions.

An alternative coding method is arithmetic coding. In contrast to the global model, a sequence of source symbols for this local model has a reasonable interpretation. It is simply a sequence of the pels of an image. In turn, each of the 3,740,000 pels in the image discussed above causes the encoder to chose one of the two defined subintervals and to divide the chosen subinterval into two subintervals in proportion to the given probabilities for the two source symbols. This method also yields an average bits per source symbol that approaches the entropy as the number of source-symbol outputs increases. For a finite image, the source-symbol outputs eventually cease, and the actual average bits per source symbol is generally somewhat greater than the entropy.

While both entropy codings in this case are simple to implement, the compressions are disappointing. For example, even at the entropy at 0.469 bits per pel, the encoded bits for the entire image are still 1,754,060. Transmission of this many bits at 4800 bits per second takes over 6 minutes for an image. In fact, the model chosen is not a very good model of image generation, specifically because it assumes that successive source symbols are independent. To achieve better compressions requires an assumption of dependence, as is made in the coding methods described below.

Coding Based on Joint-Probability Models

Block Coding

The first model discussed that assumes dependence of successive source symbols is called block coding (73). It is based on a joint-probability system, described above. We divide the image into nonoverlapping blocks of N pels each. These pels may be on a single image line or they can constitute a two-dimensional block. Each source symbol corresponds to one of the 2^N binary values that N pels may assume. Probabilities of the source symbols are estimated from frequency counts made on an ensemble of typical images. Then an image is encoded with Huffman coding or arithmetic coding. A disadvantage of block coding is the need for large codebooks.

To illustrate the potential of block coding, consider the example given above for $N = 2$. The joint-probability statistics are shown in matrix form below, with X denoting the first random variable with values x_1 and x_2, and Y denoting the second random variable with values y_1 and y_2. For images, x_1 and y_1 represent white pels, and x_2 and y_2 represent black pels.

$$P(x_1,y_1) = 0.855 \qquad P(x_1,y_2) = 0.045$$

$$P(x_2,y_1) = 0.045 \qquad P(x_2,y_2) = 0.055$$

As indicated above, the resulting marginal probabilities are $P(x_1) = P(y_1) = 0.9$, and $P(x_2) = P(y_2) = 0.1$. The per-symbol joint entropy is 0.413 bits. This is not much less than the entropy of the first-order distribution, which is equal to 0.469 bits. However, the per-symbol entropy can be made to approach the entropy of the source by taking increasingly larger blocks and, particularly, blocks defined in two dimensions to capture two-dimensional coherence. But, this is done at the expense of increased difficulty in obtaining source statistics and at the expense of increased implementation complexity.

Suboptimal Block Coding

A number of coding methods are based on the idea of block coding but have simpler implementations (88). These methods give average code bits per symbol that are suboptimal, but still encode images exactly. For example, one method divides the image into blocks of $J \times K$ pels, a block with all white pels is coded with a 0, and other blocks are coded with a 1 followed by the $J \times K$ pel values of the block (73,89). A modified form of this scheme reserves the code 0 for a line consisting only of white pels, and codes a nonwhite line as before but preceded by a 1 (90). Other related methods use hierarchies of different block sizes (91–93).

As discussed above, under the assumption of stationarity, the per-symbol joint entropy is always greater than or equal to the conditional entropy. As coding methods based on conditional models are likely to give better compressions than those based on joint models, conditional coding methods are considered next.

Coding Based on Conditional-Probability Models

In general, a source-model output may be dependent or conditioned on any number of previous outputs. However, use of high-order conditional-probability models in coding methods is sparse because of difficulties in implementation and in collection of the conditional-probability statistics. In this section, we discuss two coding approaches based on conditional-probability models where conditioning is held to orders that are relatively small.

First-Order Markov Source and Run-Length Coding

In the simplest conditional-probability model, a source-symbol output is assumed to be dependent only on the preceding output

$$P(X_N \mid X_1, X_2, \ldots, X_{N-1}) = P(X_N \mid X_{N-1}) \tag{39}$$

This is called a *first-order Markov model.* For binary random variables X and Y, the entropy of this source is shown below. This is as developed above, but for images we now denote the values of X and Y by W for a white pel and B for a black pel.

$$
\begin{aligned}
P(X \mid Y) = \ &- P(W)\,P(W \mid W) \log_2 (PW \mid W) \\
&- P(W)\,P(B \mid W) \log_2 P(B \mid W) \\
&- P(B)\,P(W \mid B) \log_2 P(W \mid B) \\
&- P(B)\,P(B \mid B) \log_2 P(B \mid B)
\end{aligned} \tag{40}
$$

A method for coding according to a conditional-probability model is based on properties of runs of pels that result as successive outputs from the indicated source. A *run* is a set of successive like values, followed by an occurrence of the other value. The number of like values in the run is called the *run length*. For images, a run is usually taken to be a series of pels of one tone followed by a pel of the other tone or by the end of the line. White runs and black runs alternate with each other. In the following paragraphs, the properties of the runs and how they lead to a coding method are summarized first, and then a more formal statement of those properties is given.

In the theoretical model, run lengths may vary from 1 to infinity. Capon showed that successive run lengths are independent of each other (94). This enables us to consider the alternating white and black runs as being produced by two separate, independent sources, (i.e., a white-run-lengths source and a black-run-lengths source). The probabilities of different run lengths produced by these two sources are linked to the underlying Markov source, as is indicated below. An entropy is defined for each of the two run-lengths sources. We may also consider the runs as being produced by a single source when each output of this source is a pair of run lengths, a white one and a black one. This is called a *compound source*, and it has an entropy that is the sum of the entropies of the separate white-run-lengths source and the black-run-lengths source. Capon also showed that the entropy of this compound source is equal to the entropy of the original Markov source. Therefore, an efficient coding of the run lengths is equivalent to a coding of individual pels based on the conditional Markov model. Since the run lengths are independent, the coding may be done by an entropy coding, such as Huffman coding with extensions or arithmetic coding. As described, the theory calls for separate coding of the white-run lengths and of the black-run lengths so that two distinct codebooks must be used.

A crucial point in the discussion above is the equality of the entropies of the pels of the Markov source and of the compound run-lengths source. This is established as follows. Since a white run of length N consists of $N - 1$ transitions from a white pel to a white pel, and one transition from a white pel to a black pel, the probability $P_W(N)$ of a white run of length N is

$$P_W(N) = P(W \mid W)^{N-1} \ P(B \mid W) \tag{41}$$

Similarly, the probability $P_B(N)$ of a black run of length N is

$$P_B(N) = P(B \mid B)^{N-1} \ P(W \mid B) \tag{42}$$

Each of these defines a geometric distribution. The average white-run length N_W and the average black-run-length N_B are represented as

$$N_W = \frac{1}{P(B \mid W)} \qquad N_B = \frac{1}{P(W \mid B)} \tag{43}$$

The entropy H_W of the white-run-lengths source is derived by substituting $P_W(N)$ in the function defining the entropy. The entropy H_B of the black-run-lengths source is obtained similarly. The resulting expressions are shown below.

$$H_W = -\log_2 \frac{P(B \mid W)}{P(W \mid W)} - \frac{\log_2 P(W \mid W)}{P(B \mid W)} \tag{44}$$

$$H_B = -\log_2 \frac{P(W \mid B)}{P(B \mid B)} - \frac{\log_2 P(B \mid B)}{P(W \mid B)} \tag{45}$$

The entropy of the white-run, black-run pair compound source is the sum of H_W and H_B. Dividing this by the average number of pels in a white-run, black-run pair gives the entropy H_{WB} on a per-pel basis. The average number of pels in a white-run, black-run pair is the sum of the average white-run length N_W and the average black run length N_B, which yields the following for the per-pel compound entropy:

$$H_{WB} = \frac{H_W + H_B}{N_W + N_B} \tag{46}$$

Simplifying the numerator first, we obtain

$$H_W + H_B = -\log_2 P(B \mid W) + \log_2 P(W \mid W) - \frac{\log_2 P(W \mid W)}{P(B \mid W)}$$

$$-\log_2 P(W \mid B) + \log_2 P(B \mid B) - \frac{\log_2 P(B \mid B)}{P(W \mid B)}$$

$$= -\log_2 P(B \mid W) - \frac{P(W \mid W)}{P(B \mid W)} \log_2 P(W \mid W) \tag{47}$$

$$-\log_2 P(W \mid B) - \frac{P(B \mid B)}{P(W \mid B)} \log_2 P(B \mid B)$$

Next, we note that, for a stationary source as is assumed here, solving the state equations gives the following marginal probabilities (shown above):

$$P(B) = \frac{P(B \mid W)}{P(B \mid W) + P(W \mid B)} \qquad P(W) = \frac{P(W \mid B)}{P(W \mid B) + P(B \mid W)} \tag{48}$$

Using these, we obtain

$$\frac{1}{N_W + N_B} = \frac{1}{[1/P(B \mid W)] + [1/P(W \mid B)]} = \frac{P(B \mid W) \, P(W \mid B)}{P(W \mid B) + P(B \mid W)}$$

$$= P(B) \, P(W \mid B) = P(W) \, P(B \mid W) \tag{49}$$

Finally, combining the results above gives

$$H_{WB} = \frac{H_W + H_B}{N_W + N_B}$$

$$= - [P(W) \, P(B \mid W)] \log_2 P(B \mid W)$$

$$- [P(W) \, P(B \mid W)] \frac{P(W \mid W)}{P(B \mid W)} \log_2 P(W \mid W)$$

$$- [P(B) \, P(W \mid B)] \log_2 P(W \mid B) \tag{50}$$

$$- [P(B) \, P(W \mid B)] \frac{P(B \mid B)}{P(W \mid B)} \log_2 P(B \mid B)$$

$$= -P(W) \, P(B \mid W) \log_2 P(B \mid W) - P(W) \, P(W \mid W) \log_2 P(W \mid W)$$

$$-P(B) \, P(W \mid B) \log_2 P(W \mid B) - P(B) \, P(B \mid B) \log_2 P(B \mid B).$$

This completes the proof, as the equation above is seen to be the defining expression for the entropy of the first-order Markov source.

Numerous run-length encoding schemes have been devised and studied, and some have been implemented in facsimile devices (95–99).

Higher-Order Markov Sources and Predictive and Extended
Run-Length Coding

As indicated above, a run-length pair source equates to a first-order Markov process, with the associated run-length coding accounting for dependencies of a given output on the immediately prior single output. Other approaches can be defined that account for higher orders of dependency using conditional-source encoding, and, furthermore, that are two-dimensional. One method, contributed by Preuss, is extended run-length coding (100). In the paragraphs below, the model upon which this method is based is described first, and then the method itself is described.

The model to be discussed is called a two-dimensional, Nth-order Markov model. In this model, it is assumed that a current pel is dependent on N previous pels in the neighborhood of the current pel. An example is shown in Fig. 9-a, where the current pel x_4 is dependent on the three pels x_1, x_2, and x_3. In general, the current pel may be dependent on any number of previous pels, which may be on the same image line as the current pel or on any number of preceding image lines. The per-pel entropy is defined as

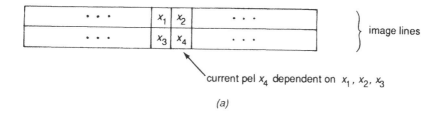

(a)

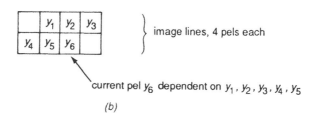

(b)

FIG. 9 Two-dimensional, higher-order Markov models: *a*, 3rd-order Markov model; *b*, 5th-order Markov model.

$$H(X_N \mid X_1, X_2, \ldots, X_{N-1})$$

$$= - \sum_{x_1, x_2, \ldots, x_N} P(x_1, x_2, \ldots, x_N) \log_2 P(x_N \mid x_1, x_2, \ldots, x_{N-1})$$

$$- \sum_{x_1, x_2, \ldots, x_N} P(x_1, x_2, \ldots, x_{N-1}) P(x_N \mid x_1, x_2, \ldots, x_{N-1}) \qquad (51)$$

$$\log_2 P(x_N \mid x_1, x_2, \ldots, x_{N-1}).$$

These definitions agree with the definition of conditional entropy given above. However, in Eq. 51, the previous pels upon which the current pel is conditioned are not all immediately preceding pels in the presumed left-to-right, top-to-bottom production of the pels by an image-generation model. This situation is shown in Fig. 9-*b*, where, for illustrative purposes, each image line is assumed to contain only four pels. The current pel, now labeled y_6, is dependent on pels y_1, y_2, y_3, y_4, and y_5. The pels y_1, y_2, y_5, and y_6 in Fig. 9-*b* correspond, respectively, to the pels labeled x_1, x_2, x_3, and x_4 in Fig. 9-*a*. The process implied in Fig. 7-*b* may be formulated as a 5th-order Markov chain, with 2^5 or 32 states corresponding to the combinations of bit values for $y_1, \ldots, y_5$. As pointed out by Arps, it is necessary to include the intermediate pels, in this instance y_3 and y_4, for proper definition of state transitions (40). Each state in the 5th-order chain may make a transition to only two states because of the overlap of pels in successive states. For example, the state based on pels y_1, y_2, y_3, y_4, and y_5 makes a transition to the state based on pels y_2, y_3, y_4, y_5, and y_6. If $y_1 y_2 y_3 y_4 y_5$ has a value of 00000, then $y_2 y_3 y_4 y_5 y_6$ may have the values 00000 or 00001. Thus, in this case, there are a maximum of 64 nonzero state-transition probabilities.

Based on the 5th-order Markov chain, a 3rd-order Markov chain corresponding to the two-dimensional 3rd-order Markov model may be derived. This is done by combining several of the states in the 5th-order chain to represent a single state of the 3rd-order chain. As an example, consider the four states for pels $y_1 y_2 y_3 y_4 y_5$ with the values 00000, 00010, 00100, and 00110. These comprise all the states with $y_1 y_2 y_5$ equal to 000, and with $y_3 y_4$ assuming all the combinations of two binary digits. The four states combined correspond to the single state 000 for pels $y_1 y_2 y_5$ in the 3rd-order chain. The derived 3rd-order chain has eight states, each being a combination of four of the states of the 5th-order chain.

Likewise, the transition probabilities of the 3rd-order chain may be derived from the 5th-order chain. Consider again the 5th-order chain states 00000, 00010, 00100, and 00110, corresponding to the single 3rd-order chain state 000. Each of the 5th-order chain states may make a transition to two distinct 5th-order chain states, yielding a total of eight possible transitions for the indicated four states. Of these, only the transitions from state 00000 to state 00000, and from state 00010 to state 00100 correspond to the 3rd-order chain transition from state 000 to state 000. Thus, the sum of the probabilities of the transitions from state 00000 to state 00000, and from state 00010 to state 00100 in the 5th-order chain equals the probability of transition from state 000 to state 000 in the 3rd-order chain. (Note that some of the 5th-order chain state transitions do not correspond to proper 3rd-order chain state transitions. For example, state 00100 may make a transition to 01000, but the derived transition from 000 to 010 is not defined.) Assuming the derived state-transition probabilities for the 3rd-order chain are well defined, then, since the chain is ergodic, the steady-state probabilities exist. Finally, each transition probability of the 3rd-order chain is equal to a conditional probability (e.g., $P(0 \mid 001)$ equals the probability of a transition from state 001 to state 010). Accordingly, both the conditional probabilities and the state probabilities can be derived from the 5th-order chain, and may be used to obtain the conditional entropy using Eq. 20. Thus, the defined two-dimensional, Nth-order Markov model is seen to be grounded in the classic Nth-order Markov chain.

The method suggested by this model is attributed to Preuss (100). This is a prediction method in which, first, an error bit map is derived from the image bit map, and then the error bit map is coded. There is a one-to-one correspondence between the elements of the error bit map and the elements of the image bit map. To construct the error bit map, each image pel is compared to a predicted value. If the two values agree, the corresponding bit of the error bit map is 0; otherwise, it is 1. The prediction of each pel is made in the following manner. A pattern of previous neighboring pels is chosen, and, for each state the pattern may assume, estimates are made of the conditional probability that the pel is 0 (white) and the probability that the pel is 1 (black). For each state, the predicted value 0 or 1 is the one with maximum likelihood (i.e., with the largest conditional probability). Then, an image is processed a pel at a time, comparing each pel to the value predicted by the state of the neighboring pels, and setting the associated bit in the error bit map.

To encode the error bit map, simple run-length coding may be used, but better compressions are possible by capturing the two-dimensional correlation in the error bit map. Consider tracing through the error bit map, and noting for each bit both the state of the neighboring pel pattern and if an error occurs in the prediction based on the state. Preuss noted that such prediction errors occur more frequently for some states than for others. Taking this into account, his method codes the position of these errors separately for each state. For a predictor pattern of N pels, there are 2^N states. For any given state, consider only the successive occurrences of that state in a given line of the bit map. Of these occurrences, those with associated prediction errors demarcate the end of runs of successive occurrences of the state. It is these successive run lengths that are encoded. As indicated, such a run-length coding is done separately for each state. Since the run lengths of each state are independent of those of other states, the runs of each state may be considered as being produced by an independent source. It has been shown that the compound-error source, consisting of the 2^N separate run-length sources, has the same per-pel entropy as the entropy of the conditional Nth-order Markov process defined above (40,100).

Generally, the advantages of predictive coding over run-length coding are simpler implementation, equal compression of negative and positive images, and less compression penalty for changes in resolution and character stroke width (40). For other predictive-coding studies and experiments, see Refs. 72, 73, and 101–105. In particular, see Ref. 106 for information concerning predictive coding with adaptive predictor probabilities.

Contour and Structural Coding

Coding Considerations

Several of the methods described above are two-dimensional in nature, including block coding, the method of prediction, and extended run-length coding. These methods capture two-dimensional dependencies by considering the relation of each pel to a set of neighboring pels, but no overt attempt is made to identify structural elements in an image. In another approach, the image is viewed as consisting of areas of one tone, called the *image tone*, placed on a single background area of the other tone, called the *background tone*. For convenience, in the discussion that follows the image tone is referred to as black and the background tone as white, although, in a negative image, the tones are reversed. An image consists of a set of separate black subareas or structural elements. The coding of each subarea consists of a coded form of the shape or contour of the subarea, along with an encoding of the position of the subarea within the image. The coding of the contour may be given in terms of the actual pels comprising the contour. This is called a *contour coding*. Or the coding of the contour may be given in terms of parameters from which the pels comprising the contour may be derived. For differentiation, this is called *structural coding*. Several examples of contour codings are given in the sections below, including coordinate coding, chain coding, predictive-differential-quantization (PDQ) coding, and blob coding. These are followed by two examples of structural coding: medial-axis-transform coding and geometric-pattern coding.

An important property of contour and structural codings is that they are feature oriented. An advantage of such codings is that they permit operations to be performed on image structural elements. Several such operations for image processing are cited below in the discussion on chain coding. Other advantages are the ability to convert exact encodings to approximate or lossy forms easily and to design lossy encodings with controlled effects on image quality. Also, feature-oriented data may be input to higher-level pattern-recognition codings, resulting in considerably higher compressions. There is also some evidence that feature-oriented codings recover more successfully from transmission errors than a number of other types of codings while maintaining comparable compressions (107). With the emerging trend of migration of the facsimile functions into computer systems, codings that are feature oriented can prove to be very useful.

Another important consideration in facsimile coding is the amount of image data that must be available to the encoder and decoder. In some methods, the entire image must be available, and in other methods, only two or just a few image lines are needed. In general, a requirement for fewer image lines results in lower memory costs and simpler logic. Optimally, a coding should be both feature oriented and have small memory requirements. Chain coding is feature oriented, but requires memory to hold the entire image. Another approach partitions an image into subgroups of lines and encodes each subgroup separately; these have been found to have relatively poor compression performance. An alternative is to code each line in reference to the previous line. This is called a *line-by-line* coding method, and obviously requires memory for only two image lines. There are many line-by-line coding methods, but only some of them are distinctly feature oriented, such as the PDQ and blob coding methods. These codings, furthermore, allow for easy conversions between data ordered in a line-by-line sequence and data ordered in a structural-element sequence. In contrast, several line-by-line codings code the pel transitions on an image line relative to pel transitions on the previous line, as well as on the same line. Both white-to-black and black-to-white transitions are coded. While these codings capture some contour information, they do not overtly identify and code structural elements, and it is not clear that they readily lend themselves to conversions between line-by-line and structural-element forms. They are referred to as contour-related codings. In this category are the two-dimensional coding method of the CCITT Group 3 and Group 4 Recommendations and several methods from which the CCITT coding method was derived. These methods are covered in a section below.

Contour coding consists of coding the contour pels of the black areas, as well as the positions of these areas within the image. Neither the white background pels nor the pels inside the black areas are coded explicitly. The decoding process may be thought of in the following manner. First, the entire image is set to white. Then the encoded contour pels are set to black. If the contour describes a solid black area, the entire internal part of the area is filled in with black pels. It is possible for contours to be nested within contours, thereby creating holes, such as the hole within the letter *O*. In this case, some provision must be made to pair appropriate contours and to fill in black pels only between each designated pair of contours. This mechanism may be provided by building into the coding scheme an identification of the pairs of contours. An example of this is indicated in the discussion of the chain coding method below. It may also be accomplished by having the encoder break up a black area with holes into a set of solid black subareas and coding the contours of each of these separately. For example, the letter *O* may be divided into four areas, consisting of a top cap, a bottom cap, a left leg, and a right leg. Other examples are discussed in greater detail below.

Coordinate Coding

A simple way of encoding a contour is to code the coordinates of each contour pel, where the coordinates are relative to a fixed point, or to the previously coded contour pel. As the number of contour pels increases, a point is reached beyond which the number of code bits exceeds the number of bits in the image and transmitting the image without coding is more efficient than with coding.

Chain Coding

An improvement over coordinate coding is possible with chain coding (108). This process traces around a contour, at each step moving from the current pel in one of eight directions to one of the eight neighboring pels. Each of the eight directions is assigned a codeword. If constant-length codewords are used, then three bits are required for each codeword. Variable-length codewords achieve greater compressions.

A unique chain encoding that provides for nested contours is given by Morrin (109). In this method, each object within an image is coded separately. An object is an image area that is enclosed by a black contour but itself is not enclosed by another black contour, although it may contain nested white areas and black areas. To code the object, first the outermost black contour is traced in a clockwise direction and chain coded. Then the outermost layer of black pels is shrunk by tracing around the object, several times if necessary, and each time changing black pels to white. If, during this shrinking, a white pel occurs to the immediate right of the tracing path, the shrinking process is suspended and the inner white area is explored. If it is a hole, then it is traced in a counterclockwise direction, chain coded, and then shrunk. Again, the shrinking may be suspended to handle any black holes within the white area. The process of coding, shrinking, and suspension of shrinking to handle nested layers continues until the innermost layer is coded, and then the suspended shrinking processes are resumed, progressing from the innermost to the outermost layers until the entire object is coded and erased (i.e., changed to white). The process also handles an object with disconnected holes.

It is seen that coding of a nested object consists of a single chain of coded outer and inner contour pels, and that each such pel is coded relative to the prior pel in the chain. This property enables a variety of useful image-processing operations to be performed directly on the encoded data, including image rotation by 90°, mirroring, translation, character generation, scaling, smoothing, and insertion of objects. Additionally, useful image measurements are obtainable as byproducts of the encoding process, including perimeter, area, center of mass, and the enclosing box of the object. (See also Ref. 110.)

Predictive-Differential-Quantization Coding and Blob Coding

Two independently developed methods that are feature oriented and are line-by-line codings are predictive-differential-quantization coding (PDQ) (96) and blob coding (2,16,111). PDQ was designed to code weather maps and newspapers. In actuality, it is not a predictive-coding method as described above. It codes differences between corresponding run lengths on successive image lines. Figure 10 illustrates the PDQ coding parameters. To code a run relative to a run in the previous line, two parameters, d' and d'', are used. The parameter d' is the signed difference in the number of pels between the left end of the run being coded and the left end of the run in the previous line. The parameter d'' is the signed difference in the lengths of the two runs, shown in the figure as $L_2 - L_1$. Additionally, a start code is used for the start of a new black area, and a merge code is used for the end of a black area. For each start code, two other parameters are coded, consisting of one run length to indicate the position of the left-most pel in the starting run, and a second run length to indicate the length of the run.

Blob coding was originally designed to code display advertisements for Yellow Pages Directories. Subsequently, it was also used as the coding method in several prototype facsimile machines, and was proposed by AT&T to the CCITT for Group 3 two-dimensional facsimile coding (110,111). As only two image lines are referred to at any given time, blob coding is also a line-by-line coding method. Blob coding has two easily interchanged forms for the coded data. One of these is a line-by-line form, in which the codes appear in the encoded bit stream in the order in which they arise in a line-by-line encoding process. The other is a blob-by-blob form, in which the codes for a given blob are contiguous in the encoded bit stream.

The definition of a blob is illustrated in Fig. 11-*a*. The blob consists of a blob head run and connecting runs, if any, ending in a blob tail run. Runs on two successive lines may become part of the same blob if their left ends are no more than three pels apart, and if their right ends are no more than three apart. To code a blob run relative to a run in the previous line, a single connection code is used to indicate the connectivity on both the left and right ends. As illustrated by Fig. 11-*b*, since the left end of the run being coded may be no more than three pels from the left end of the run on the previous line, there are

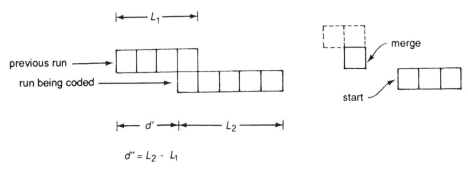

$$d'' = L_2 - L_1$$

FIG. 10 PDQ coding run-length difference parameters.

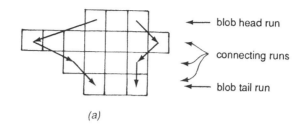

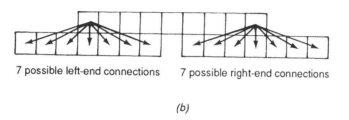

FIG. 11 Blob coding parameters: *a*, blob definition; *b*, connection codes.

seven possible left-end connections. Similarly, there are seven possible right-end connections. In combination, there are 49 possible left/right end connections, and thus there are only 49 connection codes. In associated studies on typical facsimile images, the connections on the left side of black areas were shown to be correlated strongly with the connections on the right side. This correlation is captured by using the combined left/right end-connection codes with variable-length codewords. Also, as shown in associated studies, a displacement value of three proved to be optimal (111). A maximum displacement of three permits a maximum contour slope angle of 71.6° in either direction, giving a range typically found with high probability in facsimile images. However, in PDQ coding, no apparent restriction is placed on d' or d'', and the contour slope range is assumed to be 90° in either direction, with uniform distribution in the range (96). The maximum displacement of three was also specified in the Group 3 coding method proposed to the CCITT by the British Post Office (112). It was adopted as a feature of the CCITT Group 3 Recommendation.

Another difference between PDQ coding and blob coding is in the identification of coded parameters. In PDQ coding, the parameters are d', d'', a merge code, and a start code. Following each start code are two run lengths giving the position of the new run and its length. In blob coding, the 49 connection codes replace d' and d''. The blob end code is the same as the merge code. However, in blob coding there is no distinct, single-head code, but rather there are 43 head-length codes; 40 of these are for heads of length 1 through 40, and 3 are for larger heads and require a suffix to identify the actual length. As with PDQ coding, the position of the head is an appendage to the head code. In total, the blob codebook consists of 93 codewords for which a Huffman coding is defined.

Of importance is the fact that both coding methods are feature oriented. PDQ coding codes run-length differences; blob coding simultaneously codes the left and right contours of a blob. This feature-oriented capability enables blob coding to be used as a preprocessor to obtain vector representations of engineering documents (14) and to derive features for character recognition (113).

Information-theoretic models may be defined for PDQ coding and blob coding. For PDQ coding, the model has four source outputs, namely, d', d'', a merge code, and a start code. An entropy formulation for this source must also account for the two run lengths appended to each start code. For blob coding, there are 93 source outputs, as defined above. An entropy formulation here must account for the position run length appended to each head code. In either case, probabilities of the source outputs may be approximated using frequency counts of the indicated parameters on typical images, or they may be obtained by analysis based on various simplifying assumptions (96).

Medial-Axis Transform Coding

A coding method related to PDQ and blob coding is medial-axis transform coding, also called symmetric-axis transform coding or grassfire-process coding. This approach was originally formulated by Blum to describe biological growth and visual-form-perception processes (114–120).

In Blum's approach, a given shape (or grass field) is defined by a continuous contour. In this method, consider igniting a fire simultaneously along each contour point and assume the fire wavefronts spread uniformly in all directions. Where two or more wavefronts meet, the fire is quenched. The locus of the quench points define the *medial axis* (or symmetric axis, or skeleton) of the shape. Associated with each quench point is the time from ignition of the fire to the quenching at that point, or, equivalently, the distance from the contour to the quench point. The time from ignition to the total quenching of the fire within the grass field also defines the maximum distance from the contour to the medial axis. Alternatively, one may ignite the points on the medial axis in reverse order to their sequence of quenching and at times that are in reverse to the successive times of quenching. At any given subsequent time, the advancing circular wavefronts are of varying radii and their outer envelope defines the contour of a shape. At the time the maximum distance between the medial axis and the contour is reached, the envelope defines the contour of the given shape. Equivalent definitions of the medial axis are (1) that it is the locus of the center of all maximal circles (i.e., of all circles inscribed within a shape but not contained within other inscribed circles), and (2) that it is the locus of all internal points that are equally distant from two or more contour points.

For digitized rather than continuous images, the circles of the medial-axis transform are replaced by squares or by maximal neighborhoods. Numerous methods exist to find the medial axis of a digitized shape (121–126), including one that is parallel (127). The medial-axis pels, along with the distances to the contour or the associated square sizes, actually comprise an exact encoding equivalent to contour codings. Compressions are comparable to those of chain coding (123). The digitized form of this coding is useful in pattern recognition for thinning and shading processes, and also for finding such properties relating to image subareas as status relative to convexity, intersections of two or more subareas, and whether or not a given pel is inside or outside a subarea (123,125). A generalization of this approach is growth-geometry coding, which permits any kind of propagation process in addition to the symmetric one that propagates at an equal rate in all directions (128).

Geometric-Pattern Coding

As indicated above, with medial-axis transform coding a structural element may be encoded as a set of overlapping squares. Based on this representation, a set of nonoverlapping squares may be derived, and, when applied to an entire image, it is an exact encoding (129). The structural elements of an image also may be decomposed into a set of nonoverlapping rectangles, with the coding consisting of the size of each rectangle and its placement within the image (129). Such a coding may also be obtained using a form of the blob coding method in which only direct vertical connections are permitted between the left ends and between the right ends of two successive runs in a blob (113). Greater compressions are achievable with methods that identify the set of largest inscribed rectangles in each structural element (129).

Universal Coding

The coding methods discussed above rely heavily on providing the coder and decoder with estimates of image parameter statistics. Methods in which such statistics are not provided initially but which are acquired during the encoding process are called universal coding methods (69,130–136,). They usually require complex logic and large memories, but they relieve us of having to obtain image-parameter statistics. Also, they provide a single coding solution for varying kinds of data, a property that may be of particular use in a multimedia environment.

There are numerous versions of universal coding. The basic idea is to process the elements of a string in a sequential manner, parsing the string into segments of elements called *phrases*. During the process, a dictionary of distinct phrases is constructed. Each phrase in the dictionary consists of some previously occurring phrase plus one terminal element. To parse a string, successive elements are concatenated until they form a new phrase P_n, which is entered into the dictionary. Also, since by definition the initial elements of P_n comprise a phrase P_j that is already in the dictionary, the encoder outputs the dictionary index, or a code assigned to the index, of the phrase P_j and restarts parsing with the terminal element of P_n. The final phrase may be an exception to this process.

In an early study that applied universal coding to images, experiments were made using four different versions of the coding process in combination with three different ways of defining the string relative to the image data (131). The best compression was obtained when the string was taken to be the successive bits of a bit map derived from the original image by taking the binary differences between every two adjacent image lines. This enabled the capture of some of the two-dimensional correlation in an image. Several images were encoded using universal coding and using blob coding. In all cases, the blob coding gave superior compression. The improvement in the compression factor of blob coding over universal coding ranged from 65% to 152%. Subsequent work in universal coding partitions an image into rectangular subimages that are analogous to the phrases defined for the linear case (136).

CCITT Facsimile Coding Recommendations

Overview of Recommendations

The CCITT has issued several recommendations for facsimile coding (137). Included among these are the earlier T.2 and T.3 Recommendations for analog facsimile apparatus. Equipment running under these recommendations are referred to as Group 1 and Group 2 machines, respectively. These recommendations, as well as subsequent ones, refer to an A4-size document with nominal dimensions of 215 mm (8.46 in) width and 308 mm (12.1 in) length. (Given dimensions vary somewhat in the literature.) The T.2 Recommendation permits A4-size documents scanned at 3.85 lines/mm (97.8 lines/in) to be transmitted over a voice-grade channel in the general switched telephone network in 6 minutes. The T.3 Recommendation halves this time.

Adopted subsequently was the T.4 Recommendation for digital facsimile apparatus. Equipment running under this recommendation are referred to as Group 3 machines. This recommendation assumes two-tone digitization and includes two different resolutions, referred to as standard resolution (or normal resolution) and high resolution. For both standard and high resolutions, the horizontal resolution along a scan line is specified to be 1728 samples/line, which equates to about 8.0 samples/mm or 204.2 samples/in (also called dots per inch, or dpi). For standard resolution, the vertical resolution is specified to be 3.85 lines/mm (97.8 lines/in), and for high resolution, it is 7.7 lines/mm (195.6 lines/in). The tests sponsored by the CCITT on proposed Group 3 coding methods used digitizations with 1188 lines and 1728 pels/line (or 2,052,864 total pels/image) for standard resolution, and digitizations with 2376 lines and 1728 pels/line (or 4,105,728 pels/image) for high resolution. The basic coding method in the T.4 Recommendation is a form of one-dimensional run-length coding. With this method, A4-size documents are transmitted over voice-grade channels at 4800 bits per second (bps) in one minute or less on the average. Also adopted by the CCITT is an optional extension to the T.4 Recommendation comprising a two-dimensional coding method. The one-dimensional method and the two-dimensional method are summarized below.

The T.5 Recommendation for Group 4 machines is intended to operate on digital data networks, such as the ISDN. Included in this recommendation is a wider choice of scanning densities, with horizontal resolutions of 1728, 2074, 2592, and 3456 samples/line, and vertical resolutions of 200, 240, 300, and 400 lines/inch. The coding method for Group 4 machines is specified in the T.6 Recommendation. It is similar to the two-dimensional method of Group 3, but with increased efficiencies obtainable with the error-correction ability of data networks. This improvement, coupled with higher data-transmission rates (e.g.,

64 kilobits per second (kbps) for an ISDN B-channel) permit transmission times of 1 to 5 seconds for an A4-size document.

Subsequent extensions have been made to the CCITT Recommendations, providing, for example, error-correction capability on Group 3 machines. CCITT study activity is ongoing in the areas of multilevel, color, and mixed-mode coding, and also for progressive transmission and multimedia presentation.

One-Dimensional Coding Recommendations

The Group 3 one-dimensional coding method is a form of run-length coding. It was originally formulated by the Plessey Company (137), and in modified form was later proposed by two industry associations (138). The method uses two codebooks, one for white runs and one for black runs. Run lengths from 0 to 1728 can be encoded. For a scan line starting with a black run, a white run of length 0 is inserted prior to coding the black run. Each codebook contains only 92 codewords, and uses a modified Huffman coding (MHC). In each codebook, the first 64 codewords represent run lengths from 0 to 63, inclusive. The rest of each codebook consists of a single end-of-line (EOL) codeword and 27 make-up codewords (MUC). The MUCs represent multiples of 64, ranging from 1 to 27. The output code for a run length greater than 63 consists of two codewords from the codebook (i.e., the MUC representing the highest multiple of 64 less than or equal to the run length, and that codeword from the first 64 codewords that is the difference between the run length and the value represented by the indicated MUC). Also defined is an extension of the codebook to handle lines longer than 1728 pels. For further details and properties of the code and the actual codebook values, see Refs. 8 and 137.

Two-Dimensional Coding Recommendations

The Group 3 optional two-dimensional coding method is called the modified relative element address designate (Modified READ) coding method. It is based on the relative element address designate (READ) coding method, which was proposed by Japan as the Group 3 two-dimensional coding method (11,139). The READ coding method itself is an amalgam of two earlier methods, the relative address coding (RAC) method (140–142) and the edge difference coding (EDIC) method (143,144). It has been noted that the RAC, EDIC, READ, and Modified READ coding methods can be said to be modified versions of the PDQ coding method (11). All of these methods, and the blob coding method as well, are line-by-line coding methods. However, of these, only the PDQ and blob coding methods are feature oriented in the sense defined above.

The idea of the Modified READ coding method is as follows. In a line to be coded, the method identifies each *transition pel*, defined as the second pel of a white-to-black transition or of a black-to-white transition. For this purpose, the pel previous to the first pel in the line is assumed to be white. Each transition pel is coded relative to a corresponding transition pel of the same tone on the preceding line or relative to a reference pel on the same image line as the transition pel and situated to the left of the transition pel. Additional codes are required to identify an appropriate corresponding transition pel on the preceding line.

More specifically, the coding method traverses the line to be coded and, at any point in the process, it identifies three particular pels, called a_0, a_1, and a_2, on the line being coded, and also two pels, b_1 and b_2, on the previous line, in the manner indicated below.

a_0 is the reference pel occurring to the left of a_1. Initially, it is set on an imaginary white pel to the left of the line. Subsequently, it is moved to the right during the coding process.

a_1 is the next transition pel to the right of a_0. It is the transition pel to be coded next.

a_2 is the next transition pel to the right of a_1. It has the tone opposite to that of a_1.

b_1 is the next transition pel on the preceding line to the right of a_0 and of the same tone as a_1.

b_2 is the next transition pel on the previous line to the right of b_1. It has the tone opposite to that of b_1.

At any time during the coding process, given a_0, the pels a_1, a_2, b_1, and b_2 are determined in accordance with the specifications above. Any of the pels a_1, a_2, b_1, or b_2 that cannot be found are set to an imaginary pel to the right of the indicated line. To start the process, a_0 is set to the imaginary white pel to the left of the line, and a_1, a_2, b_1, and b_2 are set in relation to a_0. Then the process continues as indicated below. The processing of a line ends with coding an imaginary pel to the right of the line being coded, followed by an EOL code. Figure 12 illustrates major aspects of the process, tracing several successive positions of the pels a_0, a_1, a_2, b_1, and b_2.

1. If b_2 is to the left of a_1, then output a codeword, called the pass-mode codeword, and set a_0 to be below b_2. In Fig. 12-a, it is seen that the pass mode results in bypassing a possible reference pel that is considered to be too far to the left of a_1. Then reset b_1 and b_2 in relation to a_0, and repeat the coding process.

2. If b_2 is not to the left of a_1, and if $|b_1 - a_1| \leq 3$, then output one of seven codewords, called vertical-mode codewords, and in particular that one corresponding to the displacement of a_1 relative to b_1. (See Fig. 12-b.) Then move a_0 to a_1 and reset a_1, a_2, b_1, and b_2 in relation to a_0. Then repeat the coding process.

3. If b_2 is not to the left of a_1, and if $|b_1 - a_1| > 3$, then output a codeword, called the horizontal-mode flag codeword. Append to this codeword two other fields representing two run lengths. The first field represents the length of the run from the transition pel back to the reference pel (i.e., $a_1 - a_0$). The second field represents the length of the run starting at the

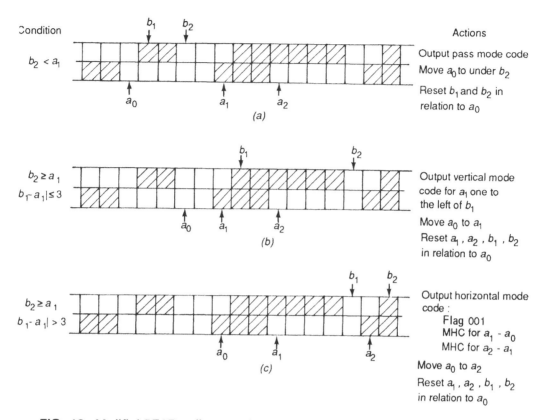

FIG. 12 Modified READ coding example: a, pass-mode coding; b, vertical-mode coding; c, horizontal-mode coding.

transition pel (i.e., $a_2 - a_1$). Thus, if the transition pel is white (black), then the first run is black (white) and the second run is white (black). The representation of the two run lengths are not codewords in the two-dimensional codebook that contains the pass-mode, vertical-mode, and horizontal-mode codewords. Instead they are codewords from the appropriate (white or black) one-dimensional MHC codebook. (See Fig. 12-c.) After outputting the horizontal-mode codeword and the two appended fields, move a_0 to a_2 and reset a_1, a_2, b_1, and b_2 in relation to a_0. Then repeat the coding process. If the first pel on the line to be coded is black, it is de facto a transition pel. For this case, if a horizontal-mode codeword is called for, then $a_1 - a_0$ is forced to be 0.

Several other features are provided in the Group 3 coding method. One of these provides for the containment of the effect of transmission errors by permitting no more than $K - 1$ sequential lines to be two-dimensionally coded. The value of K is called the K-factor. It is usually set to 2 for standard resolution and to 4 for high resolution. Each EOL code is appended with a tag bit, which has a value of 1 or 0 to indicate that the next line is one- or two-dimensionally coded, respectively. This permits the option for a facsimile device to determine the code bits for a one-dimensional and a two-dimensional coding of any given line, and to choose the more economical one, but still with the restriction that no more than $K - 1$ successive lines be two-dimensionally coded. Another feature permits a line to be coded in uncompressed form to achieve economy where the coded bits exceed the number of pels in the line. This is done by inserting a 1-D extension codeword before a one-dimensionally coded line and a 2-D extension codeword before a two-dimensionally coded line. In total, the two-dimensional codebook contains 12 codewords, consisting of codewords for the pass mode, the horizontal mode, 7 vertical-mode displacements, the EOL codeword, the 1-D extension, and the 2-D extension. For further details and codebook values, see Ref. 137.

For the Group 4 coding method, K is set to infinity, there are no EOL codes, and provision is made for image lines to have more than 2623 pels.

Figures of Merit and Compression Performance

Figures of Merit

In comparing coding methods, a primary focus often has been placed on coding compression. Other factors are also important, particularly in the evolving systems environment calling for multiple uses and multiple operations on images, including manipulation, recognition, storage, and display, in addition to the traditional transmission function. For overall optimal performance, the choice of a coding method depends on an appropriate balance of all factors for a given environment. In the discussion that follows, a number of figures of merit for comparing coding methods are summarized first, and then several results relating to compression performance are discussed.

Of major consideration is implementation cost, including that of both logic and memory, and tradeoffs between them. In some situations, more complex encoding logic is warranted to minimize memory requirements. This applies, for example, in facsimile operations with source imagery in massive, long-term image libraries. In other situations, where the data are of a more transitory nature, logic simplicity may dominate over other factors. This is the case if the facsimile function is to be built into an inexpensive peripheral primarily to service immediate hardcopy input. With respect to cost, another important factor is the extent of the design and testing effort, particularly where it involves integrated circuits and/or complex software and firmware. This has an impact on both the nonrecurring engineering cost, reflected in the market price of a product, and also the time to market.

Several other factors are also important in considering candidate coding methods. Among these are the speed of encoding and the speed of decoding. For applications with source images drawn from image libraries, very slow encoding may be tolerated inasmuch as it is performed but once in creating the library entry. However, at the receiver, very fast decoding may be required to drive an interactive, high-resolution soft-facsimile display. Alternatively, relatively slow decoding may be acceptable in a computer networked system with a facsimile server that stores incoming encoded images and, subsequently, decodes and produces nonpriority hardcopy during off-hours.

Relative to transmission, two other factors of importance are sensitivity to channel errors and code synchronizability. In simulations made on the seven coding methods proposed as Group 3 standards to the CCITT, errors were introduced into the encoded bit streams (107). Measurements were made of an *error-sensitivity factor* (ESF), defined to be the number of reconstructed pels in error divided by the number of bit errors introduced. The factor was seen to differ by as much as 24% over the seven coding methods. Additionally, an error may cause the decoder to misalign with respect to parsing the coded bit stream. The average and maximum number of bits passed until the decoder can resynchronize also varies depending upon the coding method. Longer resynchronization periods generally cause more severe image degradation.

Two other figures of merit, mentioned by Arps, are the processability of the compressed data and the robustness of the coding method (40). *Processability* is defined as the ability to perform operations directly on the encoded data. As discussed above, Morrin's method provides this ability (109). *Robustness* refers to the ability of the method to handle data variations without compression penalty, including changes in resolution, character stroke width, and polarity. For example, the compression ability of predictive coding is the same for positive and negative images, and it is more effective than other methods in handling different resolutions and stroke widths.

Another figure of merit is *compatibility*, which is the ability of a coding method to assume alternative forms, as provided uniquely by feature-oriented codings. One aspect of compatibility is the ability to define simple mappings from a line-by-line form into a feature form. This enables image operations on features and also extensions into pattern recognition. Another aspect is the ability to extend easily to effective approximate codings (discussed below). Of course, in the case of lossy encodings, a primary figure of merit is perceived image quality.

Compression Performance

With respect to compression performance, numerous studies have been made on various coding methods. The performance is given as a *compression* factor, defined as the ratio of uncoded bits (or pels for bilevel images) to coded bits. The performance may also be given as a *compression coefficient*, which is the inverse of the compression factor and, for a bilevel image, has units of bits per pel. Usually the compression performance is taken as the average over a particular small ensemble of images and is applicable only if the ensemble is truly representative of typical images handled by a facsimile system. At the time the Group 3 CCITT facsimile coding recommendations were made, a particular set of eight images was chosen as a test ensemble; many facsimile studies base comparative results on this set. The discussion below deals with results based on ensembles.

First, it is important that comparisons be made based not only on identical ensembles, but also on data resulting from an identical scanning of the images at the same resolutions. Notably, compression factors for exact encodings have been found to differ by as much as 13% on the average in one report, and 20%

in another report. These differences result from scanning the same images with different scanning devices (137,145). For approximate codings, a difference as high as 100% has been measured (145). Also, with new and evolving uses of facsimile systems, the distribution of image types may be expected to change. Using code words based on the original distribution generally will result in an overall decrease in the compression factor. Also, since different coding methods may favor different types of images, a shift in distribution of image types may lead to different rankings of coding methods.

As stated above in the discussion of image modeling, more general comparisons of different coding approaches may be made on the entropies of the source models corresponding to the coding methods rather than on the compression performance of particular implementations. The entropy divided by the coded bits per pels achieved by a particular implementation is called the *efficiency*. In coding studies, an approximation to an entropy often is made based on the frequencies in an ensemble of image parameters defining the source symbols. These entropy approximations often are referred to simply as the entropies, but the distinction should be noted. Again, care should be taken in comparing these entropy approximations for the same reasons cited above for comparing compression factors. Additionally, evolving facsimile systems can be expected to involve a greater diversity of image types, generally resulting in flatter distributions of image parameters and correspondingly higher entropies and poorer compression performances of coding methods. One way to combat this washout effect is to segment an image into regions and use different coding methods for different regions.

Table 1 highlights some of the results on compression performance obtained for the Group 3 coding methods (137). Shown are results using the one-dimensional MHC method and the two-dimensional Modified READ coding method. For each coding method, results are given for the defined standard resolution (204.2 samples per in horizontal, 97.8 lines per in vertical), and for the defined high resolution (204.2 samples per inch horizontal, 195.6 lines per

TABLE 1   Average Compression Factors for the Eight CCITT Standard Documents Using the CCITT Group 3 Coding Methods

	Average Compression Factor			
	Standard Resolution		High Resolution	
	MSLT* = 0 K† = 2	MSLT = 20 K = 2	MSLT = 0 K = 4	MSLT = 20 K = 4
One-Dimensional (1-D)	7.70	7.07	7.70	7.07
Two-Dimensional (2-D)	9.03	7.90	11.56	9.43
Percentage Improvement 2-D over 1-D	17	12	50	33

Adapted from Ref. 137.
*MSLT = Minimum scan-line time.
†K = K factor.

inch vertical). For each resolution, results are given for values of 0 and 20 milliseconds (ms) for the minimum scan-line time (MSLT). This is a minimum amount of time for transmission of the encoded bits for each image line and is specified to keep transmitters and receivers in step with each other. If the time to transmit the encoded bits for a line is less than the specified MSLT, the encoder simply adds fill bits to reach the minimum. The recommended MSLT is 20 ms, which equates to 96 bits at a transmission speed of 4800 bps. The data in Table 1 indicate several points of interest.

First, in each case in Table 1, two-dimensional coding gives improved performance over one-dimensional coding. The maximum improvement is a 50% increase over the one-dimensional result, occurring for the case with high resolution and an MSLT of 0 ms. Additionally, for selected cases with an MSLT of 0 and a K-factor of 4, exercise of an option to use the more economical of one-dimensional or two-dimensional coding on each line results in a 7% increase in the compression factor. Note that two-dimensional coding does not always excel. At resolutions lower than those in this study (e.g., at 100 dpi and lower) one-dimensional coding gives better compressions than two-dimensional codings (12).

A second observation concerning the results in Table 1 relates to the comparative performance at the standard and high resolutions. As indicated above, a high-resolution digitization has the same number of samples per line and twice the number of lines as does a standard-resolution digitization. Accordingly, at the high resolution there are twice the number of pels and twice the number of uncoded bits as at the standard resolution. We now consider the cases for one- and two-dimensional coding separately.

For the Group 3 one-dimensional coding at standard resolution, if U is the number of uncoded bits and C is the number of coded bits, then the compression factor at standard resolution is U/C. For high resolution, an additional image line is added between each two image lines at standard resolution. It is reasonable to assume that each such added line is close in content to the previous standard-resolution line. Then, for one-dimensional coding, the number of coded bits for each added line is the same, or nearly so, as that of the previous line, giving a total of $2C$ for the total coded bits. For one-dimensional coding at high resolution, this gives a compression factor of $2U/2C$, or U/C, which is the same as at standard resolution, as indeed is indicated in Table 1.

For the Group 3 two-dimensional coding, note that the method is a contour-related coding. Again, at standard resolution U is denoted as the number of uncoded bits, and A and B as the number of coded bits for horizontal and vertical contours, respectively. The compression factor at standard resolution becomes $U/(A + B)$. The number of coded bits for the horizontal contours is the same for both the standard and high resolutions. Making the same assumptions as above for the added lines, it may be assumed that the number of coded bits for the vertical contours at high resolution is twice those at the standard resolution. The compression factor at high resolution becomes $2U/(A + 2B)$. Then the ratio R of the compression factor at high resolution to that at standard resolution is

$$R = \frac{2U/(A + 2B)}{U/(A + B)} = 1 + \frac{1}{1 + (2B/A)}. \tag{52}$$

Since the upper bound on this ratio is 2, at the very best the compression factor at high resolution can be twice that at standard resolution. The actual ratio depends upon the relative values of A and B, which may vary depending upon the image, and also on the K factor. The specific relation of A and B is not included in the reported results for the CCITT coding. As a point of comparison, in a study on blob coding, which is a contour coding, B was found to be $1.36A$ for a particular image (16). Using this relation, the value of R is 1.27. From Table 1, where the MSLT is 0 ms, R is seen to have the value of 11.56/9.03, or 1.28. However, it is noted that the K factor is different for the components of this ratio; for a clearer result, the K factor should be infinity for both components.

Another observation concerning the results in Table 1 concerns the compression penalty paid for the MSLT feature. For one-dimensional coding, the decrease in the compression ratio from 0 to 20 ms MSLT is 8.2%, which is not severe. However, for two-dimensional coding, the decrease is considerable, being 12.5% at standard resolution, and 18.4% at high resolution.

In the discussion above, some of the performance results of the CCITT Group 3 coding method are highlighted. Comparisons of these results with those of other methods are valid only if the same conditions apply, or if any differences are indicated explicitly and compensations made. Specifically, the assumptions of the Group 3 coding method are use of (1) exact encoding, (2) a K-factor of 2 for standard resolution and 4 for high resolution, (3) an MSLT that is a stated duration for a given coding study and requires fill bits as needed, and (4) EOL and tag-bit codes.

For coding studies that do not adhere to these conditions, as may be expected for systems operating in a Group 4 environment, it is useful to note the results given in Table 2. In this table, coded bits and compression factors are shown, assuming high resolution, an MSLT of 0, and a K-factor of infinity. The results are given separately for all eight of the standard CCITT facsimile documents and for their average. Two sets of results are shown, one including and the other excluding the 13-bits-per-image line used for the EOL and tag-bit codes. Permitting the K-factor to be infinity results in an average compression factor for the ensemble of 13.92, which is an increase of 20.4% over 11.56, the result in Table 1 for a high-resolution, two-dimensional compression factor with the K-factor equal to 4. Additionally, excluding the EOL and tag-bit codes increases the average compression factor for the ensemble to 15.54, which is an increase of 34.4% over 11.56. For bilevel images, the average compression factor for an ensemble is the sum of the pels for all the images, divided by the sum of the coded bits for all the images. If the images are all of the same size, the average compression factor is also equal to the pels per image divided by the average coded bits per image, which is 4,105,728 ÷ 295,020, or 13.92 for the first case above, and 4,105,728 ÷ 264,131, or 15.54 for the second case above. Note that the average compression factor for an ensemble differs from the average of the compression factors.

Of interest are the relative performances of the seven coding methods proposed for the CCITT Group 3 two-dimensional recommendation (107). Of these, there are two predictive, one contour, and four contour-related coding methods. The maximum difference in the compression factor, expressed as a

TABLE 2 Coded Bits and Compression Factors for the Eight CCITT Standard Documents Using the CCITT Group 3 Two-Dimensional Coding Method

Document Number	With EOL and Tag-Bit Codes		Without EOL and Tag-Bit Codes	
	Coded Bits	Compression Factor	Coded Bits	Compression Factor
1	175,704	23.37	144,816	28.35
2	117,304	35.00	86,416	47.51
3	260,527	15.76	229,639	17.88
4	585,074	7.02	554,186	7.41
5	288,655	14.22	257,767	15.93
6	164,085	25.02	133,197	30.82
7	585,135	7.02	554,247	7.41
8	183,674	22.35	152,786	26.87
Average	295,020	13.92	264,131	15.54

Pels/Document $=$ (2376 lines) (1728 pels/line)
$=$ 4,105,728

EOL $+$ Tag-bit code bits/document $=$ (2376 lines) (13 EOL $+$ tag-bit code bits/line)
$=$ 30,888

Adapted from Ref. 137; assumes high-resolution, MSLT $=$ 0, and a K-factor of infinity.

percentage of the lowest factor, is 8.3% at standard resolution, and 16.0% at high resolution. One of these methods is considered to be an outlier with respect to compression performance. Excluding this outlier, the maximum percentage difference in the compression factor decreases to 7.6% at standard resolution and 5.4% at high resolution. Although the outlier method has the poorest compression factor, it has significantly better immunity to channel errors. The ESF of the outlier method is 32.8. As indicated above, this means that, for each bit error introduced, there are an average of 32.8 reconstructed pels in error. In contrast, for the READ code (which, with modification, was the method adopted by the CCITT), the compression factor is among the best, but the error immunity is the poorest of all seven methods. The ESF of the READ code has a value of 52.2, which is a 59.0% increase over 32.8.

In general, the CCITT study showed no gross differences between predictive coding and contour or contour-related approaches. With respect to other exact coding methods, it has been noted that block coding does not give as good compression performance as contour or contour-related codings. This is not surprising, as the conditional entropy of a source is always less than or equal to the joint entropy under the assumption of stationarity, as indicated above. Additionally, the extended run-length coding method gives somewhat better compression performance than contour or contour-related coding methods, but at the expense of more complex logic. To achieve significantly higher compressions requires use of approximate or lossy encodings, as discussed in the next section.

Approximate Coding

Introduction

With approximate or lossy codings, changes or distortions are introduced to a digitized image with the aim of realizing greater data compression. The changes made to an image may be made by a preprocessor to the encoder or by the encoder itself. A few statistical parameters characterizing the totality of the changes may be included in the coded bits. At the receiver, the decoder may simply reproduce the changed image. Alternatively, a postprocessor to the decoder, or the decoder itself may attempt, by algorithm, to reverse the effects of the changes made. Obviously, such a restoration is not the same as the original digitized image, although the two images may be perceived to be the same or similar to each other.

Approximate encodings differ in the amount and type of image distortion, the savings in coded bits (which affects delivery speed and storage and transmission costs), and the complexity of implementation. The amount and type of image distortion that is acceptable depends upon human visual perception and also upon the cognitive judgment of the human observer relative to a particular application. Not all distortions are perceptible, and not all perceptible distortions are unacceptable. A common base requirement is retention of semantic content, such as readability of text or recognizability of image features. In some facsimile applications, fidelity to the original may not be as important as aesthetic appearance. An example of this is the presentation of graphics in advertisements or other sales vehicles.

In order to compare approximate coding methods, suitable means of comparing the distortions they introduce are needed. In the next section, aspects relating to distortion measures are discussed. This is followed by sections summarizing the major ideas behind approximate encodings. As for exact coding, this summary is not an exhaustive listing of coding methods. In comparing improvements in the compression factor achieved by various approximation methods, the reader is cautioned that the compression factor frequently depends upon image resolution. A true comparison is made only if the results are based on the same digitization of the same images.

Distortion Measures

Two kinds of distortion measures are in use today, namely, subjective distortion measures and objective distortion measures. The modeling and measurement of subjective human responses to varying stimuli and of judgments relating to these stimuli are addressed in the fields of psychometrics, scaling, and multidimensional scaling (146–151). In these techniques, subjects are tested and their responses are analyzed by various statistical methods, giving such measurement values as average opinions, isopreference curves, and differences in judgment. Most of the results obtained in the past emphasize judgment of distortion relative to semantic content (i.e., intelligibility of speech or readability of text). With the expanding use of graphics, additional studies are needed concerning judgment of distortion relative to aesthetic considerations. Also, subjective data generally are difficult and costly to obtain and apply only to a specified application.

For reasons of economy, it is highly desirable to have an objective distortion measure that correlates well to subjective results. A simple objective distortion measure that is commonly used is the mean-squared error (MSE), with the error measured between an uncoded image and the reconstructed image. Other objective distortion measures are the signal-to-noise ratio (SNR) and its associated refinements (8). The SNR is defined as the variance of the uncoded pel values divided by the variance of the difference between the uncoded and reconstructed pel values. However, such measures often correlate poorly with subjective ratings. Two coding methods may produce the same MSE or SNR on a particular image or image ensemble, but they may differ considerably in the placement of the errors within an image and in the overall perceived distortion. The study of psychophysics concerns the properties of human vision, including variations in sensitivity to contrast, chrominance, spatial frequency, and such other phenomena as spatial masking and isotropic orientation (152–156). As a simple example, errors are more noticeable in low-detail image areas than in such high-detail areas as boundaries. By capitalizing on psychophysical knowledge, a coding method can reduce the coded bits by introducing distortion in ways that have little or no effect on perceived image quality.

One approach to embody psychophysical phenomena in an objective distortion measure is to employ a model of the human visual-processing system (157–164). Both space-domain and frequency-domain models are defined, and generally consist of a sequence of operations, such as nonlinear transformations, filtering, masking, averaging, and ORing. A defined sequence of such operations is applied to both the uncoded digitized image and the reconstructed image, yielding processed versions of these images. It is hoped that the processed images represent more closely the images as processed by the eye than do the uncoded and reconstructed images. The MSE or other objective distortion measure is then evaluated for these processed-image versions. This approach results in better correlations of the objective and subjective distortion measures. However, the process is complex; also, the results are only as good as the current psychophysical knowledge, which is limited. For these reasons, the modeling approach is not used widely, although the modeling of human vision gives important insights relative to distortion measurements and design of coding methods.

A related area of theoretical interest is rate distortion theory, a branch of information theory originally formulated by Shannon, and developed further by others (165–175). This theory answers the following question: Given a value of a distortion measure, what is the minimum bit rate to code a given image source? The bit rate is the number of coded bits per source symbol or, equivalently, per sequence of source symbols comprising an image. This then, at least in principle, enables us to evaluate the performance of an actual coding method on an absolute basis. Thus, for a given approximate coding method, the average values of the distortion measure and the number of coded bits, both taken over an image ensemble, may be computed. The theory gives the minimum number of coded bits corresponding to the computed distortion-measure value. The actual average coded bits may then be compared with the theoretical minimum number of coded bits. In reality, this is seldom done for several reasons. First, computation of the theoretical minimum bit rate is complex. Also, it is assumed that the source is characterized by certain parameters of the actual image ensemble. In particular, to compute the theoretical result requires use of a representative set of conditional probabilities $P(a_j \mid b_k)$, where a_j is an uncoded symbol and b_k is the reconstructed symbol. Even for modest image ensembles, there is rarely a meaningful representative set of such probabilities because of variations among images and within each image. Additionally, computation of the theoretical value depends on the definition of an appropriate distortion measure; exactly the same issues arise with respect to defining this measure, as discussed above.

Until other approaches are defined that are more robust and more computationally tractable than those indicated above, the MSE is likely to remain the accepted distortion measure. It is also noted that most of the results concerning distortion measures have been obtained for speech and for multilevel, monochrome, and color images. Very few results apply specifically to bilevel images. One study that does define a distortion measure particularly for facsimile images is reported in Ref. 175.

Binary-Vector Quantization

Binary-vector quantization is an approximate form of the exact block coding method described above. A block of pels with each pel represented by a binary value is referred to as a *binary vector*. For a block of N pels, there are 2^N possible binary vectors. In the design phase of this method, a subset of the total 2^N binary vectors is identified. The members of this subset, called template or reproduction vectors, are stored in a table in both the encoder and the decoder. To encode an image, the encoder finds the best matching reproduction vector for each image block of N pels and outputs the corresponding table index or a variable-length codeword representing the index. The best match minimizes a specified distortion measure, usually the distance, the squared error, or the MSE between the original vector and the reproduction vector. Subjective distortion criteria also are possible. To find a best match requires considerable effort. For each received code word, the decoder looks up the associated reproduction vector and uses it in reconstructing the image. An adaptive form of this method uses several reproduction vector tables and, for any given block of pels, the encoder selects the best table depending upon the contents of the block.

Of obvious importance for encoder processing, and also for storage of the reproduction vectors in both encoder and decoder, is the size of the reproduction vector table. In one experiment on newspaper text, satisfactory quality was achieved using 8-by-8-pel blocks, and only 62 reproduction vectors out of a total of 2^{64} possible vectors (176). Other considerations are the effort required to obtain the reproduction vector values and the degree to which they achieve minimum expected distortion with respect to a set of training vectors. There are several methods to obtain the reproduction vectors, differing in computational complexity and in the degree of optimality achieved. These include approaches using clustering (177), nearest-neighbor selection (178), simulated annealing (179), and neural networks (180,181). Also of concern is the degree to which the training vectors represent an end-use ensemble of images and the possibility of the ensemble characteristics changing over time. For other aspects of vector quantization, see Ref. 182.

Curve Fitting

Curve fitting is a natural extension of the exact chain coding method. Particular boundary pels are identified as critical points, called *knots*. Segments of curves are fitted through each two successive knots. Various curves, such as straight lines, parabolas, and higher-order polynomials, superellipses, and others are used (18,183–186). This approach results in considerable economy of coded bits and, in some cases, the visual result is entirely acceptable. However, the logic to identify the knots and the curve parameters is not trivial. For a comparison of polygonal approximations, chain coding, and run-length coding, see Ref. 187.

Local Modification

In the local-modification approach, image changes are made within relatively small localities by removing or inserting pels or runs of pels or by substituting reduced representations for groups of pels. The starting point often is an exact coding method with added logic to smooth or simplify an image. Visually, the changes are generally quite moderate.

Pel Modification

Simple pel modifications include the removal of isolated pels or small groups of connected pels (11,16,188) and the bridging of isolated black pels (11,188). Somewhat more complex is the smoothing of contours. In one example, gaps or notches in a contour may occur when the image level in an area is close to and oscillates about a fixed quantizer threshold level. Such artifacts may be removed by circuitry (189) or by a software algorithm.

Predictive Coding: Error Reduction

Another application of contour modification is in the use of predictive coding to simplify the image so that the prediction errors are decreased. In one study, slant edges were changed to horizontal or vertical edges and a postprocessor was used in an attempt to restore the original pel values (190). In experiments on some images, this method resulted in an increase in the compression factor ranging from 18% to 57%. Image simplification is also used in another predictive coding system that requires a constant transmission rate (191). To prevent overflow in the output buffer in this system, the number of different types of image modification allowed is changed dynamically in accordance with the state of occupancy of the buffer. Thus, as the buffer becomes fuller, the number of permissible image modifications is increased, thereby increasing the degree of image redundancy, decreasing the number of prediction errors, and, accordingly, decreasing the rate of data flow from the encoder to the buffer. Pels are changed in accordance with a morphological criterion that insures that black pels connected in the original image remain connected and that separated black entities remain separated.

Blob Coding: Reduced Runs

As mentioned above, feature-oriented exact encodings naturally lead to approximate codings by modification of the identified features. Figure 13 illustrates this principle. Figure 13-*a* is the original digitized image, which is also the reconstructed image for exact blob encoding. Figure 13-*b* is the image resulting from an approximate coding derived from the exact blob coding method described above (16). The image in Fig. 13-*b* reflects two approximation operations. The first operation simply removes all of the blob head and tail runs with lengths less than a given amount. The total number of pels so removed is included in the coded bits. Upon reconstruction, runs are annexed to the top

FIG. 13 Comparison of images reconstructed from *a*, exact blob coding; and *b*, approximate coding using local modification.

and bottom of blobs by an algorithm that preserves blob morphology but does not exceed the number of removed pels. The result is near-exact replacement of the runs removed by the encoder. This operation removes small blobs, including isolated pels. Generally, this is not perceptible, except where the blobs are part of a repeated pattern, as is the case for the Benday pattern covering the face in the drawing. The second operation selects and encodes only a subset of the runs in a blob. For two successive selected runs, the second selected run, as well as all those skipped between the first and second selected runs, have left ends no more than three away from the left end of the first selected run, and similarly for the right ends. The second selected run is coded relative to the first selected run with a connection code and with the number of runs skipped. At reconstruction, the skipped runs are filled in by interpolation.

The percentage improvement in the compression factor for the approximate coding over the exact coding varies with resolution. Additionally, the compression factors vary depending on the extent of the background area containing the drawing. The minimum enclosing rectangle for the original artwork in Fig. 13 is 2.1 in by 3.6 in. For this area, at horizontal and vertical resolutions of 192 dpi, the compression factors are 4.12 for the exact coding and 5.88 for the approximate coding. This gives an improvement of the approximate over the exact coding of 43%. At resolutions of 96 dpi, the compression factors are 2.29 for the exact coding and 3.83 for the approximate coding, giving an improvement of 67%. If the drawing is embedded in a larger area of white background such that the larger area contains K times the number of pels of the minimum enclosing rectangle, then the compression factors increase by a factor of K but the percentage improvements of the approximate over the exact coding remain the same. Thus, if the drawing is embedded in a rectangle with each side twice the size of the corresponding side of the minimum enclosing rectangle, then the compression factor increases by a factor of 4. In this case, the compression factors of 4.12 for the exact encoding and 5.88 for the approximate coding cited above become 16.48 and 23.52, respectively.

Pattern Processing

Several coding methods decompose the black pels of an image into distinct subareas or patterns and achieve compression by encoding the patterns efficiently. The patterns chosen for the various coding methods vary in complexity from elementary figures to more complex structures specific to a particular image.

Square and Rectangle Decomposition

The simplest decomposition uses geometric figures. As described in discussing exact codings, a structural element of an image may be decomposed into a set of nonoverlapping inscribed squares or rectangles. Approximate codings may be derived from these representations using a coarse quantization of the sides of the squares or rectangles, and also by retaining only those squares or rectangles with sizes above a given threshold (129). The inscribed rectangles may be found by scanning the pels of an image line by line, from left to right. At each particular pel in the scanning process, the largest rectangle with its upper-left corner positioned on the particular pel is identified, coded, and deleted from the image. The linear scan continues until the entire structure is coded. Algorithms that search for the largest inscribed rectangles at other contour points result in a smaller number of rectangles needed to attain the same quality. In one study, 92% fewer rectangles were required using the extended search (129).

Convex Blob Coding

A structural element also may be approximated by structures called convex blobs. In one definition, a *convex blob* is a blob in which all the minimum-length paths between any pair of pels in the blob are entirely contained in the blob (129). The convex blobs are identified in a region-growing process, in which successive runs in the structural element are attached to a growing convex blob. A run is attached if the convexity condition is satisfied or if it is satisfied by adding a limited number of pels to the run. The degree of approximation, and, inversely, the compression factor, may be varied by changing the ratio of the total number of pels in the convex blobs to the number of added pels. The study also includes a method for representing a structural element by means of a hierarchy of minimum convex hulls, with possible application in progressive image transmission. (See also Ref. 192.)

Protoblob Coding

Another approximation is a version of blob coding in which the blob connectivity conditions are changed in order to represent the structural elements by a set of elementary patterns called protoblobs (113). There are three types of protoblobs: trapezoids, parallelograms, and horizontal straight lines. For the trapezoid, the parameters coded are the length of the top run, the number of runs, and the two angles made by each side of the trapezoid and the vertical. For the parallelogram, the parameters are the same as for the trapezoid, except only one angle is required. For the horizontal line, the parameters are the length of the top run and a line thickness of 1 or 2 pels. A protoblob with five or less pels is discarded. Upon reconstruction, the length of the bottom run for a trapezoid is determined by algorithm. Intervening runs of trapezoids and parallelograms are obtained by linear interpolation.

The degree of approximation, and, inversely, the compression factor, may be changed by varying the quantization of the coded parameters. Other possible variations include relaxing or tightening the connectivity conditions, and coding another parameter to indicate convexity or concavity of the sides of trapezoids and parallelograms so that nonlinear interpolations may be used.

Figure 14 presents a comparison of images to illustrate the improvement of protoblob approximation over exact blob coding and approximate coding using local modification. Figure 14-*a* shows the original digitized image, which is also

FIG. 14 Comparison of images reconstructed from *a*, exact blob coding; *b*, approximate coding using local modification; and *c*, approximate protoblob coding.

the reconstructed image for exact blob coding. For comparison, Fig. 14-*b* shows the image resulting from the local modification discussed above. Figure 14-*c* shows the image resulting from the protoblob coding using a given level of parameter quantizations, and also using linear interpolation. For an image area equal to the minimum enclosing rectangle, the compression factors are 10, 14, and 33 for Figs.14-*a*, 14*b*, and 14*c*, respectively. The improvement of the local modification approximation over the exact coding is 40% and the improvement of the protoblob approximation over the exact coding is 230% (113).

Thinning and Thickening

An approximation that is applicable to images consisting of characters or symbols made up of strokes of nearly the same thickness is to use a thinning operation prior to encoding (193). For each character, outer pels are stripped away, leaving a skeleton, as in the medial-axis transform discussed above in the section on exact coding. For an approximate coding, only the positions of the skeleton pels are encoded. At the receiver, the skeletons are thickened uniformly to obtain the reconstructed image. For additional information concerning thinning processes, see also Refs. 194–196.

Symbol and Pattern Matching

An approximate coding approach that is considerably more complex than those above identifies characters, symbols, or patterns within an image. In general, these methods use a sequence of operations for segmenting, blocking, screening, library updating and management, and coding, as described below.

The earliest version of this method handles images containing only text (197). It scans the image from top to bottom, segmenting lines of text and, within each line, isolating or blocking individual characters or symbols. It then compares each such blocked symbol to bit patterns of prototype symbols in a prototype library. If a prototype is found that matches the blocked symbol within a specified tolerance, then the encoder transmits an identification code of the prototype. If no match is found, the encoder transmits a coded form of the blocked symbol and its position within the image, and also adds the bit pattern to the prototype library. At the start of processing the image, the prototype library is empty, so that the library is populated with symbols obtained from the image itself. At the receiver, any transmitted bit pattern is output to the display device and also is added to a prototype library at the receiver. Subsequent transmissions of the identification code of the symbol cause the decoder to look up the associated bit pattern in its prototype library. For this method, an initial estimate is made of a compression factor of 40.

A subsequent version of this method made several improvements (198). One of these is to segment an image into strips containing either characters or other kinds of material. The strips with characters are encoded in the same manner as indicated above. The strips containing other material are encoded by a run-length coding method. Additionally, the size of the prototype library is limited. When the library is filled to capacity, new symbols replace least-used elements.

Another set of improvements to this method was made subsequently (199). First, a reduction is made in the effort required to determine if a blocked symbol matches one of the prototypes in the prototype library. This is done by screening the prototype library to identify a relatively small subset of prototypes as candidates for matching a blocked symbol. To do this, a set of scalar features is determined for each prototype. The features of the blocked symbol are likewise computed, and a feature-space distance between the blocked symbol and each prototype is defined. Only those prototypes within a given feature-space distance of the blocked symbol are actually compared to the blocked symbol. Furthermore, the candidate prototypes are compared in the order of increasing distance. An improvement also was made to the matching algorithm. Conventional matchers identify the pels, called error pels, that differ between the blocked symbol and the prototype. A match is declared if the sum of the error pels is less than a given threshold. In the improved algorithm, the pattern of the error pels also is taken into account. For each error pel, a count is made of all the error pels, including itself, in its 3×3 pel neighborhood. The count is taken as a weighted value of the error pel. A match is declared if the sum of the weighted values of all the error pels is less than a given threshold. Error patterns that have dense clumps of error pels result in larger weighted-value sums than error patterns with sparsely distributed error pels. Thus, densely clumped error patterns less frequently result in declared matches, which is desirable because dense clumps signal gross feature mismatches. Finally, areas within a line of print that cannot be blocked into symbols are left as residue to be encoded by the run-length coding method. Simulation results indicate that, for images containing primarily text, this version of the symbol and pattern matching method has an improvement in the compression factor of about 100% over the method for an exact, two-dimensional, run-length method. For images containing primarily graphics, the compression factor is about the same for the two cases. Pratt et al. also defined an extension of the method that uses a preloaded prototype library, such as the alphabetic characters of a given font. In effect, this is a character-recognition scheme. For one image, the compression factor achieved is 257, giving a 424% increase in compression for the approximate coding method over the exact coding method.

Several other improvements have also been made to this method (145). Included is the ability for a prototype to be either a character or a small graphical symbol, or a graphical element, such as a line segment or a fraction of a larger figure. All black areas in an image are decomposed into library elements, leaving no residue to be coded in a run-length mode. Another improvement is that the screening process takes into account both the feature-space distance and the probability of occurrence of a library element. Also, the matching process uses local error patterns in making decisions. An implementation of a real-time coder for this version of the method has been made (200). Taken over all eight CCITT documents at high resolution, this version of the method gives an improvement of 92% over the exact two-dimensional CCITT code. For two of the documents, which primarily have graphical content, the compression factor is somewhat less than for the two-dimensional CCITT code; for one document, which has primarily text content, the improvement is 378%. In comparing this version of the method with the one in Ref. 199, this newer version shows an improvement ranging from 19% to 81%.

Another study on the CCITT images estimates the compressions for four variations of the symbol-and-pattern-matching approach (201). The improvement of these variations over the two-dimensional CCITT coding ranges from 200% to 290%. Finally, other experiments using a symbol-matching method on a set of images that are not the CCITT images show an improvement over two-dimensional coding of up to 185% for typewritten material, but a slight loss on graphical material (202).

Region Coding

As described above, the symbol-and-pattern-matching methods segment an image into a set of subareas, each of which generally contains either a symbol or residue. The expectation is that an image contains a large component of text, so that many subareas contain characters, and improved compressions result due to using short codewords for characters. As noted above, facsimile transmissions increasingly may contain complex material, including line drawings, photographic illustrations, and computer-generated imagery, with or without text. This points to a more general segmentation of an image into regions of varying content, with the coding of each region optimally matched to the content of the region (see Ref. 203).

Another approach extends this concept into segmenting an image into regions of varying texture and characterizing each region by texture features. Although much of the work on texture characterization has been done for scene analysis in pattern recognition, it may also be used in image coding. A texture encoding consists of the region boundaries and the texture features for each region. Several approaches for texture description are used in image processing (4,18,24,41,204). One of these uses statistical parameters, such as moments of a gray-level histogram, and descriptors of gray-level co-occurrence matrixes. Another approach defines complex texture patterns by means of combinations of simple texture primitives. A third approach uses descriptors based on the Fourier spectrum of the image. The receiver uses the texture features to synthesize the texture within a region (205–207). An important concern of texture coding is to define a mathematical measure of the differences and similarities between texture fields that agrees with human judgment (208). In one study, perceived differences between textures are characterized by a distance function based on textural descriptors (209). Of interest, experiments show that humans cannot discriminate between random texture fields that have the same second-order statistics (210).

A final coding approach using region coding is based on fractal geometry (211–214). In this method, each region is matched to a fractal image that resembles it, that is, the Hausdorff distance between the two is small. The fractal image is defined by a repetition of a set of contractive affine transformations. Given a point (x, y) in a plane, an affine transformation defines a new point (x', y') as follows:

$$x' = ax + by + e$$
$$y' = cx + dy + f$$

where a, b, c, d, e, and f are numbers. This mapping can scale, translate, rotate, and distort images. It is a contraction if, given the Euclidean distances D between any two points and D' between the transformed points, then $D' \leq kD$ for some $k < 1$. In this case, the receiver generates an output image by applying the chosen set of affine transformations many times. The set of affine transformations and their repeated application is called an *iterated function system*. Compression factors achieved with this method are reported to be about 100 for gray-level images (215) and 543 for color images (216). (See also Refs. 217–221.)

Bibliography

Books

Baxes, G. A., *Digital Image Processing*, Prentice-Hall, Englewood Cliffs, NJ, 1984.

Costigan, D. M., *Electronic Delivery of Documents and Graphics*, Van Nostrand Reinhold, New York, 1978.

Costigan, D. M., *FAX — The Principles & Practice of Facsimile Communication*, Chilton Books, Radnor, PA, 1971.

Ekstrom, M. P. (ed.), *Digital Image Processing Techniques*, Academic Press, New York, 1984.

Gray, R. M., and Davisson, L. D., *Data Compression*, Dowden, Hutchinson, & Ross, Academic Press, New York, 1976.

Green, W. B., *Digital Image Processing, A Systems Approach*, Van Nostrand Reinhold, New York, 1983.

Held, G., *Data Compression Techniques and Applications: Hardware and Software Considerations*, John Wiley & Sons, New York, 1983.

Huang, T. S., *Picture Processing and Digital Filtering*, Springer-Verlag, New York, 1975.

Jain, A. K., *Fundamentals of Digital Image Processing*, Prentice-Hall, Englewood Cliffs, NJ, 1989.

Kazan, B. (ed.), *Advances in Image Pickup and Display*, Academic Press, Orlando, FL, 1974 (Vol. 1); 1975 (Vol. 2); 1977 (Vol. 3); 1981 (Vol. 4).

Pearson, D. E., *Transmission and Display of Pictorial Information*, Halsted Press, New York, 1975.

Rosenfeld, A. (ed.), *Image Modelling*, Academic Press, New York, 1981.

Rosenfeld, A. (ed.), *Multiresolution Image Processing and Analysis*, Springer-Verlag, Berlin, 1984.

Rosenfeld, A., and Kak, A. C., *Digital Image Processing*, 2d ed., Academic Press, Orlando, FL, 1982.

Sakrison, D. J., *Notes on Analog Communication*, Van Nostrand Reinhold, New York, 1970.

Stoffel, J. C., *Graphical and Binary Image Processing and Applications*, Artech House, Dedham, MA, 1982.

Tou, J. T., and Gonzalez, R. C., *Pattern Recognition Principles*, Addison-Wesley, Reading, MA, 1974.

Surveys, Bibliographies

Algazi, V. R., (ed.), *Image Coding*, Report of a Workshop Held at Boulder, CO, December 14–16, 1977, National Science Foundation, NSF/RA-790300, 1979.

Andrews, H. C., Tescher, A. G., and Kruger, R. P., Image Processing by Digital Computer, *IEEE Spectrum*, 9:20–32 (1972).

Arps, R. B., Bibliography on Binary Image Compression, *Proc. IEEE*, 68:922–924 (1980).

Arps, R. B., Bibliography on Digital Graphic Image Compression and Quality, *IEEE Trans. Inform. Theory*, IT-20:120–122 (1974).

Habibi, A., and Robinson, G. S., A Survey of Digital Picture Coding, *Computer*, 7:22–34 (1974).

Hunt, B. R., Digital Image Processing, *Proc. IEEE*, 63:693–708 (1975).

Jain, A. K., Image Data Compression, a Review, *Proc. IEEE*, 69:349–389 (1981).

Jarvis, J. F., Judice, C. N., and Ninke, W. H., A Survey of Techniques for the Display of Continuous Tone Pictures on Bilevel Displays, *Computer Graphics and Image Processing*, 5:13–40 (1976).

Rosenfeld, A., Picture Processing, *Computer Graphics and Image Processing*, 1:394–416 (1972).

Rosenfeld, A., Picture Processing: 1973, *Computer Graphics and Image Processing*, 3: 178–194 (1974).

Rosenfeld, A., Progress in Picture Processing: 1969–71, *Comput. Surv.*, 5:81–108 (1973).

Schreiber, W. F., Picture Coding, *Proc. IEEE*, 55:320–330 (1967).

Stoffel, J. C., and Moreland, J. F., A Survey of Electronic Techniques for Pictorial Image Representation, *IEEE Trans. Commun.*, 29:1898–1925 (1981).

Yasuda, Y., et al., Advances in Fax, *Proc. IEEE*, 73:706–729 (1985).

Other Articles

Arps, R. B., *Entropy of Printed Matter at the Threshold of Legibility for Efficient Coding in Digital Image Processing*, Report No. 31, Stanford Electronics Laboratory, CA, 1969.

Bodson, D., and Randall, N. C., Analysis of Group 4 Facsimile Throughput, *IEEE Trans. Commun.*, COM-34:849–861 (1986).

Burt, P. J., and Adelson, E. H., The Laplacian Pyramid as a Compact Image Code, *IEEE Trans. Commun.*, COM-31:532–540 (1983).

Chaudhuri, B. B., and Kundu, M. K., Digital Line Segment Coding: A New Efficient Contour Coding Scheme, *IEE Proc.*, Part E, *Computers and Digital Techniques*, 131:143–147 (1984).

Coco, D., CRT Makers Looking beyond High Resolution, *EDN*, June 14, 1990.

Deutsch, S., A Note on Some Statistics Concerning Typewritten or Printed Material, *IRE Trans. Inform. Theory*, IT-3:147–148 (1957).

Fisher, S., Future of Fax Machines is Very Bright (and Diverse) Indeed, *Infoworld*, Supplement, February 26, 1990.

Glass, B., and Needleman, R., Network Fax Servers, *Infoworld*, Supplement, February 26, 1990.

Golomb, S. W., Run-Length Encoding, *IEEE Trans. Inform. Theory*, IT-12:399–401 (1966).

Grallert, H. J., Means of Picture Encoding in the Broadband ISDN, *Proc., GLOBECOM '86*, pp. 356–360 (1986).

Knowlton, K., Progressive Transmission of Grey-Scale and Binary Pictures by Simple, Efficient, and Lossless Encoding Schemes, *Proc. IEEE*, 68:885–896 (1980).

Kunt, M. Statistical Models and Information Measurements for Two-Level Digital Facsimile, *Information and Control*, 33:333–350 (1977).

Mosley, J. D., Widespread Graphics Use Spawns Diversity in Data-Compression Devices, *EDN*, 87–94 (September 15, 1988).

Robinson, J. A., Low-Data-Rate Visual Communication Using Cartoons: A Comparison of Data Compression Techniques, *IEE Proc.*, Part F, *Commun. Radar Signal Proc.*, 133:236–256 (1986).

Rooney, P., Sharp Moves Ahead with Color Fax, *EDN*, June 14, 1990.

Wilson, R., Image-Compression Chips Advance on Three Fronts, *Computer Design*, 49–51 (November 1, 1990).

References

1. Buerger, D. J., The Challenge of 1990: Bridging E-Mail, Fax, and Voice Mail, *Infoworld*, (January 22, 1990).

2. Frank, A. J., and Groff, R. H., On Statistical Coding of Two-Tone Image Ensembles, *Proc. Soc. Inform. Display*, 17:102–110 (1976).

3. Williams, T., Document Processing Moves to the Desktop, *Computer Design*, 63–70 (July 1, 1989).

4. Schalkoff, R. J., *Digital Image Processing and Computer Vision*, John Wiley & Sons, New York, 1989.

5. Garvey, J. M., Scanners Put Design On Line, *High Technology*, 63–65 (October 1985).

6. Andrews, W., Processor Board Brings Line-Scan to PC/ATs, *Computer Design*, (June 1, 1990), 123.

7. Nyquist, H., Certain Topics in Telegraph Transmission Theory, *Trans. AIEE*, 617–644 (1946).

8. Jayant, N. S., and Noll, P., *Digital Coding of Waveforms*, Prentice-Hall, Englewood Cliffs, NJ, 1984.

9. Max, N. L., Antialiasing, *IEEE Computer Graphics and Applications*, 10:18–30 (1990).

10. Arps, R. B., et al., Character Legibility vs. Resolution in Image Processing of Printed Matter, *IEEE Trans. Man-Machine Systems*, MMS-10:66–71 (1969).

11. Yasuda, Y., Overview of Digital Facsimile Coding Techniques in Japan, *Proc. IEEE*, 68:830–845 (1980).

12. Huang, T. S., Coding of Two-Tone Images, *IEEE Trans. Commun.*, COM-25: 1406–1424 (1977).

13. Warner, W. C., Compression Algorithms Reduce Digitized Images to Manageable Size, *EDN*, 203–212 (June 21, 1990).

14. Ramachandran, K., Coding Method for Vector Representation of Engineering Drawings, *Proc. IEEE*, 68:813–817 (1980).

15. American Newspaper Publisher's Association (ANPA) Research Institute Bulletin 94, Washington, DC, March 18, 1968.

16. Frank, A. J., High Fidelity Encoding of Two-Level, High Resolution Images, *Conf. Record*, IEEE International Conf. Commun., Session 26, 5–10 (June 1973). Also published as *Bell Laboratories Report*, Murray Hill, NJ, September 8, 1970.

17. Frank, A. J., and Groff, R. H., *Analysis of Field Tests to Determine Resolution Quality Standards of Logos for Automated Yellow Pages Printing*, Bell Laboratories Report, Murray Hill, NJ, 1975.

18. Pavlidis, T., *Algorithms for Graphics and Image Processing*, Computer Science Press, Rockville, MD, 1982.

19. Schwartz, M., *Information, Transmission, Modulation, and Noise*, McGraw-Hill, New York, 1970.

20. Huang, T. S., PCM Picture Transmission, *IEEE Spectrum*, 57–63 (December 1965).

21. Frank, A. J., and Schilling, J. M., Coding Ultrasound Images, *Proc. 1977 IEEE Workshop, Picture Data Description and Management*, pp. 172–181 (April 1977).

22. Roberts, L. G., Picture Coding Using Pseudo-Random Noise, *IRE Trans. Inform. Theory*, IT-8:145–154 (1962).

23. Lippel, B., and Kurland, M., The Effect of Dither on Luminance Quantization of Pictures, *IEEE Trans. Commun.*, COM-19:879–888 (1971).

24. Gonzalez R. C., and Wintz, P., *Digital Image Processing*, 2d ed., Addison-Wesley, Reading, MA, 1987.

25. Bruce, R. A., Optimum Pre-emphasis and De-emphasis Networks for Transmission of Television by PCM, *IEEE Trans. Commun. Tech.*, COM-12:91–96 (1964).

26. Spang, H. A., and Schultheiss, P. R., Reduction of Quantizing Noise by Use of Feedback, *IRE Trans. Commun. Systems*, 373–380 (1962).

27. Wezska, J. S., A Survey of Threshold Selection Techniques, *Computer Graphics and Image Processing*, 7:259–265 (1978).

28. Ting, D., and Prasada, B., Digital Processing Techniques for Encoding of Graphics, *Proc. IEEE*, 68:757–769 (1980).

29. Coale, K., Xerox Flatbed Scanner Works with Mac SE, II, *Infoworld*, (April 23, 1990).

30. Nakamura, R., AVR Unveils 300-dpi HP Scanjet Plus Clone with Picture Publisher, *Infoworld*, (May 14, 1990).

31. Coale, K., Hand-held Scanners: A Portable Alternative, *Infoworld*, (April 23, 1990).

32. Brownstein, M., Hand-held Scanner Offers "Trainable" OCR, *Infoworld*, (January 29, 1990).

33. Wilson, A., Image Scanners Speed Data Acquisition, *Electronic Imaging*, 46–49 (November 1984).

34. Pane, P. J., Sharp Introduces 386 Portable, Scanner, Desktop Laser Printer, *Infoworld*, (November 27, 1989).

35. Darrow, B., Nikon Film Scanner Teams with PCs for High Resolution Art, *Infoworld*, (January 29, 1990).

36. Helgerson, L., Scanners Present Maze of Options, *Mini-Micro Systems*, 71–81 (January 1987).

37. Kolmogorov, A., Logical Basis for Information Theory and Probability Theory, *IEEE Trans. Inform. Theory*, IT-14:662–664 (1968).

38. Fine, T. L., *Theories of Probability*, Academic Press, New York, 1973.

39. Ziv, J., Coding Theorems for Individual Sequences, *IEEE Trans. Inform. Theory*, IT-24:405–412 (1978).

40. Arps, R. B., Binary Image Compression. In: *Image Transmission Techniques* (W. K. Pratt, ed.), Academic Press, San Francisco, 1979, pp. 219–274.

41. Pratt, W. K. (ed.), *Image Transmission Techniques*, Academic Press, San Francisco, 1979.

42. Shannon, C. E., A Mathematical Theory of Communication, *Bell Sys. Tech. J.*, 27, 379–423 (Part I), 623–656 (Part II) (1948).

43. Shannon, C. E., and Weaver, W., *The Mathematical Theory of Communication*, University of Illinois Press, Urbana, IL, 1949.

44. Abramson, N., *Information Theory and Coding*, McGraw-Hill, New York, 1963.

45. Gallagher, R. G., *Information Theory and Reliable Communication*, John Wiley & Sons, New York, 1968.

46. Slepian, D. (ed.), *Key Papers in the Development of Information Theory*, IEEE Press, New York, 1973.

47. Berlekamp, E. R. (ed.), *Key Papers in the Development of Coding Theory*, IEEE Press, New York, 1974.

48. Hamming, R., *Coding and Information Theory*, Prentice-Hall, Englewood Cliffs, NJ, 1980.

49. Viterbi, A. J., and Omura, J. K., *Principles of Digital Communication and Coding*, McGraw-Hill, New York, 1979.

50. Reza, F. M., *An Introduction to Information Theory*, McGraw-Hill, New York, 1961.

51. Ash, R. B., *Information Theory*, John Wiley & Sons, New York, 1965.

52. Wyner, A. D., Fundamental Limits in Information Theory, *Proc. IEEE*, 69:239–251 (1981).

53. Papoulis, A., *Probability, Random Variables, and Stochastic Processes*, McGraw-Hill, New York, 1965.

54. Cox, D. R., and Miller, H. D., *The Theory of Stochastic Processes*, John Wiley & Sons, New York, 1965.

55. Parzen, E., *Probability, Random Variables, and Stochastic Processes*, McGraw-Hill, New York, 1965.

56. Lathi, B. P., *An Introduction to Random Signals and Communication Theory*, International Textbook Company, Scranton, PA, 1968.

57. Feller, W., *An Introduction to Probability Theory and Its Applications*, John Wiley & Sons, New York, 1950.

58. Berger, T., Minimum Entropy Quantizers and Permutation Codes, *IEEE Trans. Inform. Theory*, IT-28:149–157 (1982).

59. Huffman, D., A Method for the Construction of Minimum Redundancy Codes, *Proc. IRE*, 40:1098–1101 (1952).

60. Frank, A. J., Fast, Uniform Decoding of Minimum Redundancy Codes, *Proc. Eighth Annual Princeton Conf. Inform. Sciences and Systems*, 393–397 (March 1974). Also, Frank, A. J., U. S. Patent No. 3,883,847, issued May 13, 1975.

61. Faller, N., An Adaptive System for Data Compression, *Conf. Record Seventh IEEE Asilomar Conf. Circuits and Sys.*, 593–597 (1973).

62. Cormack, G. V., and Horspool, R. N., Algorithms for Adaptive Huffman Codes, *Inform. Proc. Letters*, 18:159–165 (1984).

63. Gallagher, R. G., Variations on a Theme by Huffman, *IEEE Trans. Inform. Theory*, IT-24:668–674 (1978).

64. Knuth, D. E., Dynamic Huffman Coding, *J. Algorithms*, 6:163–180 (1985).

65. Vitter, J. S., Design and Analysis of Dynamic Huffman Coding, *J. ACM*, 34:825–845 (1987).

66. Lelewer, D. A., and Hirschberg, D. S., *Data Compression*, Technical Report No. 87-10, Dept. Inform. and Computer Science, University of California, Irvine, CA, 1987.

67. Larmore, L. L., and Hirschberg, D. S., A Fast Algorithm for Optimal Length-Limited Huffman Codes, *J. ACM*, 37:464–473 (1990).

68. Mukherjee, A., and Bassiouni, M. A., On-the-Fly Algorithms for Data Compression, *Proc. ACM/IEEE Fall Joint Computer Conf.*, (1987).

69. Storer, J. A., *Data Compression: Methods and Theory*, Computer Science Press, Rockville, MD, 1988.

70. McIntyre, D. R., and Pechura, M. A., Data Compression Using Static Huffman Code-Decode Tables, *J. ACM*, 28:612–616 (1985).

71. Huang, T. S., Easily Implementable Suboptimum Run Length Codes, *Proc. International Conf. Commun.*, Session 7:8–11 (1975).

72. Elias, P., Predictive Coding, *IRE Trans. Inform. Theory*, IT-1:16–33 (1955).

73. Laemmel, A. E., *Coding Processes for Bandwidth Reduction in Picture Transmission*, Report R-246-51, PIB-187, Microwave Research Institute, Polytechnic Institute of Brooklyn, New York, August 30, 1951.

74. Meyer, H., Rosdolsky, H. G., and Huang, T. S., Optimum Run Length Codes, *IEEE Trans. Commun.*, COM-21:826–835 (1973).

75. Jelinek, F., *Probabilistic Information Theory*, McGraw-Hill, New York, 1968.

76. Schalkwijk, J. P. M., An Algorithm for Source Coding, *IEEE Trans. Inform. Theory*, IT-18:395–399 (1972).

77. Cover, T. M., Enumerative Source Coding, *IEEE Trans. Inform. Theory*, IT-19:73–77 (1973).

78. Elias, P., On Binary Representations of Monotone Sequences, *Proc. Sixth Annual Princeton Conf. Inform. Sciences and Systems*, 54–57 (March 1972).

79. Rissanen, J. J., Generalized Kraft Inequality and Arithmetic Coding, *IBM J. Research and Development*, 20:198 (1976); first published as IBM Research Report RJ-1591, June 1975.

80. Pasco, R. C., *Source Coding Algorithms for Fast Data Compression*, Ph.D. thesis, Department of Electrical Engineering, Stanford University, Stanford, CA, May 1976.

81. Rubin, F., Arithmetic Stream Coding Using Fixed Precision Registers, *IEEE Trans. Inform. Theory*, IT-25:672–675 (1979).

82. Jones, C. B., An Efficient Coding System for Long Source Sequences, *IEEE Trans. Inform. Theory*, IT-27:280–291 (1981).

83. Langdon, G. G., Jr., and Rissanen, J. J., Compression of Black-White Images with Arithmetic Coding, *IEEE Trans. Commun.*, COM-29:858–867 (1981).

84. Arps, R. B., et al., A Multipurpose VLSI Chip for Adaptive Data Compression of Bilevel Images, *IBM J. Research and Development*, 32:775–795 (1988).

85. Langdon, G. G., Jr., An Introduction to Arithmetic Coding, *IBM J. Research and Development*, 28:135–149 (1984).

86. Witten, I. H., Neal, R. M., and Cleary, J. G., Arithmetic Coding for Data Compression, *Commun. ACM*, 30:520–540 (1987).

87. Bentley, J. L., et al., A Locally Adaptive Data Compression Scheme, *Commun. ACM*, 29:320–330 (1986).

88. Kunt, M., and Johnsen, O., Block Coding of Graphics: A Tutorial Review, *Proc. IEEE*, 68:770–786 (1980).

89. de Coulon, F., and Kunt, M., An Alternative to Run Length Coding for Black-and-White Facsimile, *Proc. 1974 International Zurich Seminar on Digital Commun.*, C4:1–4 (1974).

90. Huang, T. S., and Shahid-Hussain, A. B., Facsimile Coding by Skipping White, *IEEE Trans. Commun.*, COM-23:1452–1466 (1975).

91. de Coulon, F., and Johnsen, O., Adaptive Block Scheme for Source Coding of Black-and-White Facsimile, *Electronics Letters*, 12:61–62 (1976), and erratum, 12 (March 1976).

92. Johnsen, O., Étude de Strategies Adaptives pour la Transmission d'Images Facsimile a Deux Niveaux, *AGEN-Mitteilungen*, 20:41–53 (June 1976).

93. Cheung, W. N., Calculation of Conditional Probabilities in Hierarchical Block Coding Systems, *IEEE Trans. Commun.*, COM-38:747–748 (1990).

94. Capon, J., A Probabilistic Model for Run-Length Coding of Pictures, *IRE Trans. Inform. Theory*, IT-5:157–163 (1959).

95. Rosenheck, B., FASTFAX, a Second Generation Facsimile System Employing Redundancy Reduction Techniques, *IEEE Trans. Commun. Tech.*, COM-18:772–779 (1970).

96. Huang, T. S., Run-Length Coding and Its Extensions. In: *Picture Bandwidth Compression* (T. S. Huang and O. J. Tretiak, eds.), Gordon and Breach, New York, 1972, pp. 231–266.

97. Huang, T. S., and Tretiak, O. J. (eds.), *Picture Bandwidth Compression*, Gordon and Breach, New York, 1972.

98. Zamperoni, P., Verbesserte Runlänge-Codierung fur Text und Zeichnungen, *Nachrichtentech. Z.*, 25:204–206 (April 1972).

99. Zamperoni, P., Quelques Methodes de Codage par Plages d'Images a Deux Niveaux, *L'Onde Electrique*, 53:379–383 (November 1973).

100. Preuss, D., Comparison of Two-Dimensional Facsimile Coding Schemes, *Proc. International Conf. Commun.*, 7:12–16 (June 1975).

101. Wholey, J. S., The Coding of Pictorial Data, *IRE Trans. Inform. Theory*, IT-7:99–104 (1961).

102. White, H. E., Lippman, M. D., and Powers, K. H., Dictionary Look-up Encoding of Graphical Data. In: *Picture Bandwidth Compression* (T. S. Huang and O. J. Tretiak, eds.), Gordon and Breach, New York, 1972, pp. 267–281.

103. Takagi, M., and Tsuda, T., Bandwidth Compression for Facsimile Using Two-Dimensional Prediction, *J. Japan*, 56-D:170 (1973).

104. Stern, P. A., and Heinlein, W. E., Facsimile Coding Using Two-Dimensional Run Length Prediction, *1974 International Zurich Seminar on Digital Commun.*, C5:1–5 (March 1974).

105. Musmann, H. G., and Pruess, D., Comparison of Redundancy Reducing Codes for Facsimile Transmission of Documents, *IEEE Trans. Commun.*, COM-25:1425–1433 (1977).

106. Kobayashi, H., and Bahl, L. R., Image Data Compression by Predictive Coding, I and II, *IBM J. Research Development*, 18:164–179 (1974).

107. Bodson, D., and Schaphorst, R., Compression and Error Sensitivity of Two-Dimensional Facsimile Coding Techniques, *Proc. IEEE*, 68:846–853 (1980).

108. Freeman, H., On the Encoding of Arbitrary Geometric Configurations, *IRE Trans. Electronic Computers*, EC-10:260–268 (1961).

109. Morrin, T. H., Chain-Link Compression of Arbitrary Black-White Images, *Computer Graphics and Image Processing*, 5:172–189 (1976).

110. Sobel, I., Neighborhood Coding of Binary Images for Fast Contour Following and General Binary Array Processing, *Computer Graphics and Image Processing*, 8:127–135 (1978).
111. Frank, A. J., Groff, R. H., and Schmidt, A. C., Jr., *Proposal for Two-Dimensional Facsimile Coding Scheme*, submitted by AT&T as Contribution No. 81 to CCITT COM Study Group XIV (March 1979).
112. British Post Office, *Proposal for Optional Two-Dimensional Coding Scheme for Group 3 Facsimile Apparatus*, submitted as Contribution No. 77 to CCITT COM Study Group XIV (March 1979).
113. Frank, A. J., Two-Dimensional Coding for Graphical Data Compression, *Proc. Bell Laboratories Graphics Processing Conference*, 1–29 (January 1980).
114. Blum, H., An Associative Machine for Dealing with the Visual Field and Some of Its Biological Implications. In: *Biological Prototypes and Synthetic Systems*, Vol. 1 (E. E. Bernard and M. R. Kare, eds.), Plenum Press, New York, 1962, pp. 244–260.
115. Bernard, E. E., and Kare, M. R. (eds.), *Biological Prototypes and Synthetic Systems*, Vol. 1, Plenum Press, New York, 1962.
116. Blum, H., A Transformation for Extracting New Descriptors of Shape, In: *Models for the Perception of Speech and Visual Form* (W. Wathen-Dunn, ed.), M.I.T. Press, Cambridge, MA, 1967, pp. 153–171.
117. Wathen-Dunn, W. (ed.), *Models for the Perception of Speech and Visual Form*, M.I.T. Press, Cambridge, MA, 1967.
118. Blum, H., Biological Shape and Visual Science (Part I), *J. Theo. Biology*, 38: 205–287 (1973).
119. Kotelly, J. A., *A Mathematical Model of Blum's Theory of Pattern Recognition*, Research Report 63-164, Air Force Cambridge Research Laboratories, Boston, 1963.
120. Agin, G. J., and Binford, T. O., Computer Description of Curved Objects, *IEEE Trans. Computers*, 25:439–440 (1976).
121. Philbrick, O., *A Study of Shape Recognition Using the Medial Axis Transform*, Air Force Cambridge Research Laboratories, Boston, 1966. Also in *Pictorial Pattern Recognition*, Thompson Book Company, Washington, DC, 1968, pp. 395–407.
122. Rosenfeld, A., and Pfaltz, J. L., Sequential Operations in Digital Picture Processing, *J. ACM*, 13:471–494 (1966).
123. Rosenfeld A., and Pfaltz, J. L., Computer Representation of Planar Regions by their Skeletons, *Commun. ACM*, 10:119–125 (1967).
124. Montanari, U., A Method for Obtaining Skeletons Using a Quasi-Euclidean Distance, *J. ACM*, 15:600–624 (1968).
125. Hilditch, C. J., Linear Skeletons from Square Cupboards, In: *Machine Intelligence 4* (B. Meltzer and D. Michie, eds.), American Elsevier, New York, 1969, pp. 403–420.
126. Meltzer, B., and Michie, D. (eds.), *Machine Intelligence 4*, American Elsevier, New York, 1969.
127. Arcelli, C., Cordella, L., and Levialdi, S., A Grassfire Transformation for Binary Digital Pictures, *Second International Joint Conf. Pattern Recog.*, 152–154 (August 1974).
128. Frank, A. J., Daniels, J. D., and Unangst, D. R., Progressive Image Transmission Using a Growth-Geometry Coding, *Proc. IEEE*, 68:897–909 (1980).
129. Zamperoni, P., Analysis and Synthesis of Binary Images by Means of Convex Blobs, *Proc. International Conf. Digital Signal Processing*, 181–186 (September 1978).
130. Elias, P., Universal Codeword Sets and Representations of the Integers, *IEEE Trans. Inform. Theory*, IT-21:194–203 (1975).

131. Seery, J., and Ziv, J., *A Universal Data Compression Algorithm Description and Preliminary Results*, Bell Laboratories Report, Murray Hill, NJ, March 23, 1977.

132. Ziv, J., and Lempel, A., Compression of Individual Sequences via Variable-Rate Coding, *IEEE Trans. Inform. Theory*, IT-24:530–536 (1978).

133. Rissanen, J., A Universal Data Compression System, *IEEE Trans. Inform. Theory*, IT-29:656–664 (1983).

134. Welch, T. A., A Technique for High-Performance Data Compression, *IEEE Computer*, 14:8–19 (1984).

135. Lempel, A., and Ziv, J., Compression of Two-Dimensional Data, *IEEE Trans. Inform. Theory*, IT-32:1–8 (1986).

136. Sheinwald, D., Lempel, A., and Ziv, J., Two-Dimensional Encoding by Finite-State Encoders, *IEEE Trans. Commun.*, COM-38:341–347 (1990).

137. Hunter, R., and Robinson, A. H., International Digital Facsimile Coding Standards, *Proc. IEEE*, 68:854–867 (1980).

138. Electronic Industries Association, and the British Facsimile Industries Compatibility Committee, *Contribution to Question 2/XIV Point B1*, CCITT Contribution COM Study Group XIV, COM XIV, No. 3, March 1977.

139. Japan, *Proposal for Draft Recommendation of Two-Dimensional Coding Scheme*, CCITT Contribution COM Study Group XIV, No. 42, November 1978.

140. Wakahara, Y., et al., *Data Compression of Facsimile Signals by Relative Address Coding*, Paper of Tech. Group IECE Japan, CS 74-115, November 1974 (In Japanese).

141. Wakahara, Y., et al., Data Compression Factors of Relative Address Coding Scheme for Facsimile Signals, *J. IEEE Japan*, 5:92 (1976). (In Japanese)

142. Nakagome, Y., Teramura, H., and Hattori, N., Digital Processing and Switching of Facsimile Signals, *Proc. Technical Symposium, World Telecommunication Forum*, Session 3.3:1–6 (October 1975).

143. Yamada, T., *Edge Difference Coding for Facsimile Signals*, Paper of Tech. Group IECE Japan, IE 76-69, 1976. (In Japanese)

144. Yamada, T., Edge-Difference Coding—A New, Efficient, Redundancy Reduction Technique for Facsimile Signals, *IEEE Trans. Commun.*, COM-27:1210–1217 (1979).

145. Johnsen, O., Segen, J., and Cash, G. L., Coding of Two-Level Pictures by Pattern Matching and Substitution, *Bell Sys. Tech. J.*, 62:2513–2545 (1983).

146. Guilford, J. P., *Psychometric Methods*, McGraw-Hill, New York, 1954.

147. Torgerson, W. S., *Theory and Methods of Scaling*, John Wiley & Sons, New York, 1958.

148. Thurstone, L. L., *The Measurement of Values*, University of Chicago Press, Chicago, IL, 1959.

149. Shepherd, R. N., Romney, A. K., and Nerlove, S. B., *Multidimensional Scaling*, Vols. 1 and 2, Seminar Press, New York, 1972.

150. Carterette, E. C., and Friedman, M. P. (eds.), *Handbook of Perception*, Academic Press, New York, 1975.

151. Kruskal, J. B., and Wish, M., Multidimensional Scaling. In: *Sage University Paper No. 11, Series on Quantitative Applications in the Social Sciences* (E. M. Uslaner, ed.), Sage Publications, Beverly Hills, CA, 1978.

152. Stevens, S. S., The Psychophysics of Sensory Function. In: *Sensory Communication* (W. A. Rosenblith, ed.), John Wiley & Sons, New York, 1961, pp. 1–35.

153. Rosenblith, W. A. (ed.), *Sensory Communication*, M.I.T. Press and John Wiley & Sons, New York, 1961.

154. Graham, C. H., (ed.), *Vision and Visual Perception*, John Wiley & Sons, New York, 1965.

155. Cornsweet, T. N., *Visual Perception*, Academic Press, New York, 1970.

156. Biberman, L. M., (ed.), *Perception of Displayed Information*, Plenum Press, New York, 1973.

157. Pearson, D. E., A Realistic Model for Visual Communication Systems, *Proc. IEEE*, 55:380–389 (1967).

158. Campbell, F. W., The Human Eye as an Optical Filter, *Proc. IEEE*, 56:1009–1014 (1968).

159. Stockham, T. G., Jr., Image Processing in the Context of a Visual Model, *Proc. IEEE*, 60:828–842 (1972).

160. Budrikis, Z. L., Visual Fidelity Criterion and Modeling, *Proc. IEEE*, 60:777–779 (1972).

161. Mannos, J. L., and Sakrison, D. J., The Effects of a Visual Fidelity Criterion on the Encoding of Images, *IEEE Trans. Inform. Theory*, IT-20:525–536 (1974).

162. Sakrison, D. J., On the Role of Observer and a Distortion Measure in Image Transmission, *IEEE Trans. Commun.*, COM-25: (1977).

163. Hall, C. F., and Hall, E. L., A Nonlinear Model for the Spatial Characteristics of the Human Visual System, *IEEE Trans. Sys. Man Cybern.*, SMC-7:161–170 (1977).

164. Sakrison, D. J., Image Coding Applications of Vision Models. In: *Image Transmission Techniques* (W. K. Pratt, ed.), Academic Press, San Francisco, 1979, pp. 21–71.

165. Shannon, C. E., Coding Theorems for a Discrete Source with a Fidelity Criterion, *IRE Nat. Conv. Record*, Part 4, 142–163 (1959).

166. Berger, T., *Rate Distortion Theory, A Mathematical Basis for Data Compression*, Prentice-Hall, Englewood Cliffs, NJ, 1971.

167. Sakrison, D. J., The Rate Distortion Function for a Class of Sources, *Inform. Control*, 15:165–195 (1969).

168. Haskell, B. G., Computation and Bounding of Rate Distortion Functions, *IEEE Trans. Inform. Theory*, IT-15:525–531 (1969).

169. Hayes, J. F., Habibi, A., and Wintz, P. A., Rate Distortion Function for a Gaussian Source Model of Images, *IEEE Trans. Inform. Theory*, IT-16:507–508 (1970).

170. Sakrison, D. J., and Algazi, V. R., Comparison of Line-by-Line and Two-Dimensional Encoding of Random Images, *IEEE Trans. Inform. Theory*, IT-17:386–398 (1971).

171. Davisson, L. D., Rate Distortion Theory and Application, *Proc. IEEE*, 60:800–808 (1972).

172. Blahut, R. E., Computation of Channel Capacity and Rate-Distortion Functions, *IEEE Trans. Inform Theory*, IT-18:460–473 (1972).

173. Tasto, M., and Wintz, P. A., A Bound on the Rate Distortion Function and Application to Images, *IEEE Trans. Inform. Theory*, IT-18:150–159 (1972).

174. Gray, R. M., and Davisson, L. D., Source Coding Theorems without the Ergodic Assumption, *IEEE Trans. Inform. Theory*, IT-20:502–516 (1974).

175. Wilson, S. G., and Troxel, J. R., Facsimile Coding: Distortion Measures, Code Generation, and Tree Encoding, *Conf. Record, International Conf. Commun.*, 8.4.1–8.4.5 (June 1979).

176. Knudson, D. R., Digital Encoding of Newspaper Graphics, *Electron. Sys. Lab.*, M.I.T. report ESL-616 (August 1975).

177. Linde, Y., Buzo, A., and Gray, R. M., An Algorithm for Vector Quantizer Design, *IEEE Trans. Commun.*, COM-28:84–95 (1980).

178. Equitz, W. H., *Fast Algorithms for Vector Quantization Picture Coding*, Masters thesis, M.I.T., Cambridge, MA, June 1984.

179. Kirkpartrick, S., Gelatt, C. D., Jr., and Vecchi, M. P., Optimization by Simulated Annealing, *Science*, 220:671–680 (1983).

180. Kohonen, T., *Self Organization and Associative Memory*, Springer-Verlag, New York, 1984.

181. Kohonen, T., The Neural Phonetic Typewriter, *IEEE Computer*, 21:11–21 (1988).

182. Nasrabadi, N. M., and King, R. A., Image Coding Using Vector Quantization: A Review, *IEEE Trans. Commun.*, COM-36:957–971 (1988).

183. Montanari, H., A Note on Minimal Length Polygonal Approximation to a Digital Contour, *Commun. ACM*, 13:41–47 (1970).

184. Mergler, H. W., and Vargo, P. M., One Approach to Computer Assisted Letter Design, *J. Typographic Research*, II:299–322, (1968).

185. Frank, A. J., Parametric Font and Image Definition and Generation, *Proc. AFIPS Fall Joint Computer Conf.*, 135–144 (1971).

186. Schreiber, W. F., Huang, T. S., and Tretiak, O. J., Contour Coding of Images. In: *Picture Bandwidth Compression* (T. S. Huang and O. J. Tretiak, eds.), Gordon and Breach, New York, 1972, pp. 443–448.

187. Pavlidis, T., Techniques for Optimal Compaction of Pictures and Maps, *Computer Graphics and Image Processing*, 3:215–224 (1974).

188. Kawade, T., and Nakagawa, T., Effects of Reducing and Bridging Isolated Picture Elements in Coding of Facsimile Signal, *1976 Nat. Convention Record*, IECE Japan, No. 1027 (1976). (In Japanese)

189. Kukinuki, T., Notchless Bi-Level Quantizer for Facsimile and Its Effect on Coding Efficiency, *IEEE Trans. Commun.*, COM-26:611–618 (1978).

190. Takagi, M., and Tsuda, T., Comparison of Facsimile Bandwidth Compression Using Two-Dimensional Prediction and Signal Modification, *Conf. Record, IEEE International Conf. Commun.* (1976).

191. Margner, V., and Zamperoni, P., Transmission of Binary Pictures with Constant Bit Rate by Source Coding Using a Morphological Irrelevance Criterion, *Frequenz*, 31:221–227 (1977).

192. Wilson, G. R., and Batchelor, B. G., Convex Hull of Chain-Coded Blob, *IEE Proc.*, Part E, *Computers and Digital Techniques*, 136:530–534 (1989).

193. Usubuchi, T., Mizuno, S., and Iinuma, K., Efficient Facsimile Data Reduction by Using a Thinning Process, *Proc. National Telecom. Conf. '77*, no. 49-2 (December 1977).

194. Salari, E., and Siy, P., The Ridge-Seeking Method for Obtaining the Skeleton of Digital Images, *IEEE Trans. Syst. Man Cybern.*, SMC-14:524–528 (1984).

195. Zhang, T. Y., and Suen, C. Y., A Fast Parallel Algorithm for Thinning Digital Patterns, *Commun. ACM*, 27:236–239 (1984).

196. Lu, H. E., and Wang, P. S. P., A Comment on "A Fast Parallel Algorithm for Thinning Digital Patterns," *Commun. ACM*, 29:239–242 (1986).

197. Ascher, R. N., and Nagy, G., A Means for Achieving a High Degree of Compaction on Scan-Digitized Printed Text, *IEEE Trans. Computers*, C-23:1174–1179 (1974).

198. Chen, W., Douglas, J. L., and Widergren, R. D., Combined Symbol Matching — A New Approach to Facsimile Data Compression, *Proc. Soc. Photo-Optical Instrumentation Engineers*, 199:2–9 (1978).

199. Pratt, W. K., et al., Combined Symbol Matching Facsimile Data Compression System, *Proc. IEEE*, 68:786–796 (1980).

200. Cash, G., Johnsen, O., and Segen, J., A Real-Time Pattern Matcher Coder, *International Picture Coding Symposium*, 55–56 (March 1983).

201. Bodson, D., et al., Measurement of Data Compression in Advanced Group 4 Facsimile Systems, *Proc. IEEE*, 73:731–739 (1985).

202. Silver, D. M., and Johnson, D. A. H., Facsimile Coding Using Symbol Matching Techniques, *IEE Proc.*, Part F, 131:125–129 (1984).

203. Frank, A. J., Region Recognition Using an Adaptive Viterbi Algorithm with Application to Efficient Coding of Yellow Pages Graphics, *Bell Laboratories Report*, Murray Hill, NJ, August 1978.

204. Haralick, R. M., Statistical and Structural Approaches to Texture, *Proc. Fourth International Joint Conf. Pattern Recog.*, 45–69 (1978). Also published in *Proc. IEEE*, 67:786–804 (1979).

205. Rosenfeld, A., and Lipkin, B. S., Texture Synthesis. In: *Picture Processing and Psychopictorics*, (B. S. Lipkin and A. Rosenfeld, eds.), Academic Press, New York, 1970, pp. 309–345.

206. Lipkin, B. S., and Rosenfeld, A., (eds.), *Picture Processing and Psychopictorics*, Academic Press, New York, 1970.

207. Hassner, M., and Sklansky, J., Markov Random Fields as Models of Digitized Image Texture, *Proc. IEEE Computer Conf. 1978 Pattern Recog. and Image Processing*, 346–351 (1978).

208. Sakrison, D. J., Image Coding Applications of Vision Models. In: *Image Transmission Techniques* (W. K. Pratt, ed.), Academic Press, San Francisco, 1979, pp. 23–71.

209. Zobrist, A. L., and Thompson, W. B., Building a Distance Function for Gestalt Grouping, *IEEE Trans. Computers*, C-4:718–728 (1975).

210. Julesz, B., Experiments in the Visual Perception of Texture, *Scientific American*, 232:34–43 (1975).

211. Mandelbrot, B., *The Fractal Geometry of Nature*, W. H. Freeman, San Francisco, 1983.

212. Barnsley, M. F., *Fractals Everywhere*, Academic Press, Boston, 1988.

213. Barnsley, M. F., and Sloan, A. D., A Better Way to Compress Images, *Byte*, 215–223 (January 1988).

214. Peitgen, H. O., and Saupe, D., *The Science of Fractal Images*, Springer-Verlag, New York, 1988.

215. Cipra, B. A., Image Capture by Computer, *Science*, 243:1288–1289 (1989).

216. Coco, D., Fractals Take on Color Image Compression, *EDN*, (July 12, 1990).

217. Jurgens, H., Peitgen, H. O., and Saupe, D., The Language of Fractals, *Scientific American*, 60–67 (August 1990).

218. Frame, M., and Erdman, L., Coloring Schemes and the Dynamical Structure of Iterated Function Systems, *Computers in Physics*, 500–505 (September–October 1990).

219. Brammer, R. F., Unified Image Computing Based on Fractals and Chaos Model Techniques, *Optical Engineering*, 28:726–734 (July 1989).

220. Dennis, T. J., and Dessipris, N. G., Fractal Modelling in Image Texture Analysis, *IEE Proc.*, Part F, No. 5, 227–235 (October 1989).

221. Coco, D., Fractal Research: Payoff at Last, *EDN*, (September 20, 1990).

AMALIE J. FRANK

Colpitts, Edwin H.

Edwin H. Colpitts was born in Point de Butte, New Brunswick, Canada, in 1872. Little has been recorded about his early life, but this seems to be in keeping with the private, quiet demeanor he displayed throughout a life marked by significant accomplishment in the field of telecommunications. Dr. Colpitts distinguished himself early in his academic career when he graduated with honors from Mt. Allison University, where he received his A.B. in 1893. He went on to Harvard and graduated with both bachelor of arts and master of arts degrees in 1897. Dr. Colpitts remained at Harvard for three years while he worked as an assistant in physics. In 1926, he received an honorary LL.D. from Mount Allison.

Upon his departure from Harvard, Dr. Colpitts began his lifelong association with the Bell System (1). In 1899, he took the first of a series of positions with the telephone company when he joined the engineering department of the American Bell Telephone Company in Boston. It was in this first position that Dr. Colpitts did important research in the area of electrical cable loading. His work in verifying G. A. Campbell's (2) research provided the first conclusive demonstration of the practicability of loading. This research led to improved long-distance service by upgrading transmission efficiency, thereby extending the range of long-distance telephone conversations beyond the then-existing 1000-mile limit. In 1907, Dr. Colpitts transferred to the Western Electric Company in New York to head the physical laboratory in its department of engineering. This and subsequent rapid promotions demonstrated, in a very tangible manner, the importance of his scientific contributions and his capacity for applied research. His research during this period produced 24 patents on which he was either the primary or co-investigator.

Dr. Colpitts's work can be divided into three major categories: magnetic coils for loading, transformers, and filter purposes; studies of cross talk and capacity unbalance between adjacent telephone circuits; and the application of the vacuum tube to long-distance telephone communications.

Dr. Colpitts's work with magnetic coils led to two important achievements. The first breakthrough resulted in an improved battery-supply repeating coil for the common battery system. The second important achievement resulted in the design of retardation coils for use in composite telegraph circuits. Basic features of both the repeating and retardation coils were used to improve voice telecommunications techniques throughout the first half of this century. Colpitts's work with cross-talk interference between adjacent circuits led to the first effective method of determining the capacity unbalance between telephone wires on an actual pole line. This laid a foundation for the design of voice transportation systems and for the balancing of long-distance cables. Research into the application of the vacuum tube to telephony led to patents that protected the vacuum tube as oscillator, modulator, and amplifier. In fact, his design of a vacuum tube oscillator became popularly known as the Colpitts oscillator due to its wide use.

Dr. Colpitts also played a significant role in the achievement of transcontinental telephony and in the transmission of speech across the Atlantic Ocean.

His paper, "Carrier Current Telephony and Telegraphy," with O. B. Blackwell (3), was a key contribution in accomplishing transcontinental telephony. Likewise, his paper "Radio Telephony," with E. B. Craft (4) contributed significantly to the achievement of trans-Atlantic communication. These papers, though written in the 1910s and 1920s, are still of technical and historical interest to the student of telephony.

After his retirement from Bell Laboratories in January 1937, Dr. Colpitts continued to work diligently to further telecommunications engineering. In 1937, he received the Fourth Order of Merit of the Sacred Treasure award from the Japanese government. The award was presented to him by Emperor Hirohito in recognition of his promotion of electrical engineering in Japan. Colpitts had traveled to Japan after his retirement to deliver a series of lectures, sponsored by the Japanese Institute of Electrical Engineers. In 1941, he was elected director of the Engineering Foundation of New York, an affiliation that remained active until his death in March 1949. In 1948, he was awarded the Elliot Cresson Medal by the Franklin Institute. He was a Fellow of the American Institute of Electrical Engineers, the Institute of Radio Engineers, the American Physical Society, the Acoustical Society of America, the American Association for the Advancement of Science, and was a member of the American Chemical Society.

Bibliography

Edwin H. Colpitts, *Bell Laboratories Record*, Vol. 27, January 1949–December 1949.

Edwin H. Colpitts, Phone Pioneer, *New York Times*, 21:4 (March 7, 1949).

Heads Engineering Unit, Dr. Colpitts is Elected Director of Research Bureau, *New York Times*, 17:8 (October 1, 1947).

Japan Honors Colpitts, U.S. Telephone Pioneer, *New York Times*, 31:5 (April 25, 1937).

Johnson, K. S., *Transmission Circuits for Telephone Communication,* D. Van Nostrand Company, New York, 1925.

Work on Communications Wins 1948 Cresson Medal, *New York Times*, 46:5 (September 7, 1948).

References

1.　Brown, C. L., The Bell System. In: *The Froehlich/Kent Encyclopedia of Telecommunications*, Vol. 2 (F. E. Froehlich and A. Kent, eds.), Marcel Dekker, New York, 1991, 65–86.

2.　Brittain, J. E., Campbell, George Ashley. In: *The Froehlich/Kent Encyclopedia of*

Telecommunications, Vol. 2 (F. E. Froehlich and A. Kent, eds.), Marcel Dekker, New York, 1991, 241–247.

3. Colpitts, E. H., and Blackwell, O. B., Carrier Current Telephony and Telegraphy, *Transactions of the AIEE*, XL:205–296 (1921).

4. Colpitts, E. H., and Craft, E. B., Radio Telephony, *Transactions of the AIEE,* col. 38, part 1, 305–343 (1919).

MONICA KRUEGER

Comma-Free and Synchronizable Codes

Synchronization of a signal can be maintained using synch pulses or by reference to a particular frequency component of the signal. It is possible, however, to design a code so that synchronization may be obtained from the message itself without the use of any such devices. Such a code is called a *synchronizable code*. (A more precise definition follows.)

Example 1. Suppose that we set up a code with words of length 3 using the letters {*A, B, C*}. In the codeword dictionary, if we allow only those words in which all three letters are not the same and in which alphabetical order is not violated, (e.g., let our dictionary $\mathcal{C}$ = {*AAB, AAC, ABB, ABC, ACC, BBC, BCC*}, we obtain a synchronizable code. That is, the word boundaries in any sufficiently long string of letters taken from a message can be determined. Suppose we consider the following string of letters

$$B\,B\,C\,C\,A\,B$$

under the assumption that it is a randomly chosen segment of some long message in the code. Clearly, the word boundaries must be

$$B\,BCC\,AB,$$

since the other two possibilities,

$$BB\,CCA\,B$$

and

$$BBC\,CAB$$

both yield nonwords: neither *CCA* nor *CAB* occurs in the dictionary $\mathcal{C}$. We are now in a position to make a more careful statement of the definition.

Suppose that by a code we mean a set of words $\mathcal{C}$ whose letters are taken from an alphabet $\mathcal{I}$. By a word, we mean a non-empty finite string. $\mathcal{C}$ is synchronizable if and only if, given any sequence of letters from $\mathcal{C}$ (infinite in both directions)

$$\ldots a_{-2}\,a_{-1}\,a_0\,a_1\,a_2\,\ldots$$

there is at most one way to group this sequence into words (i.e., there is at most one partitioning of the integers into intervals such that all the corresponding strings of letters in the given sequence belong to $\mathcal{C}$.)

A simple example of a code that is not synchronizable is any code that contains a word that is a cyclic permutation of another codeword. The example below illustrates this point.

Example 2. Suppose that $\mathcal{C} = \{ABC, BCA\}$. Then we can set up a sequence

$$a_{-3} = A$$
$$a_{-2} = B$$
$$a_{-1} = C$$
$$a_0 = A$$
$$a_1 = B$$
$$a_2 = C \qquad \text{and so on}$$

which could be grouped $[\ \dots\ (a_{-2}\ a_{-1}\ a_0)\ (a_1\ a_2\ a_3)\ \dots\ ]$ or as $[\ \dots\ (a_{-3}\ a_{-2}\ a_{-1})\ (a_0\ a_1\ a_2)\ \dots\ ]$. Cyclic codes, which by definition include all cyclic permutations of any codeword, are consequently not synchronizable.

In Example 1 above, we were able to obtain synchronization given a message fragment of only 6 symbols. That this was not merely a coincidence we see below, for in this particular code any 6 consecutive symbols of any message is enough to determine the synchronization, although 5 may not suffice. Accordingly, this value 6 is referred to as the synchronization delay of this code. Although it is not the case that every synchronizable code has a synchronization delay, the only synchronizable codes that do not have a synchronization delay have an infinite number of codewords.

Example 3. Consider the 11-character code for the natural numbers using their ordinary base-10 representations each followed by a comma character __

$$\mathcal{C} = \{1\_, 2\_, 3\_, 4\_, 5\_, 6\_, 7\_, 8\_, 9\_, 10\_, 11\_, \dots\}$$

Synchronization would be uniquely determined by the comma characters, but there could be no a priori bound on the length of indeterminate segments. Since essentially all practical codes consist of a finite number of codewords, any practical synchronizable code will have a synchronization delay. A code that has a synchronization delay could be described as effectively synchronizable.

Block Codes

A block code is a code whose words all have the same length. The codes in Examples 1 and 2 are examples of block codes. The code in Example 3 is not a block code, but instead a variable-length code. Any block code has a finite number of code words. If n is the number of letters in the alphabet $\mathcal{I}$ and k is the length of each word in $\mathcal{C}$, then there are at most n^k words in $\mathcal{C}$.

Prefix/Suffix Graphs

To determine whether a given code is synchronizable and, if it is, the synchronization delay, it is useful to construct a prefix/suffix graph (1,2). A *prefix* of a

code word is a proper initial segment. For example, the prefixes of the word
AAB are A and AA. Similarly, the suffixes of AAB are B and AB. The nodes
of a prefix/suffix graph of a code $\mathcal{C}$ are the prefixes and suffixes of all the
words of $\mathcal{C}$. Two nodes x and y are connected by an edge (or arrow since this is
to be a directed graph), whenever x is a prefix, y is a suffix, and, taken together,
they form a word of the code. Thus, the 7 words of the code in Example 1 give
rise to 14 edges as follows (see Fig. 1 for the entire prefix/suffix graph of these
codewords):

1.	AAB	$A \rightarrow AB$	$AA \rightarrow B$
2.	AAC	$A \rightarrow AC$	$AA \rightarrow C$
3.	ABB	$A \rightarrow BB$	$AB \rightarrow B$
4.	ABC	$A \rightarrow BC$	$AB \rightarrow C$
5.	ACC	$A \rightarrow CC$	$AC \rightarrow C$
6.	BBC	$B \rightarrow BC$	$BB \rightarrow C$
7.	BCC	$B \rightarrow CC$	$BC \rightarrow C$

Paths through this graph yield strings of indeterminate synchronization. For
example, the path

$$A \rightarrow AB \rightarrow B \rightarrow CC$$

yields the indeterminate string ABB. Two conflicting groupings of symbols in
which this string occurs may be obtained from the path by taking the first node
of the path, A in this case, as a prefix and grouping the nodes as prefix-suffix,

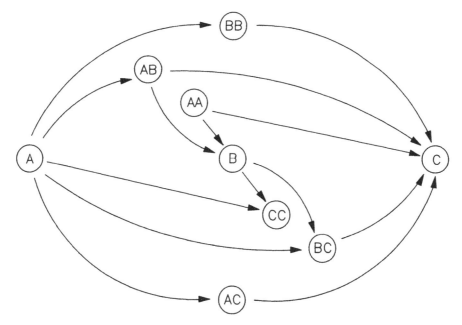

FIG. 1 Prefix/suffix graph for Example 1.

prefix-suffix, and so on from that point, and then doing the same thing starting from the second node of the path, AB in this case. Here we get ABB imbedded in the two groupings

$$(A \ A \ B) \quad (B \ C \ C)$$
$$(A \ B \ B)$$

The indeterminate segment ABB itself is just the string that remains after deleting the first and last nodes A and CC from the path. To get indeterminate segments that are as long as possible, we find paths through the graph that cannot be extended further; in other words, we find all maximal paths and then determine the lengths of the corresponding indeterminate segments. The maximal paths through the graph in Fig. 1 are shown below.

Path	Indeterminate Segment	Length
$A \ \ \to AC \ \to C$	$A \ C$	2
$A \ \ \to BB \ \to C$	$B \ B$	2
$A \ \ \to BC \ \to C$	$B \ C$	2
$A \ \ \to CC$		0
$A \ \ \to AB \ \to B \ \ \to BC \ \to C$	$A \ B \ B \ B \ C$	5
$A \ \ \to AB \ \to B \ \ \to CC$	$A \ B \ B$	3
$A \ \ \to AB \ \to C$	$A \ B$	2
$AA \to B \ \ \to CC$	B	1
$AA \to B \ \ \to BC \ \to C$	$B \ B \ C$	3
$AA \to C$		0

In the table above, note that the longest indeterminate segment in this code is the segment of length 5. Thus, any segment of length 6 or more can be imbedded in at most one way in a string of code words. The synchronization delay of this code is therefore 6.

A code that is not synchronizable has indeterminate segments that are infinitely long. For a finite code, this can only occur if some path in the prefix/suffix graph contains a cycle. In the graph of the code in Example 2 (Fig. 2), we observe the cycle $A \to BC \to A$. Conversely, any such cycle in a prefix/suffix graph corresponds to an infinitely long indeterminate segment. Thus, any code with a finite set of words is synchronizable if and only if its graph has no cycles.

Returning now to Example 1, we note that in using the graph in Fig. 1 to determine the synchronization delay for this code we ignored the end points of

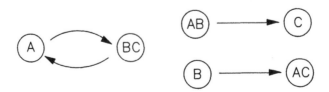

FIG. 2 Prefix/suffix graph for Example 2.

the maximal paths (e.g., *A* and *C* in the case of the first maximal path listed in the table above). Sometimes, however, the end points may not be ignored, as in the following example.

Example 4. Suppose we consider the code $\mathcal{C} = \{ARSTU,\ ABRST,\ VWXCD,\ UVWXD\}$, where $\mathcal{I} = \{A,\ B,\ C,\ D,\ R,\ S,\ T,\ U,\ V,\ W,\ X\}$. If we list all maximal paths for the prefix/suffix graph for this code we find one path

$$ARST \rightarrow U \rightarrow VWXD,$$

which gives us an indeterminate segment *U* with the end points omitted. This indeterminate segment, however, can be extended if we consider the end points of the path. Note that the segment *R S T U V W X* occurs in both

$$(A\ R\ S\ T\ U)\,(V\ W\ X\ C\ D)$$

and

$$(A\ B\ R\ S\ T)\,(U\ V\ W\ X\ D)$$

and is consequently indeterminate. The end point *VWXD* contributes one of its prefixes *VWX* to the indeterminate segment because *VWX* is also a prefix of a codeword namely *VWXCD*. To take into account the end points, then, each right end point of a maximal path contributes the longest of its prefixes, which is also the prefix of some codeword. Similarly, a left end point contributes to an indeterminate segment the longest of its suffixes that is also the suffix of some codeword. The left end point *ARST* of the path under consideration has a suffix *RST* that is the suffix of the codeword *ABRST*. If we list all the maximal paths, *RSTUVWX* is the only indeterminate segment that arises. Hence the synchronization delay of this code is 8.

In the table for the code of Example 1, the first right end point *C* has no prefixes. The only prefix of the other right end point *CC* (namely *C*) is not a prefix of any codeword and thus cannot contribute to any indeterminate segment. Similar considerations apply to the left end points. That is why the correct value for the synchronization delay can be obtained even though the end points are neglected. One further example is given to clarify the procedure.

Example 5. Consider the code $\mathcal{C} = \{ABC,\ BCB,\ BAC\}$. The prefix/suffix graph of this code is provided in Fig. 3. All the maximal paths, together with their indeterminate segments, are given in the table below.

Path							Indeterminate Segment	Length
A	→	*BC*	→	*B*	→	*CB*	*B C B*	3
A	→	*BC*	→	*B*	→	*AC*	*B C B A*	4
AB	→	*C*					*B*	1
BA	→	*C*						0

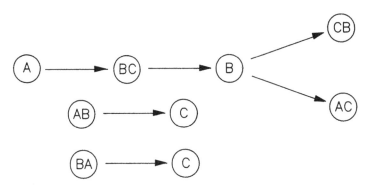

FIG. 3 Prefix/suffix graph for Example 5.

Hence the synchronization delay of this code is 5.

This method finds all indeterminate segments that cross or abut a message boundary, hence all indeterminate segments are of a length greater than or equal to $k - 2$, where k is the word length. An example of a code where this is not the case follows.

Example 6. Suppose $\mathcal{C} = \{ABXYZD, AXYZCD\}$. A search of the prefix/suffix graph turns up no indeterminate segments at all, but XYZ occurs in both

$$\ldots (A\ B\ X\ Y\ Z\ D) \ldots$$

and

$$\ldots (A\ X\ Y\ Z\ C\ D) \ldots$$

Hence the synchronization delay of this code is 4.

Thus, when the longest indeterminate segment found by the graph search is less than $k - 2$, one can either remain content with $k - 2$ as a conservative estimate of the synchronization delay, or continue the search in the manner below.

   If some letter occurs at more than one position in a word, then that letter is an indeterminate segment. In Example 6, X occurs in the second and third positions and hence is an indeterminate segment, as are Y and Z.

   If some doublet (e.g., XY or YZ) occurs at more than one position, then it is an indeterminate segment.

   Continue the search until the longest such segment is found. The synchronization delay is then one more than the longest segment found in both the letter and doublet searches. (Alternately, the synchronization delay of a code is sometimes defined in terms of word splitting so that it is always at least k.)

Comma-Free Codes

The concept of a synchronizable code first arose as a generalization of comma-free codes. A *comma-free code* is a code in which no two words taken in succession contain as an overlap a string that is also a codeword (3). For instance, the code of Example 2 is not comma-free

$$(A\ B\ C)\ (A\ B\ C)$$
$$(B\ C\ A)$$

It is not difficult to see that a comma-free code (1) is a code whose prefix/suffix graph has no paths of length greater than or equal to 3 (3 edges) and (2) is a special case of a synchronizable code. Recall that a block code is synchronizable if and only if its prefix/suffix graph has no paths of infinite length.

Comma-free codes were first proposed as a possible solution to the problem of how the genetic code synchronized itself at a time when the alphabet of the code was known but the code itself was not (4). It was assumed that none of the four known code letters functioned as a "comma," hence the name comma-free code. The genetic code turned out not to be comma free, although it is the case that none of the four letters of the code functions as a comma. The usage "comma-less" and sometimes even "comma-free," unfortunately, has survived as a description of the genetic code.

Conversely, it is also unfortunate from the standpoint of terminology that the existence of a comma character in a code does not keep that code from being comma free. It is as though arbitrary quadrilaterals were referred to as "right-angle free," leading to the problem that rectangles, being quadrilaterals, would be "right-angle free." This phenomenon is illustrated in Example 7.

Example 7. Suppose that $\mathcal{C} = \{AAC, ABC, BAC, BBC\}$. Clearly, C is functioning as a comma character even though $\mathcal{C}$ is a comma-free code.

Variable-Length Codes

In block codes, synchronization means determining the placement of word boundaries; determining one boundary determines them all. However, in variable-length codes it is not clear a priori which word boundaries need to be determined. Determining one is not the same as determining all of them. If we were to require the determination of all word boundaries, then there would exist synchronizable codes with only a finite number of codewords that would not possess a synchronization delay.

Example 8.   Consider the code $\mathcal{C} = \{ABC, ABCDE, DEF, FGH, GHI\}$. In this code, one can construct arbitrarily long segments

$$(A\ B\ C)\,(A\ B\ C)\,(A\ B\ C)\,(A\ B\ C)\,(A\ B\ C)\,(A\ B\ C\ D\ E)\,(F\ G\ H)$$
$$(A\ B\ C)\,(A\ B\ C)\,(A\ B\ C)\,(A\ B\ C)\,(A\ B\ C)\,(A\ B\ C)\,(D\ E\ F)\,(G\ H$$

where the synchronization would be indeterminate if by synchronization one meant determining all the word boundaries of the segment. This code is, however, both synchronizable and finite. Consequently, it is preferable to define synchronization as the determination of at least one word boundary.

By a word boundary, we mean a partition $\{(-\infty, i],[i + 1, \infty)\}$ of the integers into two intervals. To say that a given sequence (finite or infinite)

$$\ldots a_{-2}\, a_{-1}\, a_0\, a_1\, a_2 \ldots$$

determines a word boundary means that there exists a word boundary such that no grouping of symbols into words

$$\ldots (a_{-2}\, a_{-1})\, (a_0\, a_1)\, a_2 \ldots$$

of any sequence (which is infinite in both directions)

$$\ldots a_{-2}\, a_{-1}\, a_0\, a_1\, a_2 \ldots$$

containing the given sequence includes an interval that crosses the word boundary. For example, if the letter X never occurs in any codeword except as the second letter, then any string containing the letter X determines a word boundary.

A code has a synchronization delay q if and only if every sequence of symbols of length q determines a word boundary and this is not the case in relation to any $q' < q$. (Note that any sequence of symbols that cannot be imbedded in some string of words determines a word boundary vacuously.)

Although, as Example 8 shows, a segment that determines one word boundary does not necessarily determine all the word boundaries, a sequence that determines two word boundaries in a synchronizable code does determine all word boundaries in between. In other words, a finite sequence of symbols cannot be grouped into words of a synchronizable code in more than one way. This property is sometimes called *unique decipherability* and follows directly from the definition of synchronizability (5,6). Conversely, any uniquely decipherable code that has a synchronization delay q is synchronizable.

To find the indeterminate segments of a variable-length code, prefix/suffix graphs as defined above are not sufficiently general. Capocelli, however, found a neat way to generalize these graphs in such a manner that the indeterminate sequences may still be obtained from paths in the graph (5). Each step in a path $x_0, x_1, \ldots, x_n$ of a prefix/suffix graph (as previously defined) is a word $x_{i-1}x_i \in \mathcal{C}$. Suppose we define ab^{-1} to be the unique string of symbols c that, when concatenated with b, yields a, expressed as $cb = a$. For example, $(ABCD)$

$(CD)^{-1} = (AB)$. Therefore, ab^{-1} exists only if b is a suffix of a. Let us now connect by an arrow two vertices x and y of the graph if either $xy \in \mathcal{C}$ or $xy^{-1} \in \mathcal{C}$. Capocelli called paths in such a graph *linked sequences*.

Example 9. Suppose $\mathcal{C} = \{ABC, BCDEFG, DE\}$. Then a path through the graph of this code is

$$A \to BC \to DEFG \to FG$$

Notice that the third arrow corresponds to the fact that $(DEFG)(FG)^{-1} = DE \in \mathcal{C}$. The indeterminate segment represented by this path is $BCDE$. It can be grouped both as

$$\dots (A\,B\,C)(D\,E)\dots$$

and

$$\dots (B\,C\,D\,E\,F\,G)\dots$$

It is also the longest indeterminate segment in this code, hence its synchronization delay is 5.

As with block codes, end points of a path can contribute to an indeterminate segment. Unlike block codes, however, it can also happen that the initial or final node of a path may itself be a word of the code.

Example 10. Suppose $\mathcal{C} = \{ABC, ABCDE, DEFGHI, FG, HGFD\}$. In this case, the indeterminate segment corresponding to the path

$$ABC \to DE \to FGHI \to HI$$

is $D\,E\,F\,G\,H$. This segment comes from the groupings

$$\dots (A\,B\,C\,D\,E)\,(F\,G)\,(H\,G\,F\,D)\dots$$

and

$$\dots (D\,E\,F\,G\,H)\dots$$

Although the second reading can be extended

$$\dots (A\,B\,C\,D\,E)(F\,G)(H\,G\,F\,D)\dots$$

$$\dots (A\,B\,C)(D\,E\,F\,G\,H)\dots$$

the segment $A\,B\,C\,D\,E\,F\,G\,H$ is no longer indeterminate even though it is consistent with two distinct readings, because the two readings do determine one word boundary (the one preceding the letter A). Hence the synchronization delay of this code is 6.

Paths such as this one that begin with a word of the code are related to a problem that arises only in connection with variable-length codes, namely, given the beginning of a message, how long it is necessary to wait until one can determine the first word boundary. The number of symbols necessary to determine the delay is called the delay of the code, and could also be described as the synch-maintenance delay. This delay is found by considering paths like the one in this example that begin with a word. The indeterminate segments used in finding the synch-maintenance delay consequently begin at the given word boundary. The required delay is one greater than the length of all such segments. In this example, the synch-maintenance delay is 9, one more than the length of $ABCDEFGH$.

Clearly, any code with a synchronization delay q will have a synch-maintenance delay q', which is at most $q + \ell$ where ℓ is the maximum of the lengths of all words in the code. The converse is not the case. A code with a synch-maintenance delay need not be synchronizable. Clearly, any block code has a synch-maintenance delay of 0, but not all block codes are synchronizable.

Paths that both begin and end with a word of the code yield indeterminate segments bounded at each end by a word boundary.

Example 12. Suppose $\mathcal{C} = \{ABCDE, FG, ABC, DEFG\}$. The path

$$ABC \rightarrow DE \rightarrow FG$$

corresponds to the readings

$$\ldots (A\,B\,C\,D\,E)(F\,G) \ldots$$

$$\ldots (A\,B\,C)(D\,E\,F\,G) \ldots$$

Codes whose graphs contain such paths are therefore not uniquely decipherable and not synchronizable.

There are many codes that, although not synchronizable in the deterministic sense discussed here, may yet be synchronizable to within a high degree of probability. Codes in which the probability of a segment having a determinate synchronization that tends to 1 as the length of the segment tends to infinity are called *statistically synchronizable* (7). Examples of statistically synchronizable codes intended for practical application are discussed in Refs. 8 and 9.

Glossary

COMMA-FREE CODE. A code in which there do not exist words x, y, and z such that z is a segment of the concatenation xy that includes the boundary between x and y.

DELAY (of a code). See SYNCH-MAINTENANCE DELAY.

EFFECTIVELY SYNCHRONIZABLE CODE. A code that has a (finite) synchronization delay

FINITE CODE. A code with only a finite number of words.

INDETERMINATE SEGMENT. A string of letters that does not uniquely determine any word boundaries.

PREFIX (of a word). An initial segment of length ℓ, where $0 < \ell < n$ and n is the length of the word.

SUFFIX (of a word). A final segment of length ℓ, where $0 < \ell < n$ and n is the length of the word.

SYNCH-MAINTENANCE DELAY (of a code). The least number n such that any string of words of length n (symbols), whose initial word boundary is known to be at the beginning, uniquely determines another word boundary. The synch-maintenance delay is usually referred to as the *delay*.

SYNCHRONIZABLE CODE. A code in which any infinitely long string (infinitely long in both the positive and negative directions) can be grouped into words of the code in, at most, one way.

SYNCHRONIZATION DELAY (of a code). The least number n such that no string of letters of length n is an indeterminate segment (i.e., any string of length n must uniquely determine at least one word boundary).

UNIQUELY DECIPHERABLE CODE. A code in which no finite string of symbols can be grouped into codewords in more than one way.

Bibliography

Golomb, S. W., and Gordon, B., Codes with Bounded Synchronization Delay, *Information and Control*, 8:355–372 (1965).

References

1. Ball, A. H., and Cummings, L. J., Extremal Digraphs and Comma-Free Codes, *Ars Combinatoria*, 1:239–251 (1976).
2. Levenshtein, V. I., Certain Properties of Code Systems, *Soviet Physics–Doklady*, 6(10):858–860 (1962).
3. Golomb, S. W., Gordon, B., and Welch, L. R., Comma-Free Codes, *Canadian J. Math.*, 15:202–209 (1958).
4. Crick, F. H. C., Griffith, J. S., and Orgel, L. E., Codes without Commas, *Proc. Nat. Acad. Sci.* 43:416–421 (1957).
5. Capocelli, R. M., A Note on Uniquely Decipherable Codes, *IEEE Trans. Inform. Theory*, IT-25(1):90–94 (1979).
6. McMillan, B., Two Inequalities Implied by Unique Decipherability, *I.R.E. Trans. Inform. Theory*, (2):115–116 (1956).

7. Wei, V. K., and Scholtz, R. A., On the Characterization of Statistically Synchronizable Codes, *IEEE Trans. Inform. Theory.*, IT-26(6):733–735 (1980).
8. Rudner, B., Construction of Minimum-Redundancy Codes with an Optimum Synchronizing Property, *IEEE Trans. Inform. Theory*, IT-17(4):478–487 (1971).
9. Titchener, M. R., Construction and Properties of the Augmented and Binary-Depletion Codes, *IEEE Proc.*, 132(31):163–169 (1985).

BOB NEVELN

Command, Control, and Communications Systems

Introduction

The term *command, control, and communications* (1) (often articulated as "C-cubed," or C^3) is used, most often in national defense applications (2), to describe equipment complexes that assist managers to direct widespread resources performing critical, time-urgent tasks. The alternative *command, control, communications, and intelligence* (C^3I) suggests systems used to acquire, analyze, or make use of military intelligence information about the enemy. In this article, we use the term C^3 systems to describe systems performing any or all of these functions. Large C^3 systems are usually comprised of large numbers of separately designed and constructed subsystems, all of which their designers and users refer to as C^3 systems; this terminology is also used in this article.

While the most complex and expensive of these systems have been used for national defense purposes (3), C^3 systems are also found in civil air-traffic control and are important parts of city or state police and fire safety operations.

Characteristics of a Typical C^3 System Application

Information-Action Loop

A typical C^3 system application is embodied in an information-action loop, a control loop that performs data analysis and decision making (Fig. 1). A situation, change, or specific event is detected. Information about it is collected, sometimes automatically, and transmitted to an analysis site. There, the information is examined for importance and credibility in light of other knowledge, perhaps including historical data. Information that appears sufficiently important is forwarded to the point at which responsive actions can be directed. Alternative responses may be examined, together with those resources available to be committed. A course of action is then decided upon. Directives are issued in an appropriate form. These are forwarded to those directly controlling the resources that will respond. Later, when directed actions are completed or abandoned, the C^3 system may again be used to observe and analyze the results. Follow-up may need to be directed, or the system may await another event.

In principle, such an information-action loop is not conceptually different from the operation of a furnace and thermostat to stabilize the temperature in a building. In practice, in a military control center or the control tower of a busy airport, the situation is enormously more complex. It may be made especially difficult by the limited information on which actions must be based, and can be strongly influenced by the personalities and moods of humans in the loop.

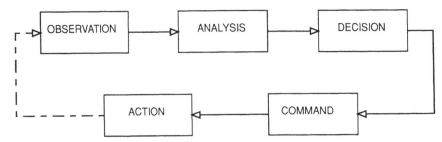

FIG. 1 An information-action loop. The principal functions of many C³ systems can be described as control loops involving complex data analysis and decision making, parts of which are carried out by humans. Each step has complexities: analysis of the observations requires recognition functions; the decision to take action requires examination of the situation and the responses possible to meet it; and the command must be interpreted in terms of feasible action in light of resource constraints.

Most C³ systems are perceived by their users as merely aids for humans who actually analyze events, available resources, and possible courses of action. The presence of the humans in the loop makes C³ systems very different from automatic control systems used for most repetitive industrial processes. In the latter, humans are generally viewed as monitors should the automated system malfunction. Humans "program" automated control systems, predefining in detail their responses to various combinations of inputs. The situations managed using C³ systems are usually too complex to program fully in advance.

C³ systems are used in applications where events are highly unpredictable. Though there is no firmly agreed-upon definition, presence of humans in management, analysis, decision, and/or implementation functions of the system distinguishes C³ systems from many other control systems, including supervisory control systems, where the role of humans is to approve or disapprove system-proposed actions.

Command and Control Activities

The term *command and control* is predominantly a military one, with parallels in nonmilitary organizations. Command is the management function, exercised hierarchically from top commanders to the lowest working levels (4). Just as in large industries, there are many levels between top and bottom in the U.S. military, where formally, there are as many as 19 levels; in practice, there are considerably fewer.

Control does not imply command. A combat air controller or civil-air traffic controller is in no way the manager of the aircraft flight crews whose flights he or she controls. The controller provides guidance and assistance based on information that may not be fully available to the controlled resources, and a set of rules that structure how the controlling operation is to be carried out (5). A harbor pilot for a large vessel, whose job it is to assist the vessel through the confined waters of the harbor to safe docking, is analogous to an air-traffic controller.

A C^3 system may provide support for either command or control and, quite often, for both (Fig. 2). Control functions may involve no large-scale or long-term issues; instead, controllers assist air traffic or vector military fighters toward potentially hostile aircraft. Command functions, however, include long-range planning, analysis of intelligence or market data, allocation of all sorts of resources, logistical support, coordination with cooperating organizations, alternative plans, and the like. Competent high-level commanders and managers leave their subordinates opportunity for initiative, and recognize the longer time-scales and greater uncertainties of the environments in which strategic commands are interpreted and executed.

A C^3 system supporting command functions may address one or more managerial issues. Large organizations with many subordinate headquarters operations may use dozens of automated systems to support their management activities. Some must deal with short-fused, real-time issues; though most issues have, on average, less urgency, managers are often unwilling to grant permanent low priority to tasks that could become critical. It makes good sense for a system designer not to clutter critical real-time systems with high volumes of low-priority work. Since some of the same data and personnel are involved with both high- and low-priority work, users often have a strong desire to unify all functions within a common framework. In practice, a tremendous variety of design concepts can be found in C^3 systems, matching the variety of organizations that they are designed to serve.

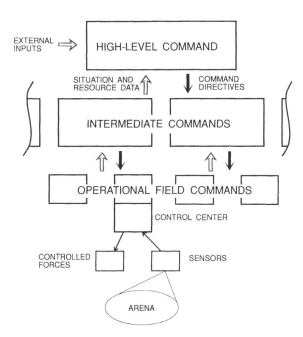

FIG. 2 Command and control functions. Both command and control functions are supported by C^3 systems. Both involve the examination of situation data and available resources to determine the most appropriate course of action. High-level command deals with aggregated resources and must take future possibilities more heavily into account. Control functions provide central coordination to distributed, independently operated resources.

Distributed Character of C³ Systems

The need for extensive telecommunication is a major consequence of the distributed nature of C³ systems. In some instances, many megabytes of bandwidth must connect remote terminals (e.g., sensor platforms in space) to command nodes. Not only the usual natural barriers to communications integrity, but also intentional disruptions are often expected. Those tasks to which C³ systems are applied, moreover, may pose users with no simpler but effective alternative to telecommunications.

From an operational point of view, the widespread distribution of a C³ system leads to problems of lengthy response times, use of noncurrent information as a basis for future actions, and unpredictable errors in interpretation of directives, along with the possibility of duplicative responses or failure to respond at all.

On the positive side of the ledger, a well-designed C³ system often includes many redundant elements: parallel communications links using different phenomena or routings; multiple nodes, each capable of carrying out the duties of some other node; and intranodal redundancy, which guards against single-thread failures within nodes.

While many automatic control systems have a single mode of operation, by its very nature a C³ system may operate in a variety of modes determined by its users' reactions to events and by commandable resources (6). An important dimension is that of global versus local control (Fig. 3). A complex national C³ network may require days to close certain information-action loops. The time required for a multilayer management hierarchy to respond formally to a situation reported from operational levels is often so great that action could never be taken in a timely way. Imagine an airline where the corporate chief executive had to be consulted before every landing! A part of the solution to this dilemma is to authorize and encourage lower levels to respond with minimal delay to urgent local ("tactical") situations, within the overall guidance of a global ("strategic") plan and/or a set of "rules" that guides their actions. Higher levels, using aggregated local-situation data, can concentrate their attention on the overall situation: moving resources to where they are or will be needed and planning needed shifts in resources or future coordinated action. For certain situations, it may be most reasonable to bypass intermediate-level delays, permitting high-level authority to pass certain crucial decisions directly to operational personnel (and concurrently to intervening levels). Thus, use of nuclear weapons is made a determination for the chief executive, but is given priority handling by steering implementation directly to operations.

Operationally, C³ system users often have both too little and too much information on which to base decisions. Seldom can C³ systems provide all the information that might be thought pertinent to decisions, especially in a competitive arena where the opponent's plans and moves may be hidden from view. At the same time, top-level C³ center(s) for a large operation are often awash in data and without the time to review and interpret it all. Evolution of computer-based decision aids is one of the most important recent technological trends that will affect C³ system development (7). A paucity of pertinent information might be addressed by using information fusion to combine sparse information from many sources into a more complete, coherent (and, hopefully,

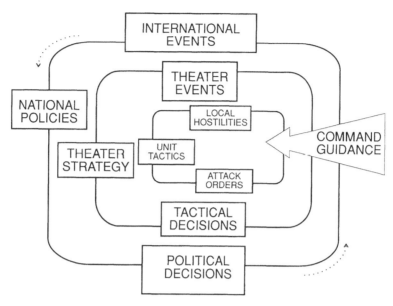

FIG. 3 Global versus local control illustrated by military command and control at three levels: personnel at the national level pass long-term direction to theater-level commanders, who in turn issue commands to units under their control. Each operates in a different information domain and different context. One or more coupled C³ systems pass commands downward and forward situation information to higher levels. Each level also obtains outside information that, when pertinent to other levels, should be transmitted to those who need to know it.

correct) situation assessment. The problem of too much information is another side of the same coin, susceptible to similar but more mature techniques; database management systems provide tools for allocation of resources, though only if the data can be rapidly updated.

Classes of C³ Systems and Their Applications

The majority of existing C³ systems have been developed for military purposes. During the 40-year cold war between the United States and the Soviet Union, both nations dedicated much attention to offensive and defensive C³ systems for nuclear warfare. High-flying military reconnaissance and communications satellites (8), oceanic acoustic-sensor networks, ultra-low-frequency global radio systems, launch-control bunkers, and computerized command centers are components of these systems. Prior to initiation of nuclear attacks, communications can be carried out largely via encrypted ground-line communications. In tactical warfare, uncertainty about the location of possible future military actions largely requires mobile or transportable C³ components using radio communications.

Military C³ Systems

Battle- and Force-Management Systems

Electrical communications networks have been used to couple commanding officers to operational troops and to one another in a battle area since World War I, when field telephones were in use. Telephones were supplemented by radio in World War II. Only in the past two decades has progress been made in converting these ad hoc communications networks into something that could properly be called a system.

A major problem in military communications is uncertainty of reaching units when they are on the move or heavily involved in combat. Though this is certainly true when using telephones, beyond line of sight it is often a problem when using very high frequency (VHF) and ultra-high frequency (UHF) radios. Compounding the problem is the limited availability of radio bandwidth for the many users who wish to communicate simultaneously. Even in the UHF range, where wavelength is relatively plentiful, the problem of establishing contact can be complicated by complex frequency-assignment strategies that must often be implemented manually. Some recently developed military spread-spectrum communications systems (9) employ multiple-carrier frequencies that are frequency-hopped using security codes that are frequently changed. In time division multiple access (TDMA) versions of these systems, each transmitter sends periodic but infrequent composite pulses that contain encrypted data or voice signals.

Battle-management communications systems are typically based on VHF or UHF radio for short ranges of a few miles, either high-frequency (HF) or tropospheric scatter propagation at distances of a few hundred miles, and satellite communications for communicating anywhere on the globe. While early military communication was either point-to-point or broadcast, some modern field communications systems incorporate transportable relaying and switching nodes supporting subordinate-unit communications. These nodes connect high-level commanders and staffs to subordinate units and support communication between them. Though the nodes may communicate via radio, switching functions may be similar to those in the central exchanges of commercial telephony.

Surveillance and Early Warning Systems

In the information-action loops referred to above, the triggering event for a response may be detection of enemy activity by a sensor connected into the C³ system. Sensors operate using a variety of phenomena: electromagnetic reflection is the controlling principle of all radars and velocity-dependent frequency shift is also used by Doppler radars. Sensors may detect electromagnetic emissions from radars that are intended targets or capture visible-light or infrared images of potential target areas. Some sensors are intended to detect enemy communications that reveal the presence, approximate location, numbers, and perhaps the identity of enemy units (10).

A surveillance or warning system may be based on one or many sensors. A single sensor might be used to determine the presence of intruders at a single site, such as a naval vessel or an airport (called *point defense*). Multiple sensors of the same type might be used to protect a perimeter, as in the use of radar to warn of aircraft approaching a national frontier or closed-circuit television cameras to guard a building's entrances (termed *area defense*).

Sensors of the same or different types may also be used in combination to increase probability of detection. Multiple coverage of the same area by two or more sensors provides tolerance to failure of sensors. Sensors of different types (e.g., radar plus infrared-emission sensors) provide improved detection probability because they are intended to sense different features of a target's signature.

Combining multiple sensors into a single system may be as simple as grouping TV monitors at a guard station. A netted radar system, however, is representative of the more powerful real-time warning systems: an aircraft detected by one radar is "handed off" using computer automation as it moves into an area scanned by another radar in the net. Measuring the speed and direction of the aircraft electronically allows prediction of its destination. (This technique was first used in the Battle of Britain in 1940 with entirely manual netting.)

Sensing aircraft by the use of radar is one of the easier military surveillance tasks, at least in peacetime. If one has full information regarding friendly and commercial aircraft operations, it is, in principle, possible to determine which aircraft are potentially hostile. However, aircraft may sometimes avoid clear detection by flying low enough to be out of sight most of the time.

Sensing naval vessels within visual range is relatively simple, so long as they are on the high seas. However, naval vessels in confined waters pose more difficulties because of competition from other objects. Isolated ground targets, such as armored vehicles, may be mistaken for civilian vehicles or even billboards; however, in practice they can be identified by their participation in formations.

A key factor in C^3 systems intended for early warning of attack is timeliness. If attackers cannot be detected until they are close and they take advantage of terrain to reduce their visibility, locating them is made much more difficult; directing attacks against them may result in attackers arriving too late to establish contact. Good C^3 systems should enable a military response within the limited time ("window of opportunity") available. When this time interval gets down to a few minutes, as in nuclear-missile attacks, the defense problems are indeed severe.

Electronic Warfare Systems

Countermeasures are intended to deny an adversary the ability to detect the protected target or to prevent communication of information needed to carry out an attack (11). Countermeasures are most often directed at sensors, but may be directed against the communications channels that link sensors, controllers, and their controlled forces. Countermeasures may be *passive* (i.e., do not radiate electromagnetic radiation), *active* (radiate energy to counter the enemy's sensors), or *lethal* (physical attacks against the sensors, communications, or nodes). Electronic countermeasures have been exercised since World War II (12,13), when metal foil strips ("chaff") were used to attenuate and reflect radar signals, and transmitters were used to "jam" radar receivers (see Fig. 4). Because the reflected energy returned to the radar depends on the inverse fourth power of the distance between the radar and its target, while the target's detection of radar energy depends on the inverse square of the distance, a radar-illuminated target can usually detect the radar's presence long before the radar detects the target. When an attack is being made against a defended target, passive countermeasures (including "stealth" treatment to reduce radar reflections from the attacker) and chaff are preferred, since they may prevent detection of the attacker. Once the enemy becomes aware that the attacker is nearby, however, active countermeasures may be employed by the attacker without further increasing its risk.

The most common active countermeasure for either radar or communications is *jamming*. Radar jamming usually attempts to confuse the radar operator or electronic data processor. The most effective jamming, even more so than that which inhibits detection or communication, is that which can cause the adversary's system operators to distrust their own sensors or communications equipment.

Deceptive radar or communications signals attempt to convince an enemy that the situation is different from that which his sensors suggest. Radar deception, for example, may imitate radar echoes from a larger flight of aircraft or make signals appear to come from other than the true direction. Sensors or their supporting information analyzers can be "spoofed" if the situation is made to appear different from the real one.

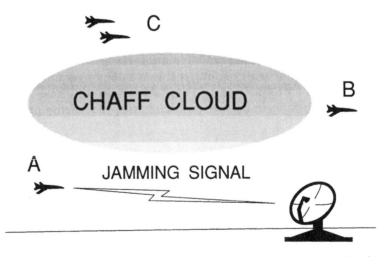

FIG. 4 Electronic warfare countermeasures. Aircraft A is transmitting a jamming signal (active countermeasure) toward the radar as it approaches at low altitude. Aircraft B has distributed a mass of chaff (passive countermeasure) so that Aircraft C can approach without being detected by the radar.

Radionavigation Systems

Radionavigation is construed to be part of the C³ field because electronic signals are used to guide aircraft or ships. It is not, however, part of the information-action loop so important in C³ applications. The simplest radionavigation aids are small transmitters (*beacons*) whose direction is inferred by receiving their signals via a small directional antenna. The most complex radionavigation system thus far is the U.S. NAVSTAR GPS (Global Positioning System—see Fig. 5). A constellation (nominally including 18 satellites) of satellite-based, time-synchronized transmitters is spaced around the earth in orbits about one-half the diameter of the geosynchronous orbit. By measuring the differences in arrival times of signals from four different satellites, the three spatial coordinates of the receiver location can be determined quickly and with remarkable accuracy. Like GPS, most modern radionavigation systems use time-difference techniques, which can be highly accurate because of the great stability of modem frequency standards based on molecular spectra.

Civil C³ Systems

Air-Traffic-Control Systems

In the United States, civil air-traffic control is a responsibility of the Federal Aviation Administration (FAA). Its nationwide network of radar and com-

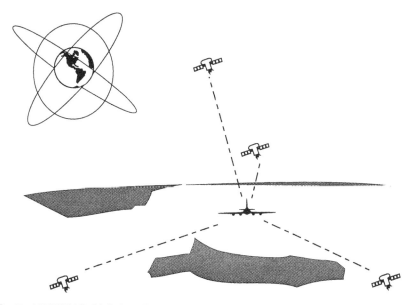

FIG. 5 NAVSTAR Global Positioning System, comprised of a space component with satellites spaced uniformly within three inclined orbits (shown at upper left of the illustration), showing a user in an ideal receiving position for time-difference position estimating: three satellites are at 120° spacings close to the horizon and one is almost directly overhead.

munications-equipped control centers (including military facilities within the network) monitors passage of private, commercial, and military aircraft along designated "airways," and into and out of the nation's major airports. The growth of commercial aviation worldwide has necessitated the evolution of air-traffic standards, many based on systems or practices initiated in the United States because of its early development of commercial aviation.

Air-traffic-control systems fulfill our definition of C^3 systems. Aircraft crews report their positions and intentions, are given directions by air-traffic controllers, and correct their flight trajectories accordingly. The human element is important in this control loop because of uncertainties due to weather, queueing of landings and takeoffs at busy airports, and nonscheduled private or military flying. The civil air-traffic application appears to be amenable to increased automation; there have been serious proposals to automate flight-space allocation. Subjective passenger safety considerations, however, have thus far discouraged efforts toward greater automation of flight-control operations. However, radar-controlled approach procedures, used at some airports in bad weather, have been successfully automated at a few locations.

Physical Security Systems

Physical security systems are intended to discourage unauthorized intrusions onto property by enabling rapid response to such intrusions. (Prison escape-prevention systems present the other side of the same coin.) Industrial security systems often possess the characteristic information-action loop. They sense intrusion through any of a variety of sensors, alert security personnel using alarms and displays, and may also be relied upon to display the results of the responsive actions. Most commercial security systems protect an area having a closed perimeter and usually involve less sophisticated sensors and communications than those used for national defense against intruding bombers or missiles. The primary purpose of both systems is, of course, deterrency.

Timeliness is important as well. If security response forces are unable to react in time to reach the intruders to prevent their carrying out their objectives, the security system's only possible value is as a deterrent. For a deterrent to be successful, a potential intruder must believe that it will function effectively.

Many security systems have never been realistically tested. If stakes are high enough, the performance of such systems and the complementary response mechanisms should be tested regularly by arranging for real and/or simulated intrusion attempts at unpredictable intervals, with the attempts being made by persons or organizations adept at this activity.

Fire and Police Emergency Systems

The most numerous nonmilitary command and control systems are those that control police and fire responses. As with other C^3 systems, situations potentially requiring action are alerted by sensors. Such systems may initiate a phone call to an emergency number and play a prerecorded message. Then police may respond directly to the intrusion alarm systems activated at the intrusion site.

In the past century, many large cities have installed fire-alarm devices that are mounted in various locations throughout the city. Though many such devices are typically connected to a single signal circuit, each alarming device signals its location uniquely by using a simple telegraphic code. Police and fire-safety organizations make use of human dispatchers whose function it is to interpret alarms and other input information and then to transmit directives to pre-positioned or mobile vehicular units. Telephone, facsimile, and/or teletype-writer communications over wire or radio channels may be used, depending on the urgency of the transmission and the type of unit or location that is to receive the information. It is not uncommon in police work to use both wire and radio relay of signals, such as by cable from the central control node to a remote transceiver and then by UHF radio to a patrolling vehicle. Satellite communications links are now employed in some civil emergency situations (e.g., to detect and locate an emergency transmitter emitting from a downed aircraft). However, the majority of civil emergency systems cover small geographic regions and therefore can make good use of short-range radios with commercial communications media.

Other Nonmilitary Systems

A familiar C^3-like application is radio dispatching of taxi fleets. Requests for service are received by phone, and in the case of future dates or times, they must be recorded, later recalled, and the service scheduled at the appropriate time. Unlike most C^3 applications, this system's objective is a simple statistical one: keep as much of the fleet operating as much of the time as possible. In practice, most such systems rely on UHF transceivers in each taxi, with fixed transceiver(s) serving the dispatcher(s).

Management control systems in business and industry have some of the features by which we have characterized command, control, and communications systems. For example, production control systems at some manufacturing firms respond directly to order receipts and put into motion the processes necessary to produce products ordered. This is a departure from traditional manufacturing, where manufacturing planning leads order receipt by months or years. The objective is to reduce capital committed to unsold inventory, uncompleted products, and product components. Large, well-managed development and construction projects involving thousands of workers and millions (or billions) of dollars often possess certain elements of a C^3 system. Control actions may be triggered by exception (i.e., if certain prescheduled objectives are not met). Typical control processes of this type identify key completion schedules or such development parameters as product performance or estimated costs. These parameters are "tracked" via periodic review or some other monitoring method. When a tracked parameter goes beyond the range of acceptable values, an action is expected to be taken.

Many production processes, especially in materials and chemical industries, employ discrete or continuous process control as a means of maintaining product throughput and quality; human operators may "monitor" control loops, with the supposed objective of preventing systematic errors or catastrophic failures of the automated system. If humans in the loop have the necessary knowledge to alter the system constructively and can provide the required timeliness in response, the scheme makes sense. Recent disasters in nuclear-reactor control, however, have emphasized the shortcomings of human "controllers" required to recognize and act when a complex automated system fails.

Characteristics Shared by C³ Systems

Distributed-System Resources

Most C³ systems are characterized by distribution of the resources involved in carrying out system functions. This distribution may be natural if the work must be done in specific, scattered locations. Or it may be for convenience, as in an organization that distributes its operations to take advantage of land or labor availability or the proximity to markets.

Network Structure and Growth

A requisite of most C³ systems is rapid communication among sensors, control elements, and controlled elements. Because of this, a C³ system typically relies on a communications network designed to meet its needs. Such a network may include both wire/cable and radio communications links, depending on the mobility of all of the system resources. Channels may be owned or leased. They may be operated continuously or intermittently, depending on the system's peak and average communications traffic and the allowable delays. Most such systems embody one or more command and control nodes that handle and use large parts of the system communications. The system's geographic distribution, moreover, may be such that multiplex switching nodes are required to concentrate and distribute communications traveling between groups of resources remote from major control nodes or other resource groupings. Some such systems may be characterized as being built around a hierarchy of nodes and subnodes, others can be envisioned as having a *communications backbone,*, a high-capacity channel or set of channels connecting a group series of major nodes with which other system resources can communicate (see Fig. 6).

   Whatever the networking concept, one of the most important considerations is its adaptability to system growth and restructuring. Primarily for this reason, many C³ systems have been based heavily on the use of leased commercial communications. Large commercial networks enjoy economies of scale that cannot be matched by stand-alone C³ applications. If commercial communications channels are not available or for some reason are not acceptable, system network channels must be designed with growth and change in mind. In recent years, this has usually suggested digital communications, even where current signals are largely analog in character.

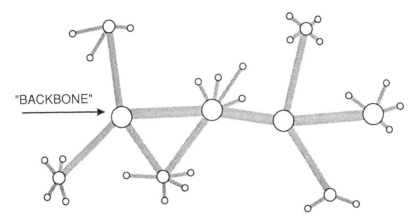

FIG. 6 Graphic representation of ground-based communications facilities of a hypothetical C³ network. Four major nodes are situated along a high-bandwidth "backbone" with sub-nodes to either side acting as communication hubs for minor nodes. Backup communications via radio could be used to bridge gaps in this permanent network, albeit perhaps operating at lower total bandwidth.

Network Control Concepts

Principal C³ network control issues relate to connectivity, timeliness, and channel availability. Terminal pairs that must communicate must be able to obtain an appropriate connecting channel. If the communication requires minimal delay, it cannot rely on shared channels with large queueing or switching delays. Traditionally, such links were supported by continuous-duty channels, often in the form of leased circuits. Packet communications are an increasingly viable alternative for many such requirements.

Channel availability in the presence of equipment failures, natural interference, or intentional hostile interruption requires redundancy and/or diversity. Essential C³ systems are designed to include one or more means to achieve every necessary communication. Cable communications may be augmented by radio relay, troposphere- or satellite-echoed communications, or point-to-point sky-wave or ground-wave radio transmission, depending on distance, geometry, and other application factors. Some C³ systems rely on special waveforms that cannot be handled by readily available alternative communications (e.g., because of bandwidth). This increases system vulnerability.

Incorporating or Interfacing with Dissimilar Subsystems

Long-established C³ systems are seldom homogeneous, often incorporating older-generation equipment of lower capability than more recently added portions. Some systems must operate across organizational or even national boundaries. This can add a variety of interface issues, from languages and character sets used by operators and their keyboards, to signal levels and formats. These issues have been faced for decades in international telegraphy and telephony.

However, international commercial communications service does not require standardization of message formats or even resolution of semantic differences, both of which may be needed for successful international C^3. Extensive interaction has taken place between the United States and other members of the North Atlantic Treaty Organization (NATO) in efforts to agree upon the C^3 standards important to all members of that military alliance. The steady advance of technological capabilities poses one difficulty in reaching agreement: a nation invested in mature and currently acceptable technology is often unwilling to endorse new standards that require premature abandonment of its investment.

Human Interfaces for Receiving Information and Promulgating Decisions

Humans are important elements in C^3 systems. Accordingly, the means by which they interact with a system should be of great concern, though this has often been overlooked by C^3 system designers. Traditional human–computer interfaces have addressed either limited applications (e.g., keying input of routine data), or complex but non-time-critical applications (e.g., programmer's workstations).

Traditional real-time human interfaces to communications systems have been largely voice and low-performance teletypewriter interfaces. Voice interfaces are too error-prone for many applications. Happily, human–machine interfaces are being attacked on a broad front in connection with the development of personal-computer and programming/engineering workstation software.

Voice, Keyboard/Printer, and Facsimile Terminals

Since the early 20th century, the telephone (or radio telephone) and teletypewriter have been the principal terminal devices for C^3 applications. As early as the U.S. Civil War, telegraphy was relied upon. Both North and South realized the strategic importance of the telegraph lines to nationwide coordination of troop movements. Both Northern and Southern army leaders were subjected to political meddling in purely military affairs, made quicker and more damaging by the speed and ubiquity of the telegraph. The telephone became the principal communications vehicle of military command and control during World War I; it was complemented by the radio telephone in World War II.

For many years, telephonic facsimile has been available that is capable of reproducing a printed page. Within little more than a decade, integrated computer chips have greatly enhanced the speed and quality of facsimile output. The chief difficulty of both voice and facsimile communications is that further, error-prone steps are required to retain the information in computer-accessible form.

The teletypewriter (a keyboard/printer combination, also equipped to write and read data on a paper tape) has a chief advantage of producing a printed record, though it is a poor choice for receiving locations in foxholes or aircraft cockpits. Continued popularity of teletypewriter communications (now often communicating directly to computer display screens) is due in large part to its archival potential.

Computer Input and Display Terminals

The cathode-ray tube (CRT) and its flat-panel display cousin represent the state of the art in human–computer interfaces. Although a standard typewriter keyboard is the most common input device, a variety of alternatives is available to fit special needs (see Fig. 7). These input devices include the light pen, a sensor that allows the user to touch a point on the screen; the more convenient mouse, whose movement over a planar surface is picked up via a rolling ball and position sensors and imitated by an electronic cursor on the screen; the trackball, an upside-down mouse whose large rolling ball is moved by hand;

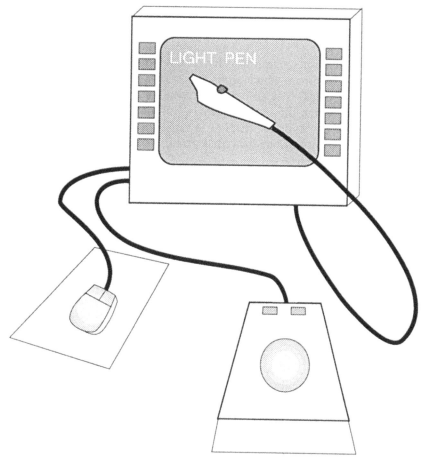

FIG. 7 A CRT display equipped with four nonkeyboard input devices. The light pen contains a tiny light sensor that detects when the CRT's electron beam impinges on the point where the light pen is aimed. If the push-button is depressed, the computer can determine where on the screen the light pen was pointed. At the left of the illustration is a computer-input mouse, which can be moved around on a nonslip pad while the corresponding position of a cursor moves on the screen. The trackball (right) is in essence a large inverted mouse whose rolling ball is moved by the user's palm while fingers are used to press one of the buttons. The rows of pushbuttons on either side of the screen can be used to select commands whose description is displayed adjacent to the buttons. They and/or the trackball are compatible with operation in a military aircraft.

and the digitizer, a calibrated equivalent of the mouse that requires a document-sized base, typically used to measure positions on scaled drawings or maps. A CRT display may be covered with a touchpanel that in effect sensitizes areas of the screen. Alternatively, a ring of push buttons may be placed around the periphery of a CRT with their functions displayed on the edges of the screen.

A microphone may also be used as a computer input device if the computer is augmented with a speech-recognition subsystem. This input scheme may be highly effective, provided the device is "trained" to the user's voice, a limited number of short words need to be distinguished, and there is little noise and distortion in the voice signal.

Intrinsic limitations of the CRT display screen are its two dimensions, the diameter of the electron beam by which it forms images, and the resolution it can give to color images. Flat-panel screens are based on a fixed array of display cells that equals the number of picture elements (*pixels*) from which an image may be formed.

In most C^3 applications, computer-driven CRT displays create a rectangular raster of discrete pixels from which alphanumeric or graphic images are formed in monochrome or color. The principal alternative to the raster is the vector graphic display, in which the beam is moved to "paint" lines connecting pairs of points. Though this type of display is superior to raster technology for displaying maplike information, it is used mainly for special applications such as wall-projected map imagery.

A prevailing trend in commercial computer display technology is use of large (20″ diagonal or more), high-resolution display screens that permit placing large amounts of text or graphic information on a single screen. The complementary software trend is "windowing," wherein a number of applications or different aspects of a single application may be represented on different or overlapping areas of a single screen. This scheme has some applicability to C^3 displays. However, in most C^3 applications, display formats need to be standardized. Operators viewing different CRTs must discuss their contents over an intercommunication network, and operators coming on duty for their 4- or 8-hour duty period should greet a familiar representation. Both situations discourage the provision of "design-it-yourself" display format and content.

The value added by color to C^3 displays has long been appreciated. In most C^3 applications, several categories of objects must be distinguished: friendly versus hostile troops, vehicles, ships, or aircraft; controlled versus uncontrolled air traffic; different varieties of radars; and the like. Less is known about optimal design of color displays, save that which relies on knowledge about color sensitivity of human vision. Still less is known about optimizing details of representation, such as the use of polygons, circles, or crosshairs, and the use of solid, dashed, and dotted lines.

Tradition still plays too great a role in human interface design. Despite longstanding knowledge of greatly improved keyboard layouts, the "QWERTY" layout is still used, even on keyboards to be used by personnel lacking keyboard training prior to learning to operate some particular C^3 system. Warnings are depicted on aircraft and other displays in the color red, despite its being a color to which human eyes have the least sensitivity, and also the color used universally in cockpit displays to maximize eye adaptation for night operations.

Also, the radio voice codes ("ALFA," "BRAVO," "CHARLIE," etc.) used for voice communications of encrypted information or in noisy environments were not designed as a maximally distinguishable set, as one would have hoped.

Computer-Based C³ Decision Aids

Because performance of humans is so important in C^3 system operations, substantial benefits can be obtained from intelligent use of decision aids (14). In a C^3 system context, a decision aid may be a computer resource that, because it is able to examine the data available to a human operator, can quickly carry out analyses to help the operator select from a set of options. Though the development of decision aids has frequently been associated with activities in artificial intelligence, this is somewhat unfortunate. There are many decision aids that need not rely on unproven technology. Simple decision aids, for example, can carry out elementary mathematical or logical operations based on selected data. Capabilities as elementary as simple and rapid computation of travel distance or time along routes marked on a display would be of great help in some C^3 systems. For any given application, candidates for decision aids may be found by examining the tasks system operators must repeatedly carry out mentally or with the aid of a scratchpad and pencil.

There is also need within the C^3 systems arena for more elaborate advanced analysis and simulation tools. These are most appropriate to planning, scheduling, and the like, rather than to the real-time resource-allocation operations for which most C^3 systems principally have been designed.

Computer and Software Intensiveness of C³ Systems

C^3 systems are strongly dependent on computers and software, not only because of the complex communications signals that they handle, but because system users at their nodes and terminals must deal with data on characteristics, numbers, and locations of the controlled resources. In the case of military C^3 systems, additional computer support is required for data on the characteristics and location of the adversary's resources.

Computer Needs

C^3 systems may impose near-real-time requirements on computer support. This may take the form of requirements that inquiries from operator terminals be responded to in a few seconds or less. Other requirements may specify the maximum time required for processing certain kinds and volumes of data. Where graphic data is converted to suitable formats for transmission (also, perhaps, including encryption), much processing is required before the data can be used. These requirements demand that the many computers associated with a C^3 node, and perhaps computers distributed throughout many nodes, should be networked (15). Real-time requirements placed on such systems should not specify delay criteria more stringent than actually needed for effective system operation. Human system operators may find computer delays of a second or two annoying, particularly if they must await the response before continuing their work. If this petty annoyance is translated into a stringent delay requirement, it could possibly render the system unaffordable or even infeasible.

In highly important applications, computer-system reliability and availability frequently become critical. This is especially so when a system must operate on a continuous, 24-hour, 7-day schedule. In such instances, there is no practical alternative to a fault-tolerant, redundant computer installation. Few commercial computer systems are designed for uninterrupted operation. However, duplicate conventional computer systems can be employed, provided reliable (and perhaps redundant) input and output connections are offered along with means for sensing computer failure and removing the failed computer from further activity until repaired. If even seconds of downtime cannot be tolerated, three or more computers may be operated in parallel, correct operation being deduced by automatic comparisons of their outputs. If a brief hiatus in operation is acceptable, however, a two-computer complex may suffice, provided it has means for continuously monitoring performance and for transferring critical processes resident on the failed computer to the still-operable computer.

Distribution of Computer Resources

C^3 systems often impose requirements for distribution of data and processing functions around a collection of remotely located computer sites. When coupled with the usual demand for real-time system responses, the combination is one of the most demanding of any computer application.

Databases are usually geographically distributed because the controlled resources and/or adversary systems they describe are distributed (perhaps globally, as in the instance of a space-based strategic missile defense system). Processing functions may be required to permit transmission of processed data from a sensor, rather than transmission of less-compact "raw" data.

The most difficult system task in widely distributed computer complexes is achieving the necessary synchronization of data and events. A processor should not, for example, aggregate stale data from some remote locations with fresh data from others. Though a generic solution is to forward new or altered data from each location to all other network locations immediately, this will likely impose excessive communications loads. A more subtle and more acceptable approach is to update quickly only information of immediate importance, waiting for periods of low system and network activity to transmit more complete updates. This data transmission paradigm should be coupled with a database layout that addresses the relative importance of data elements involved. Data of low importance, or that which is not needed instantly, can be archived on slow-access storage media; that which is needed instantly should be stored in computer memory.

Data Utility and Timeliness

A central problem in C^3 systems is dealing with data whose value and timeliness are greatly different. An air-traffic-control system, for example, may have recorded that a particular aircraft initiated a flight between two particular airports at a particular time. Later, unconnected data might reveal that some aircraft abruptly disappeared from the view of a long-range radar at a certain time and location. Put together, these two items of data might allow determination of the identity of a missing aircraft.

The problem is even more complex in some military C^3 systems. Observers on a merchant vessel may have reported seeing a large missile-carrying submarine in midocean on a certain day. It may be desirable to relate this informal information to real-time data from such detection systems as sonic detectors or surveillance satellites.

Standard Message Formats

A requirement in many C^3 systems is that data be transmitted in a standard message format, similar to a record in a database. This typically could involve transmitting position data in universal coordinates, estimated speed in knots, heading in degrees from North, time as Greenwich mean time, and identifying objects or formations uniquely or as a member of a class. Each data field may be identified by position in the message and/or by an identifying prefix. Unknown information may be left blank or merely omitted. Messages received in such formats can be rapidly machine processed and analyzed.

Despite widespread use of such formatted data structures in C^3 systems, there is as yet less attention to the idea that formats include accuracy or uncertainty estimates for their data. Certainly, these will be valuable in connection with future decision aids. For example, a brief radar observation of a far-distant aircraft flying across its field of view at a heading nearly normal to its line of sight may yield far less accurate velocity information than if the heading were parallel to line of sight. If a velocity-accuracy factor is not recorded specifically in the data, it is not possible to assess it correctly, unless the radar location, aircraft location, detection interval, and radar characteristics are all known or specified in the message. Even if all these factors were known by the message recipient, a complex search and calculation would be required. Even though expected limitations to communications bandwidth discourages inclusion of unneeded data, future automated analyses would benefit greatly from accuracy estimates.

Message Filtering

Data filtering in C^3 systems involves correcting or discarding data, typically at message level. This may be part of processing when messages are received, or later, if and when the data is selected for processing. Unlike signal filtering and signal processing, this filtering is based on semantics of the data. A message whose fields are unreadable after decryption should not be forwarded. Likewise, messages that contain redundant information deserve little or no attention. Data from known sources having known offsets or biases may be altered to remove or accommodate that bias. Incorrectly spelled values might be replaced with "closest-match" identifiers from a working set. This could be a complex process, especially if the data also contain uncorrected errors introduced at the source or in transmission. In military C^3 systems, there is the possibility that spurious messages are introduced by adversaries or their agents. In interpreting a given message, both its content and the context in which the message was received must be considered in any filtering operation that alters it.

Context- and content-based filtering can benefit from application of rule-based processing, a product of artificial-intelligence research. A set of rules may be devised for handling each message format and each field within the format. On receipt, applicable rules are applied to the data to correct or normalize them and prepare them for additional analysis. Both syntactic (form) and semantic (function) rules usually will be required.

Data Aggregation and Integration (Data Fusion)

The term *data fusion* implies a process wherein data obtained from one or more sources, possibly in a variety of formats, are selected and grouped according to specific objects or events they are believed to describe. For example, a message to the effect that a particular aircraft left a certain airport at a certain time, plus reports of the sighting of an unidentified aircraft at later times and locations consistent with that takeoff, might be "fused" in an attempt to infer the identity, destination, and objectives of the flight.

Data fusion is, or should be, an exercise in reducing uncertainty. In C^3 system application, it ranges widely in both difficulty and utility. Successive sightings of a particular aircraft by a net of radars whose fields of view overlap significantly may be fusible into a wholly valid representation of the flight path. If it is one of many aircraft that fly near the ground and on zigzagging courses to avoid continuous radar contact, any conclusion as to destination (or even the number of aircraft) might be much less certain. The difference between these two cases is uncertainty, which arises when we have fragmentary information.

Data fusion is what intelligence analysts often do, usually with extremely fragmentary information. An analyst who holds some particular viewpoint strongly may justify it using all the noncontradictory data. On the other hand, if all noncontradictory data were discarded (for any reason), the case in favor of another viewpoint would appear to be strengthened. In fact, the total body of existing data may point to a third, very different conclusion. The most closely related field of study and research to that of data fusion is probably the study of evidence (16).

Data Security

Users of most C^3 systems are often concerned with data security. Lack of it takes several forms, for example, accidental or malicious loss or alteration of data within the C^3 system itself, interception or corruption of messages by competitive or hostile organizations, and access to data by unauthorized personnel inside the system.

These breaches of data security can take place in a number of ways, principally involving computers or communications channels. Consequently, a variety of techniques is required to address the total problem. For data in transit, protection is most commonly through application of encryption (encoding). The issue here is not whether an outside party could eventually decrypt system messages, but whether it would be worth the cost to do so, and whether it could be done quickly. To decrypt quickly and correctly, one must have both the encryption algorithm and the key used for the particular encoding. If both are changed periodically and great care is taken to safeguard each, decryption by an adversary may be rendered both costly and untimely. The design of secure encryptions is a mathematical specialty that has been practiced mainly inside intelligence organizations. Modern codes, now usually applied and removed using computers, are far more complex than those that occupied large groups of technical specialists during World War II.

Data security within C^3 system nodes requires that certain data be made accessible only by the most trusted employees. Very critical data may be revealed or changed only if two or more such employees join to access it. Major strides have been made in protecting data in systems used by both highly trusted employees and others less highly cleared (termed *multilevel security*) (17). The complexity of present-day computer hardware and operating systems is so great that most have weaknesses that can be exploited to extract or alter protected data. While data security originally arose as a military concern, the growth of communications-based computing and irresponsible "hacking" have resulted in widespread concerns over data security. New systems and software increasingly show attention to these concerns. As with other attempts to reach perfection, the cost associated with improvements in computer security increases steeply as the level of required protection is increased.

Sensors

C^3 systems employ sensors that range from simple intrusion detectors to multi-million dollar radars. A sensor's purpose is to detect objects, events, or environmental changes of potential interest to its users. Large sensors, such as ground-based surveillance radars, are familiar, as are such simple detectors as magnetically operated switches that indicate when a building door has been opened. Some sensors are cloaked in secrecy, such as those on satellites that detect, from space, the ground launch of nuclear missiles.

A major subdivision of C^3 sensors is that of active and passive sensors, depending on whether or not the sensor emits radiation. Most radars are active (radiation emitting) sensors. Television or infrared cameras are usually passive (non–radiation emitting). Sensors may also be grouped into imaging and nonimaging sensors, depending on whether or not they can produce a two-dimensional photo-like representation of an area or object. Some nonimaging sensors merely register the presence of possible objects of interest in the area of coverage, while others are designed to *track* (point toward and follow motion of) a target. Outputs of most imaging sensors are intended for human observers, though computer-based automatic pattern recognizers can also be used to examine the sensor data for features of possible interest.

Sensor Parameters

Sensors differ in five critical parameters: field of view, scan rate, sensitivity, range, and resolution. The sensor's instantaneous field of view may be limited by the optics of the sensor, and the total field of view by limits to its mechanical orientation (Fig. 8). Specification of an instantaneous field of view usually suggests that electronic scanning is possible over that field. However, this scanning may or may not produce acceptable moving images, depending on the scanning rate and the response time of the sensing mechanism.

Scan rates are important because, if a scan period is too long, a target could move far enough during one scan that it is not in view during the next scan. When a target is near the maximum range at which it should be detectable, sensor detection probability should be increased if the sensor scans too slowly. For example, if the sensor is a pulsed radar, its maximum range depends on the number of pulses echoed off a target before the radar scan moves to where the target is no longer positioned in the radar beam.

Sensitivity and maximum detection range usually go hand in hand. The higher the sensitivity, the weaker the signals needed from targets to detect them. As in communications, sensitivity is not merely a function of transmitter power and receiver amplification, but must also take into account receiver noise and

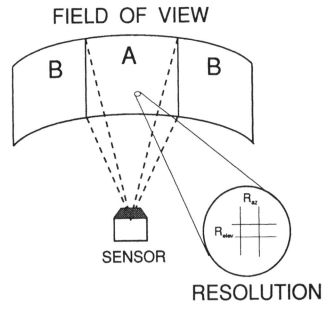

FIG. 8 Field of view, a sensor parameter. This sensor has an instantaneous field of view *A*, a total field of view (provided by mechanically scanning left to right) *BAB*, and different resolutions R_{az} and R_{elev} in azimuthal and elevation axes, respectively.

signal attenuation. If sensor signals are attenuated significantly by the atmospheric paths through which they travel (as for example, in the millimeter-wave electromagnetic domain within the normal atmosphere of the earth), useful ranges can be very short, almost independent of the gain or illumination power available.

Resolution of a sensor usually refers to the uncertainty in size or location of a target. Use of optical principles in most imaging sensors implies resolution inversely proportional to the transverse dimensions of the optics (measured in wavelengths). Radar antennas are seldom designed to produce resolutions much smaller than 1°. Conventional radar imagery is, understandably, rather coarse. Synthetic-aperture radars can be used in some applications to increase radar resolution by several orders of magnitude. In these radars, the echoes from a series of pulses are combined mathematically to produce an equivalent antenna the size of the area covered by all of the antenna positions when the pulses are emitted. (There must be relative motion between radar and target to produce the desired effect.)

These metrics are inappropriate for sensors that do not use electromagnetic radiation for detection. Trip-wires, proximity detectors, and nonimaging heat sensors, for example, may be best characterized in the parameters of the application for which they are intended.

Passive versus Active Sensors

Electromagnetic sensors operating at frequencies from a few hundred megahertz to a few tens of gigahertz are frequently active sensors, sensing reflection of radiation emitted from the sensor location (*monostatic active sensors*) or from some other location (*bistatic active sensors*). Infrared sensors are most often passive devices. In the widely used 8–12-micron (wavelength) band, warm objects such as human bodies, aircraft, or wheeled/tracked vehicles emit radiation that may sometimes be detected at distances of many miles. At shorter wavelengths, flames or heated jets from aircraft or missile engines can be readily detected. At wavelengths just longer than the infrared limit of human vision, starlight or an infrared source can be relied on for illumination.

Some passive electromagnetic sensors are designed to monitor radio transmissions or to detect distant radars. The simplest passive sensors are magnetic-field, sound, or movement detectors, any of which may be used in C^3 systems. The principal advantage of passive sensors is that operation of them does not readily divulge their presence. A small sensor on a target in the beam of a conventional radar, on the other hand, can detect the radar long before the radar can detect the target because of the inverse fourth-power range relationship obeyed by the radar echo, versus the inverse square range relationship of the illuminating signal intensity at the target.

Passive sensors may also be desirable because they use little electrical power, sometimes none at all, when in operation. However, many clever detection techniques are based on phase measurements, which are generally not meaningful with passive sensors.

Design and Application Issues

C^3 systems impose certain requirements uncommon in other communications or control systems. In this section, a few of the more important of these are addressed.

System Architectural Issues

Extensibility

The purchaser of an appliance or vehicle, even a building, is seldom concerned with the degree to which that item can evolve and be expanded. It is usually easier to merely "trade up" to another item meeting the new requirements. C^3 systems, however, may evolve over long periods of time. They become part of the essential fabric of the organization they serve, and tend to evolve along with that organization. If this is not anticipated in initial design, the using organization may soon find itself faced with the need for a complete replacement, with attendant costs and operational disruption.

Evolutionary development makes it difficult, as the evolution of telephony illustrates, to take prompt and full advantage of new technology. The U.S. military has long held that its military superiority must be based on qualitative rather than quantitative superiority, emphasizing limited quantities of new technology rather than masses of less capable, older technology.

Over the long life of a typical C^3 system, frequent growth and occasional structural change is likely. In this sort of system environment, it is usually unwise to attempt a very high degree of optimization since it will so quickly become invalid. Systems offering broad performance optima and easy extensibility are usually preferable.

Design for growth and change requires strong attention to careful selection and intelligent use of interfaces since these are normally the vehicles for change and extension. While in short-lived, quantity-produced systems such as automobiles, the major interfaces (engine transmission, drive train body, etc.) may be changed frequently, many kinds of C^3 system interfaces can outlast the technology they connect. This is in large part a matter of intelligent partitioning of the system into its subsystems (18).

Reliability, Availability, and Maintainability

The qualities of reliability, availability, and maintainability are important in most large-scale systems, and no less so in C^3 systems. They pose additional challenges when systems are widely distributed. It may not always be possible, for example, to learn why (or even if) some remote system element has failed. Satellite communications relays or sensor satellites are examples of complex and essentially unrepairable subsystems. In addition, in military use these may be exposed to jamming or direct physical attack.

Even the proper measures of reliability of such systems are subject to debate. For an individual subscriber, commercial telephone systems usually measure their integrity in terms of the likelihood that a call can be completed successfully within a given time period. This is basically an availability rather than a reliability measure, since the individual subscriber has little, if any, involvement with central-equipment or channel reliabilities.

C^3 systems typically support activities having priorities ranging from unimportant to urgent. Failure of a C^3 system to close some critical loop correctly and in a timely way could result in a disaster of major proportion.

The designer's concern is also one of availability: how much attention (and how many resources) should be dedicated to the highest priority system tasks as opposed to that given to other system activities. Networks supporting the system must recognize and respond to this range of task priorities; commercial networks are usually not designed to do so. It is often desired to design a C^3 system to carry out certain critical tasks, even though all other tasks may be delayed or aborted as a result.

Maintainability requirements for C^3 systems are similar to those for other telecommunications systems, and may call for little or no disruption of existing service during periodic maintenance, predictable types of repairs, and expansion of system capacity.

Interface Definition

Good interface design is the essence of systems that must evolve and grow. Though commercial interface standards are used in C^3 systems where applicable, use of new technology in a system often may require interfaces that cannot be implemented using existing standards. Many C^3 systems have no precedents in existing systems and are developed on demanding schedules by cost-competitive contractors to satisfy unique and exacting specifications. In this environment, there is seldom opportunity to assemble industrywide expertise and develop a thoroughly thought-out interface specification.

System Design Issues

Understanding the Application

It is the nature of a good C^3 system to reflect the mission and organization of its users. If the system is to be useful, its structure in certain ways will imitate the organization and the functional and geographical distribution of the organization(s) using it. Despite the activity the user may have engaged in to develop a system specification, it is desirable for the system developer further to study the organization, its people, and its plans in detail. Otherwise, when inevitable design decisions arise, design personnel may have no application basis for an intelligent resolution. If one does not exist, it is important to develop a representational model for communications within and outside the organization.

There are many ways to represent communications activity. One may construct graphs with nodes representing the communicators (or organizations) and arcs that are numbered to represent average, peak, and other measures of the volume of communications between each pair of nodes. While this is helpful to envision the task or describe it to others, for serious analysis these same data are best captured in tabular (or database) form, with one record per pair of nodes or communicators.

If it turns out that the new system will require changes in the way the organization's activities are carried out or managed, such a major change must be appreciated well in advance. Often, simulating certain functional elements of the proposed system helps users to visualize and design new operating procedures.

Use of Advanced Technologies

For competitive reasons, developers or users of new systems may desire to incorporate new and sometimes unproven technology into their designs. An argument in favor of this is the long system life cycle (typically, the time from onset of design to eventual system replacement). Without the latest technology, the system might soon become technologically obsolete, perhaps even before it goes into service. If the selected technology does not mature as expected or is leapfrogged by yet another technology, the results of such bold thinking may be disastrous unless the firm has little concern about competition.

Conservative system designers eschew any technology lacking a proven record of acceptability. A moderate approach often practiced is to design the system so that either a new technology or a traditional one can be incorporated to meet system requirements. With careful design, the result then can be acceptable using the traditional technology and even better with the new.

There are, of course, systems that owe their existence to new technology: microwave radar would not have prospered lacking the unproven magnetron transmitter tube. Fifty years later, we are now in an era when computer hardware and software are often pacing system technologies.

Integration of Hardware and Software Elements in C^3 Systems

C^3 systems are perhaps the most software-intensive of all classes of modem systems. By this it is meant that not only the majority of the communication signal-processing functions, but all of the display and control functions, decision support, and system control functions are defined almost completely through computer programs. This makes software, and more particularly hardware controlled by software, the central technology of these systems.

In addition, the high degree of system coordination required in near-real time places stringent requirements on software to cooperate, rapidly and without error, with other software. There is no longer sufficient time for humans to take over the interface and correct incompatibilities. When the various elements of the system are developed by different specialist teams, as is usually the case, this poses further stringency on definition of subsystem functions and interdependency and on the relevant software/hardware interfaces.

A good design approach to software-intensive computer/communications systems includes (1) careful definition of interfaces to minimize software interdependencies across the interfaces; (2) realistic simulation of the interfaces in operation (perhaps at scaled-down data rates and with certain noncritical operations simulated roughly); (3) thorough testing and evaluation of activity at the interfaces; and (4) organized overall (rather than piecemeal) refinement of design in light of these experiences. When full-scale hardware and the actual system software later become available, the efforts required to produce the design models again can be used to study and resolve exceptions discovered in the real system.

Testing Issues

Test Conditions and Criteria

C^3 systems pose some of the most difficult challenges of system testing. Though these are not unlike the problems encountered in testing communications networks, additional factors are imposed: there are often stringent requirements on real-time performance; priority tasks must be handled, with reliability and availability far above average; military C^3 systems may be required to operate in the presence of intentional interference; and portions of the system may be damaged or destroyed by sabotage or other physical attack.

A C^3 system for emergency civil or military use is expected to work well when outside events are most disruptive and with near-peak communications traffic. Because this is precisely its normal working milieu, the situation seems to suggest worst-case testing approaches. The big difficulty with this is that no one, especially the system user, is willing to suggest what that "worst case" should be. Almost any complex system can be brought to its knees by sufficient nodal problems and channel disruptions. Rather than seeking such a worst case, a more practical approach is to determine relationships among subsystem failures, system activity level, and those performance parameters of importance to users. Then it should be clear where the limitations are. More important, factors controlling system performance as it approaches ultimate limits (as illustrated in Fig. 9) need to be thoroughly understood.

Simulation of Subsystem and System Environments

Simulation can be a valuable technique at many times during a C^3 system's life cycle, including its use for design exercises, integration tests, operational maintenance, and as a mechanism for realistic operator training. In any of these situations, it is generally infeasible to use realistic system activity using actual system inputs, since, in design phases, the system or its real environment may not exist. Even later, when they do, it would be impractical to attempt to get subsystems (including remote nodes and terminals) to follow a rigid "script." In

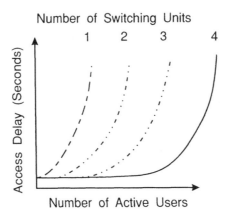

FIG. 9   Factors controlling system performance. By analysis, simulation, and/or hardware testing, performance variables of importance to the system's users can be determined as
functions of the design parameters. This curve might help a designer select number of
switching units. However, if the required access time were near or below the asymptotic
value, a change in the design concept (or the requirement, perhaps) would be called for.

addition, the entire system could be tied up merely to test some part of it. During
initial integration, this would not be feasible; for operational maintenance or
training, it would be undesirable to do so. Simulation requires additional elements (hardware and/or software drivers and performance-measuring means) to
simulate parts of the system or the external environment. In software-controlled
systems, simulations are often supported on actual system equipment. Alternatively, it may be desirable to have special hardware and software especially for
these purposes.

Representative simulations may be developed to test the system after minor
changes or reconfiguration, to evaluate performance at high activity rates or
other stressful conditions, or to take system operators or teams through scenarios similar to those they will experience in actual operation. With forethought
in the design, it may be possible to employ the system's own simulation resources
to support portions of developmental tests.

Operational Issues

Many operational issues are of little concern to system analysts or designers.
Such logistical matters as providing power, fuel, maintenance personnel, and
facilities are similar to those of other electronic systems operated in similar
environments. Personnel management is likewise similar to that in other activities. In this section, a few operational matters that have an important bearing
on system design are discussed.

Personnel Availability and Requirements

As computer-based systems have grown in capability and complexity, there has
frequently been a tendency for designers to create systems requiring personnel

having years of technical training or experience for competent operation and maintenance. In practice, since most military C^3 system operators do not have this kind of background, most of the special training should relate to the application and their roles in it rather than to technical details of the equipment. While it is not unreasonable to require some special training, a period of a few weeks to a few months is usually all that reasonably can be allowed. Similar considerations apply to nonmilitary systems, as is the case with civil air-traffic controllers in the United States, who receive about a year's special training.

Training of personnel places special constraints on system design. Competent operation of the system may not require large amounts of training. The designer must also keep in mind that a C^3 system operator may be mentally occupied with the external situation and his or her role in it. This places a premium on system operation that, while performing all necessary operations, is both simple and intuitive.

Performance Testing and Maintenance in Mature System Operation

The key roles of humans in the information-action loops of C^3 systems implies that changes of operating personnel can result in performance changes in the system. In continuously operating C^3 systems, personnel changes occur every shift or watch (i.e., every 4- or 8-hour period). Performing maintenance over the life of a complex C^3 system requires not only courses of training for system operators, but regular refreshers and continual monitoring of operator behavior patterns during regular operation. In many such systems, it is important to organize operators into teams (crews). Exercises and simulations should be scored by measures meaningful to system objectives. Team competitions and post-exercise discussion of its performance by the team help to maintain interest in what is otherwise often viewed as a dull job.

Management of System Integrity and Performance

A C^3 system may contain a wide variety of different elements with various amounts of parallelism and redundancy. As such, its users may be inclined to view it strictly from a personal perspective. Each operator could perceive it through his or her own controls and display. Its maintenance personnel might view it merely as a collection of "boxes" to be kept in repair.

Achieving overall objectives usually requires that both operational and management personnel appreciate the system in its top-down view, that is, in terms of the actual operational functions it (and they) carry out. Only by relating the hardware and software resources to the functions they provide to the users is it possible to develop a highly cooperative team with a realistic view of performance and a rationale for improvement.

The system designer can start, and later help with, this process. Because neither the human components of the system nor the processes that they carry out are controlled by the hardware and software designers, the system's users must define meaningful operational measures and set working standards or criteria for each. In a police emergency system, for example, perhaps the most important criterion is how long after a call was received did the patrol car arrive? Any system has measures of this sort that should be the bases for system management and improvement.

Management of System Growth

Seldom are a system's users able to predict fully in advance all of the system features they need or to define appropriate values or limits of all of the important operating parameters. By refining initial views with experience gained in actual system operation, the system can be made to evolve most beneficially. The establishment and regular updating of the aforementioned framework of system metrics forms a proper basis from which to define system evolution.

Even when details of later system evolution are still unformed, initial system requirements should provide for some growth. This may include planning for simple expansion of numbers of communication channels, switching circuits, computers or processing power, memory, and storage. With modern digital technology, is it often possible to provide for later expansion at little initial cost. Since every sort of expansion potential involves some initial cost, growth estimate parameters should be as carefully thought out as available data allow.

References

1. Beam, W. R., *Command, Control and Communication Systems Engineering*, McGraw-Hill, New York, 1989.
2. Boyes, J. L., and Andriole, S. J. (eds.), *Principles of Command and Control*, Armed Forces Communications and Electronics Association (AFCEA) Press, Fairfax, VA, 1987.
3. Bracken, P. J., *The Command and Control of Nuclear Forces*, Yale University Press, New Haven, CT, 1983.
4. Cardwell, T. A., *Command Structure for Theater Warfare: The Quest for Unity of Command*, Air University Press, Maxwell Air Force Base, AL, 1986. (Available from Superintendent of Documents, U.S. Government Printing Office, Washington, DC.)
5. Orr, G. E., *Combat Operations: Fundamentals and Interactions*, Air University Press, Maxwell Air Force Base, AL, 1983.
6. *The World Wide Military Command and Control System*, U.S. Government Accounting Office, Washington, DC, 1980.
7. Turban, E., *Decision Support and Expert Systems: Management Support Systems*, 2d ed., Macmillan, New York, 1990.
8. van Trees, H. (ed.), *Satellite Communications*, Institute of Electrical and Electronics Engineers (IEEE) Press, New York, 1979.
9. Dixon, R. C. (ed.), *Spread Spectrum Techniques*, IEEE Press, New York, 1976.
10. Hovanessian, S. A., *Introduction to Sensor Systems*, Artech House, Norwood, MA, 1988.

11. Schleher, D. C., *Introduction to Electronic Warfare*, Artech House, Norwood, MA, 1986.

12. Price, A., *The History of US Electronic Warfare*, Vol. 1, Association of Old Crows, Alexandria, VA, 1984.

13. Jones, R. V., *The Wizard War — British Scientific Intelligence 1939–1945*, Coward, McCann and Geoghan, New York, 1978.

14. Hopple, G. W., *The State-of-the-Art in Decision Support Systems*, QED Information Sciences, Wellesley, MA, 1988.

15. Blanc, R. P., and Cotton, I. W. (eds.), *Computer Networking*, IEEE Press, New York, 1976.

16. Schum, D. A., *Evidence and Inference for the Intelligence Analyst*, University Press of America, Lanham, MD, 1987.

17. U.S. Department of Defense, *Trusted Computer System Evaluation Criteria*, doc. no. CSC-STD-001-83, August 1983. (Available from U.S. Government Printing Office, Washington, DC.)

18. Beam, W. R., *Systems Engineering: Architecture and Design*, McGraw-Hill, New York, 1990.

WALTER R. BEAM

Common Channel Signaling

Background and Motivation for Common Channel Signaling

Common channel signaling (CCS) is a technique for telecommunications signaling that separates the physical channel used for signaling from that which is used to carry the end-user's telecommunications traffic. One signaling channel carries signaling to control a group of circuits, hence the term "common channel" signaling. CCS is generally used for interswitch signaling in such circuit-switched networks as public telephone networks and Integrated Services Digital Networks (ISDNs).

In-band Signaling Characteristics and Limitations

Historically, signaling was done "in band" — that is, the signaling information was exchanged over the same channel that eventually would be used to carry the user's traffic. In-band signaling systems have been in use in telephone networks since the inception of automatic telephony switching. Actual in-band systems include direct current (DC) signaling, single-frequency (SF) signaling, and multifrequency (MF) signaling. As these systems emerged and were implemented, they represented the then-current state of the art in terms of speed and information-carrying capacity.

In-band systems in use today have roughly the equivalent of 30 bits per second (bps) bandwidth. They are designed to signal a fairly limited set of information types, primarily strings of digits, whose significance is implicit in the context in which they are sent. Because they are in band, they are mostly restricted to signaling during the setup and teardown phases of a call, with very limited ability to signal during the active phase of a call.

Impact of Services on Signaling Capabilities

As new services are developed for telecommunications networks, they usually have some implications for interoffice signaling that require new information to be passed between switches. Either the signaling systems must be upgraded to allow these services to be implemented fully, or the services may be restricted to operate only within the context of a single switching system. As services become more complex, the limitations of existing in-band signaling systems have become more burdensome, and new signaling systems have thus been developed.

Network Architecture Trends

Two major trends in telecommunications network architectures have also placed demands on interoffice signaling. The development of the ISDN concept has

163

been based on the definition of a message-based, digital out-of-band signaling interface between the end user and the network (1). This user-network interface allows end users to request a variety of types of service over a single interface, and results in a huge increase in the amount of signaling information being received and processed by the network. Thus, the interswitch signaling system similarly is required to handle much more signaling information much faster.

The emergence of the intelligent-network concept is another trend with immense impact on signaling (2). The creation of centralized points of service control, the modularization of service functionality, and the flexibility of new services introduction, which are fundamental to the intelligent-network architecture, all place entirely new demands on the network signaling system to tie together all the pieces of this complex picture.

History of CCS

In the early days of CCS, the primary consideration was to provide faster signaling systems for call setup. In every telephone call, there is a fixed overhead required to set up a path to the terminating end through a series of switches and interswitch trunks. Even if the call is not completed, this overhead is incurred and represents a part of the holding time of the traffic offered to the network. Any reduction in this holding time represents potential savings of resources.

Common Channel Interoffice Signaling (CCIS)

In the national public telephone network operated by AT&T prior to 1984, the longest and largest trunk groups (those at the top of the switching hierarchy) offered the greatest savings from faster call setup and, in 1976, AT&T introduced common channel signaling in the long-distance network. The system was based on a signaling system recommendation of the International Telegraph and Telephone Consultative Committee (CCITT) for use in the international telephone network; since this was the sixth such recommendation, it was known as Signaling System No. 6 (3). The AT&T implementation of this protocol was known as Common Channel Interoffice Signaling (CCIS).

CCIS used modems operating over analog transmission facilities to carry its signaling information. The modems ran at either 2.4 kilobits per second (kbps) or 4.8 kbps. Rather than requiring a signaling channel between each pair of switches, a packet switch, known as a signaling transfer point (STP), was used to route signaling traffic between switches. Packets consisted of from 1 to 9 20-bit "signal units" that were encoded with the signaling information according to the rules of the CCIS protocol.

In 1980, the CCIS protocol was enhanced and the network was modified so that not only could switches exchange messages for trunk setup, but they could also send messages to databases to request translation of 800 service numbers and validation of telephone credit-card numbers. This new capability was called *direct signaling*.

Emergence of Signaling System No. 7

The protocol architecture of CCIS is very compact. Information fields are defined by bit position within a signal unit, and all protocol functionality is defined in a single protocol specification. This can be highly efficient, but leads to great complexity when new features need to be added. During the 1970s, new protocol design strategies, collectively known as open systems interconnection (OSI), were emerging that would simplify the job of augmenting protocols as new needs were identified. Octet-oriented protocols and protocol layering to separate different functions into different protocol specifications were two key elements of this new approach. In the late 1970s, the CCITT had begun applying the new concepts to develop a new signaling system, Signaling System No. 7 (SS7). The CCITT published its first SS7 recommendations in 1980 (4). The SS7 approach included much more than the latest protocol design concepts. It also embraced digital technology for its transmission channels, specifying them to run at the much higher speeds of 56 kbps and 64 kbps, and it upgraded the packet size from the CCIS limit of 180 bits to the SS7 size of 62 octets.

In 1984, the CCITT issued a revised and expanded set of SS7 recommendations, this time adding interoffice signaling to support the user-network signaling of ISDN. This is known as the ISDN User Part (ISDN-UP). As the OSI concept had been greatly refined, the 1984 SS7 recommendation also contained a new protocol layer, the Signaling Connection Control Part (SCCP), to align SS7 with OSI better (5).

AT&T also divested the Bell Operating Companies in 1984, and a new committee for drafting national standards for telecommunications was formed. This was the American National Standards Institute-accredited Committee T1. One of the first projects of the new T1 committee was to adapt CCITT SS7 standards to the United States. This led to the publication in 1987 of the first ANSI standards for SS7 (6-11).

Some of the advances in the 1987 ANSI standard, compared to the CCITT 1984 standard, were the expansion of the SS7 network addressing space from the international limit of 14 bits to a national standard of 3 octets (24 bits), and the selection of the CCITT-permitted national option to extend the packet size to 272 octets. In addition, a major new layer of protocol, the Transaction Capabilities Application Part (TCAP), was added to SS7 in the ANSI standard to support intelligent-network database applications. This new layer was subsequently added to the CCITT version that was reissued in 1988 (12).

Signaling System No. 7

Physical Architecture

Like CCIS, SS7 networks use a packet switch, or STP, to route messages between nodes on the network. The nodes are switches, service control points (SCPs), and operator services systems (OSSs) that exchange messages to carry out functions ranging from simple telephone call setup to complex database-driven services.

STPs are deployed in mated pairs in which both carry traffic under normal conditions and each acts as a backup system for the other in case of failure. Figure 1 illustrates the physical architecture of a generic CCS network. The links connecting an end node to its primary serving STP pair are known as *A-links*. If a node has links to another STP pair, these are known as *E-links*. Links between STPs that are not mates are known as *B-links* unless the STPs are arranged to use a hierarchical routing pattern, in which case links between different levels of the hierarchy are known as *D-links*. The links between two members of a mated pair of STPs are known as *C-links*. SS7 allows up to 16 links between any two points. The collection of all links between two points is called a *linkset*.

Protocol Architecture

SS7 is a suite of data-communications protocols that, taken together, support the implementation of modern telecommunications circuit-switched networks controlled by common channel signaling.

In order to accomplish such a broad objective, the SS7 protocol suite includes several protocols for various types of application-level communications between two switching systems (exchanges), or between a switching system and an SCP. These protocols are the Telephone User Part (TUP), the Data User Part (DUP), the ISDN User Part (ISDN-UP), and the TCAP. These four protocols offer implementers a choice of application-layer protocols suited to different applications. Any one signaling message will use only one of these four, depending on the application being supported by that message.

While application-level protocols are used to support end-node communications (switch-to-switch or switch-to-SCP), they do not include the functions necessary to route messages between end nodes, to concentrate message traffic via packet switches to allow economical connectivity of many end nodes, and to assure timely, accurate, and reliable delivery of messages between end nodes, within and across networks. These functions are performed by the Network Services Part of the SS7 protocol suite and are required for every message, regardless of the application. Thus, any TUP, DUP, ISDN-UP, or TCAP mes-

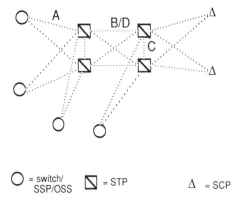

FIG. 1 Physical architecture of a generic common channel signaling network.

sage is carried through a signaling network by the protocols of the Network Services Part. The Network Services Part is divided into the Message Transfer Part (MTP), which is required for every SS7 message, and the Signaling Connection Control Part (SCCP), which is only required for some SS7 applications (primarily TCAP).

Schematically, this overall structure is captured in Fig. 2, where a single SS7 signaling message is represented by a vertical cross section. The TUP and the DUP were developed first and, in the United States, the ISDN-UP, which came later, has subsumed all applications originally intended to be supported by the TUP and the DUP. This reflects the move toward integration of voice (supported by TUP) and data (circuit-switched data is supported by the DUP) into the ISDN (both voice and circuit-switched data are supported by the ISDN-UP).

The emergence of the SCP as an alternative to switch-resident stored program control was accompanied by the development of TCAP as a part of SS7. TCAP supports service applications that use the exchange of signaling messages without the establishment of a circuit-switched connection.

Detailed discussion of ISDN-UP and TCAP and the types of applications they support, are included below. The TUP and DUP are not discussed further, as they are not widely used in the United States.

The ISDN User Part (ISDN-UP)

The ISDN-UP is commonly referred to as a protocol for circuit control and circuit-related services control. This means that ISDN-UP has two major functions. One is the supervision of circuits (trunks) between switches, both in the idle state and in the active state, as well as during the transition between these states. The other is the transfer of information between switches to allow software-based value-added services to be provided in conjunction with the establishment of a basic end-user to end-user circuit-switched connection.

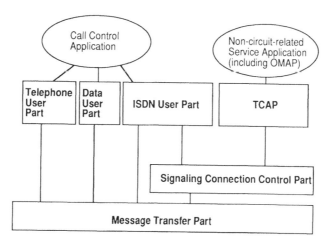

FIG. 2 Signaling System No. 7 protocol architecture.

Circuit Supervision and Control with ISDN-UP.   In traditional telephony terms, circuit supervisory signals are such things as on hook, off hook, address digits, and status (busy, idle, out of service). The ISDN-UP approach to reporting this sort of information is to define a message set, where each message type corresponds to a type of supervisory function. The ultimate goal of the message set and the procedures for using it is to allow two switches to exchange the necessary information regarding a particular trunk that connects them.

The ISDN-UP message set contains two basic classes of messages. Messages in one class come into play as part of call setup, call supervision, or call release—that is, they relay events that take place between the originating customer's call request and the final clearing of the call. The messages of the other class are call independent, relating only to the ongoing monitoring of status of circuits or circuit groups.

A Simple ISDN-UP Call Sequence.   ISDN-UP supports calls between two analog users, between two ISDN users, and between an analog and an ISDN user. It also supports a network with a mixture of SS7 and non–SS7 controlled trunks. Thus, there are dozens of different call setup scenarios possible. A few simple examples are described in detail below. For further detail, see Refs. 13–15.

The first example of a call setup is for a call between a pair of analog subscribers, using a call path through three switches, with SS7 controlling all the trunks (see Fig. 3). Several important facts about ISDN-UP are discussed here in connection with the example illustrated by Fig. 3.

First, the initial address message (IAM) causes the selected trunk to be marked busy in both switches. Trunks are set up all the way to the terminating switch before the status of the called party is checked. No "look ahead for busy" is performed in this example, although SS7 is capable of performing such a function if it is desired.

Second, no feedback is given to the originating switch regarding the progress of the call until the setup attempt reaches the terminating switch. The address complete message (ACM) is generated by the terminating switch, not the tan-

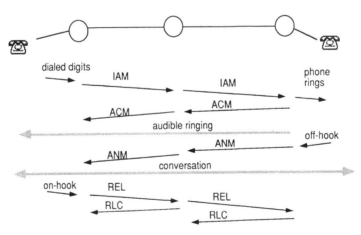

FIG. 3 Simple call control.

dem. This design allows the originating switch to attempt to recover from failures encountered anywhere along the path to the terminating switch. The answer message (ANM), also generated from the terminating switch, informs the originating switch that the called party has answered.

If the called line is busy, a release message (REL) (not an ACM) is generated. The REL carries a code informing the originating switch of the reason for release, permitting the busy signal to be generated locally, while the trunk network is idled. (With in-band signaling, the trunk network is held, and the busy signal is played by the terminating switch.)

Third, unlike the ACM, the release-complete message (RLC) is not an end-to-end acknowledgment, but a local acknowledgment. This design allows trunks to be idled as quickly as possible.

Finally, the IAM is not sent until customer dialing is complete. With traditional signaling schemes, it is commonplace to begin forward signaling prior to the completion of customer dialing (*overlap outpulsing*). The 56- (or 64-) kbps data rate available with SS7 makes this unnecessary, permitting all the call setup information to be packaged into a single message.

Variations on ISDN-UP Call Control. Because networks need to be able to evolve from traditional signaling systems to SS7, it is necessary to be able to operate in an environment of mixed signaling types. This general category of considerations is called *interworking*. If either of the trunks depicted in Fig. 3 were controlled via conventional signaling, then interworking would exist at the tandem. The principal mechanism for dealing with interworking encountered in the forward direction of call setup in ISDN-UP is the early return of the ACM from the point of interworking. This ACM carries with it a code indicating that interworking has occurred, which allows the originating switch to take into account the lack of end-to-end SS7 capability on the call.

Continuity checking on trunks adds another variation to the basic call control sequence. While state-of-the-art transmission and switching implementations monitor transmission continuity independently of the network supervisory and address signaling, earlier versions rely on the successful exchange of in-band signaling to verify that circuit continuity exists and that the call can be usefully completed using the chosen trunk circuit. A conversion to SS7 removes this in-band exchange of tones, and ISDN-UP provides substitute procedures. When a trunk is chosen that is marked in the switch memory as requiring a continuity test, this fact is indicated to the switch at the other end of the trunk via a code embedded in the IAM. The two switches then attach tone transmitting and receiving equipment to the circuit and attempt to exchange a standardized set of tones within a fixed time window. The switch that originated the test then reports the results to the other switch via the continuity test message (COT).

Timing is critical in this process since knowledge that a test is pending propagates all the way along the SS7–controlled trunk path, and the destination switch will not alert the called party until it learns, by receipt of a COT with a "success" indication, that all pending continuity checks have completed successfully.

Relationship of ISDN-UP to ISDN.   The ISDN concept was the principal driver in the move to define ISDN-UP in SS7. Prior to this, the TUP and DUP existed to serve the needs of analog telephone subscribers and subscribers to circuit-switched data service. Much of the design of the ISDN-UP message set was linked to the parallel design of the user-to-network call control signaling protocols for ISDN, termed the Digital Subscriber Signaling System No. 1 (DSS1). The ISDN-UP provides a capacity for information commensurate with that of DSS1, plus a protocol message flow that maps easily to that of DSS1. Thus, with full ISDN, including SS7, calls are controlled from end-user to end-user by the exchange of digitally encoded out-of-band signaling messages. Figure 4 illustrates a simple ISDN call flow.

Structure of ISDN-UP.   An ISDN-UP message has a structure designed to meet several objectives. Among these are (1) a logical association between signaling messages and the physical resources they control; (2) a clean separation of information elements that perform different tasks; and (3) a relatively easy evolutionary path for adding new functionality in support of end-user services needs.

The logical association function is supported by the first three mandatory information fields in every ISDN-UP message. These are the origination point code (OPC), the destination point code (DPC), and the circuit identification code (CIC). The OPC is the SS7 network address of the switch originating the ISDN-UP message. The DPC is the switch receiving the message. These two addresses taken together provide a unique context in which to assign numbers to every individual circuit between the two switches. The circuit numbers so assigned are the CICs. The CIC field is 14 bits in length, allowing a theoretical maximum of 16,384 circuits. Reserved bits in the CIC field provide a means to add additional octets if needed. The unique context of the OPC/DPC pair allows CICs to be reassigned for each pair of switches in a network. The fourth mandatory field contained in every ISDN-UP message is the message type. This

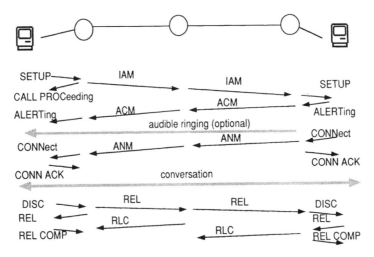

FIG. 4   ISDN call flow.

permits the basic functions of the message to be identified early in its processing.

The remainder of an ISDN-UP message depends on its type, plus the specifics of the application context in which the message is generated. The possibilities range from very flexible (the IAM) to very constrained (the RLC). A detailed look at the IAM is illustrative. An IAM has a mandatory fixed part, consisting of fields that must always be present and whose lengths are fixed, a mandatory variable part whose fields must be present but have variable lengths, and an optional part, consisting of whatever additional fields (or none) are needed to suit the application at hand.

Figure 5 illustrates the mandatory parts of the IAM. The ISDN-UP standard defines the additional fields permitted in the optional part of an IAM.

As the structure suggests, the purpose and use of these fields is to carry information at a deeper level of detail than the message set. In essence, these fields carry the detailed codings needed to indicate the varieties of situations that can exist within the context of an IAM. For example, "the nature of connection" indicators field provides codings to advise the receiving switch whether a continuity check is required on this circuit and whether a continuity check is pending at an earlier circuit in the connection.

At this point a discussion of routing is in order. The "called party number" field is mandatory in the IAM, since this is used by the receiving switch to determine whether it is the final switch in the call path and, if it is not, to determine which switch to send an IAM toward in order to further the call. This analysis of called party number is part of a routing process in the switch call-processing program. As noted above, there is an SS7 network routing function as well, supported by the Network Services Part of SS7, to enable the message actually to be delivered to the chosen switch. This is a separate routing function that acts not on the called party number, but rather on the SS7 network address, which is the DPC field of the IAM. Once the call processing identifies the next switch and the outgoing trunk circuit, then SS7 software must map that information to a DPC. As an IAM moves from switch to switch, its called party number remains constant but the DPC is replaced at every hop. Thus, the IAM is not simply relayed at a tandem switch, it is actually regenerated.

| Origination Point Code |
| Destination Point Code |
| Circuit Identification Code |
| Message Type = Initial Address Message |
| Nature Of Connection |
| Forward Call Indicators |
| Calling Party Category |
| User Service Information |
| Called Party Number |

FIG. 5 Mandatory fields of the initial address message (IAM).

ISDN-UP and End-User Services.   The list of optional parameters for the IAM is a veritable index of service possibilities. The single optional parameter "calling party number" opens the way to a host of features known as CLASS[SM] features. Delivery of calling number to the called party, distinctive ringing based on comparison of calling number to a preprogrammed list of numbers, selective rejection or forwarding, and automatic recall of the last incoming call are just some of these possibilities (16–18).

Call-forwarding service has been widely available for a number of years. With in-band signaling, calls forwarded across interswitch trunks are indistinguishable from new calls originating from the forwarding line. With ISDN-UP, optional parameters are available to carry the fact that a call is a forwarded call, why it was forwarded (busy, no answer, or unconditional forwarding), what the forwarding party number is as well as the original calling party number, and so on. This kind of information supports such powerful voice messaging applications as customized answering ("Ms. Smith is on her line" or "Mr. Jones is away from his office"). These voice messaging applications can now be more efficiently provided over a wide serving area due to the ability of ISDN-UP to concentrate call-forwarding traffic with the necessary supporting information at a single call-answering point (19–21).

The Transaction Capabilities Application Part (TCAP)

The TCAP of SS7 emerged during the late 1980s as part of the intelligent-network concept. It is an application protocol intended for use in query/response situations where there is no circuit-switched connection established between the querying and the responding nodes. Its first use was to support communications between a switching system and an SCP, where control logic and information would be stored for access by many different switching systems.

Figure 6 shows a typical TCAP message flow for 800 service using a database

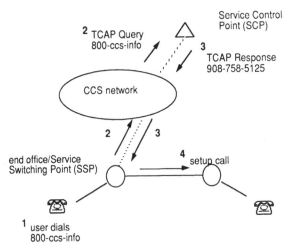

FIG. 6   Transaction Capabilities Application Part message flow for 800 service.

located in an SCP to translate the dialed 800 number into a routing number the network can use to complete the call.

Structure of TCAP. The method for coordinating applications in TCAP is through the establishment of a logical association called the *transaction*. The TCAP *query* message initiates a transaction, and the *response* message ends it. Each message carries *transaction identifiers* allowing the messages to be associated. If a transaction requires more than a pair of messages, TCAP provides a *conversation* message type that can be used to continue the transaction. If only a single message is needed to accomplish the service, a *unidirectional* message is defined in TCAP that does not need a subsequent response message to close out a transaction. Finally, to support abnormal termination of processing, an *abort* message is available to advise the remote node of a failure in a transaction.

Within the TCAP protocol, a second level of structure is provided for coordinating different processes that are parts of a single transaction. Structures at this level are called *components* in TCAP. A TCAP message can carry an *invoke* component, which initiates some processing at the remote node. Other component types are the *return result* component, the *return error* component, and the *reject* component. These last three are possible responses to an invoke, giving either successful (return result) or unsuccessful (return error) completion information on the invoked operation, or notifying the invoking node of abnormalities (reject). An added flexibility is the ability to respond to an invoke with another invoke, as for instance in the case where the first invoke requests instructions and the responding invoke answers the request with instructions to do something (hence the responding invoke rather than return result.) So that multiple processes can run concurrently as part of a single transaction, the component layer maintains its own set of identifiers to associate invoke components with their responses.

Completing the structure of TCAP are the parameters that each component can carry. For example, an invoke component requesting the 800 service application for instructions on how to handle a call will include the 800 number the customer dialed. If the TCAP response message contains a responding invoke component instructing the switch to route the call, one parameter in the responding invoke component will be the routing number. Figure 7 summarizes the general structure of a TCAP message.

Message Type
Originating Transaction Identifier
Responding Transaction Identifier
Component Type
Originating Component ID
Responding Component ID
Parameter Set

FIG. 7 Transaction Capabilities Application Part message structure.

CLASS Features Using TCAP. While the earliest applications of TCAP were for database queries to SCPs for services like 800 service, TCAP is the protocol of choice for any service application in which no circuit setup is performed between the two nodes involved. It is applied in this context to support CLASS features in which the TCAP messages are exchanged between two switching systems. Figure 8 illustrates this application.

First, a call attempt is made from one user to another that results in a busy signal. The caller then activates an automatic callback feature, which causes the switch serving the caller to launch a TCAP query message requesting that the switch of the called user monitor the line and notify caller's switch when he or she becomes free. The switch serving the called user acknowledges this request with a response message, and some time later notifies the caller's switch with another TCAP message that he or she is now free. The caller then receives a signal that the automatic callback feature can set up the desired call, and if the caller answers, the call can then be set up to the called user in the usual way.

The Message Transfer Part (MTP)

The MTP provides the physical-, link-, and network-level functions for the SS7 network. The physical layer defines a 56- or 64-kbps digital data-transmission channel, and the link layer provides error detection and correction, sequencing, and packet acknowledgments between the two ends of a link. These aspects of the physical-layer and link-layer specifications are fairly similar to other, non–SS7 physical and link layers (e.g., high-level data-link control, or HDLC, or link access protocol-balanced [LAP-B]) and are not discussed to any greater depth in this article.

The network level of the MTP, which provides routing, network traffic management, and testing and maintenance functions, is highly specialized to meet the unique availability and reliability objectives of CCSNs.

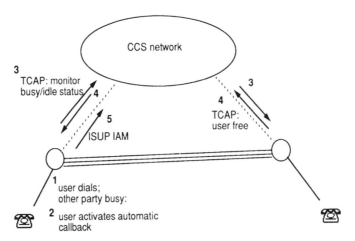

FIG. 8 Transaction Capabilities Application Part service application: automatic callback.

Routing.   Routing in the MTP is based on the following information, which is provided in every message by the MTP user part (e.g., ISDN-UP or SCCP):

1. DPC. Every node in an SS7 network is assigned an address called a *signaling point code*. The signaling point code of the intended destination node for a message becomes the DPC for that message.
2. Signaling link selection (SLS) code. The MTP provides for multiple physical links between any two nodes. A fundamental design philosophy in SS7 is that, under normal conditions, traffic is spread across all the available resources. Thus the SLS code is usually provided by the MTP user part as a random number that the MTP then uses to select one of the multiple links to the next node in the route toward the destination. Randomness of SLS codes leads to a roughly uniform spreading of traffic across the available links. Another option available to the MTP user part is to assign several messages the same SLS code. This will lead to the MTP routing these messages over the same links, which will ensure sequential delivery (in the absence of link failures) even though the MTP itself is a connectionless network-layer protocol.
3. MTP user part indication. This information is used at the destination node by the MTP function to distribute the message to the correct MTP user part at that node.
4. Priority indication. MTP does not provide priority routing unless there is congestion on a link (defined by buffer occupancy exceeding some threshold). When congestion is detected, traffic sources are notified to stop sending traffic of priority lower than the current congestion level. Then, if buffer occupancy continues to rise and exceeds a discard threshold, messages can be dropped by MTP. The priority level is used to determine which messages are dropped first. Four levels of priority are defined and three levels of congestion threshold are set, so that, at the first level of congestion, the lowest-priority messages are dropped. If congestion continues to increase, then, when the second threshhold is passed, the second-lowest-priority messages are dropped, and so on. The highest-priority messages are those used to manage MTP traffic, and thus are never to be dropped as a strategy for easing congestion.

MTP Network Management.   The MTP incorporates a number of strategies to allow it to maintain a high probability of message delivery in the presence of failures in network nodes or links. The majority of these strategies are based on built-in redundancy in network designs.

The MTP supports both node redundancy and link redundancy. Node redundancy is most often seen in SS7 network designs as the implementation of STPs in mated pairs. Thus each STP in the network has a mate that can take over the full load should the STP fail. In the absence of failure, the two STPs share the load and, from the perspective of the rest of the network, are equivalent paths to a given destination.

Link redundancy is manifested as mated sets of links to mated pairs of STPs. As above, in the absence of failure, the redundant resources share the traffic load and are viewed as equivalent paths to a given destination. When a link set fails, its traffic is diverted to the redundant links, which have been engineered with spare capacity. In addition, the multiplicity of links permits network operators to obtain greater reliability by selecting diverse routes for the various links.

Since the strategy is to move traffic around when failures occur, the MTP must provide the control information to support this coordinated activity among various nodes in the network. This is the purpose of the MTP network-management messages and procedures. There are too many MTP management procedures to review them all in this article. Two categories of procedures that are discussed in more detail to illustrate this point are procedures to react to failed links and procedures to react to failed or isolated nodes by communicating route status information between nodes.

When a link fails and there are redundant links between the same two nodes, traffic can simply be diverted to the remaining links. In order to minimize message loss, there needs to be some coordination between the two nodes to allow messages either in transit or queued on the failed link to be re-sent on a different link. Thus the two nodes exchange "changeover" messages. When the conditions that led to the changeover have been resolved, "changeback" messages are exchanged to facilitate the restoration of traffic to its normal pattern.

When a route from an STP to a destination fails—that is, all the normal links that lead toward that destination become unavailable (see Fig. 9)—then the STP is forced to reroute traffic onto the C-links to its mate, where routes to the destination still exist. Since such use of the C-links is considered a temporary, emergency measure, messages are sent to the sources of such traffic to tell them to reroute it to the mate STP over A-links. These messages are called *transfer restricted* (TFR) messages, and they carry a "concerned point code" field to indicate which destination has become unreachable via the normal routes through the STP. At the same time, to cover the possibility that the mate may also have a failure of routes to the same destination, another message, called a *transfer prohibited* (TFP) message, is sent to the mate STP. This advises the mate that traffic is being re-routed toward the concerned point code via the

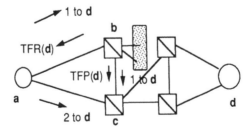

FIG. 9 Message Transfer Part network management. Upon detection of loss of primary path to *d*, the STP at *b* responds to a received message for *d* by diverting it to the backup path through its mate *c* and informing the sender *a* that, until further notice, it should not send *b* any messages bound for *d* unless no other paths exist. In addition, mate is notified not to send any messages bound for *d* since they will loop.

C-links. Thus, the mate knows not to route traffic to that same destination via the C-links (in the opposite direction), which would result in messages looping between the STPs.

The Signaling Connection Control Part (SCCP)

The SCCP was added to SS7 in 1984 as part of the ongoing effort to incorporate OSI principles into SS7. It provides extensions to the network-layer services of the MTP, including connection-oriented network services if desired. In addition, the SCCP provides a more flexible addressing scheme that allows SS7 service applications to be addressed separately from the SS7 nodes where they reside. These separate addresses are called subsystem numbers (SSNs) in SCCP. Thus applications that use SCCP are completely addressed by the combination of point code plus SSN.

The great advantage that this addressing scheme offers is that it allows modularity of network software applications. For example, if an SCP node supports 800 service and credit-card service, these two applications can be addressed separately; if backup applications are to be provided for each, then the 800 module could be backed up at one SCP while the credit card application is backed up at a different SCP.

With this flexibility comes the need to have procedures to route messages to the active copy of a given application, which can change with time. Thus SCCP has its own management messages and procedures that allow nodes that launch messages to applications using SCCP addresses to update their information as the status of a subsystem changes.

A further refinement allows these network maps to be centralized in STPs. The mechanism for doing this is an alias addressing procedure whereby the source of a message can attach a logical name to the message (e.g., "800 service"), and send the message to an STP, where the mapping from the logical name to the address of the currently active copy of that application can be made. The alias addresses in SCCP are called *global titles* and the process at the STP is called *global-title translation*.

The centralization of this information saves effort in the non-STP nodes, which far outnumber the STPs in a typical network. It also allows faster response to changes in network status by minimizing the number of nodes that must update their tables in response to the change.

The SCCP protocol supports global title translation through the structure of the SCCP calling and called party addresses. These fields contain a set of indicators as to whether the address consists of a global title, a point code, an SSN, or a combination of these. They also indicate whether the recipient of the message is expected to perform the global title translation function. Figure 10 illustrates the use of the address fields in support of global title translation. If global title translation is required, the routing label of the message contains the point code of the STP performing the translation as the DPC of the message, and the SCCP called party address field contains the global title. When the STP performs the translation, it places the new DPC into the routing label and the new SSN into the SCCP called party address field.

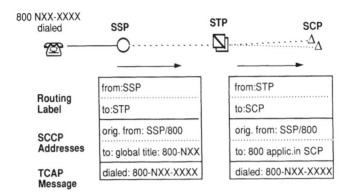

FIG. 10   Global title translation.

SCCP management messages are defined that allow notification of the failure or congestion of a subsystem at a node. This allows STPs to update their global title translation tables to route to a backup copy of a failed application, for example. These messages can also be received by the backup application itself, allowing it to prepare to receive traffic, for example, by postponing low-priority processing tasks.

Operations, Maintenance, and Administration Part

The operations, maintenance, and administration part (OMAP) of SS7 is actually an application designed to assist in the operation and administration of CCS networks by carrying out automatic testing of the validity of network routing data or determining the maintenance status of specific network resources. As a non-circuit-related application, it uses TCAP messages to carry test requests and their results through the SS7 network. The current extent of OMAP standards is to verify the validity of MTP and SCCP routing data and the status of signaling link equipment.

Summary

In summary, SS7 is a family of protocols and applications that is positioned to meet the widest range of signaling needs for modern telecommunications networks. While it is sufficiently mature to have already seen widespread deployment in networks throughout the world, including the United States, it is also sufficiently flexible to be the subject of significant ongoing standardization for new and emerging applications. At least through the early 1990s, SS7 evolution will be driven primarily by the complexities of interconnecting CCS networks, by the emergence of the advanced intelligent network concept for modularizing and accelerating new feature development, and by the drive toward such new end-user services as new CLASS and ISDN supplementary service features.

Acknowledgment: CLASS is a service mark of Bellcore.

Glossary

A-LINK. Access link from an end node in a CCS network to its home STP
ACM (*address complete message*)
ANM (*answer message*)
B-LINK. Bridge link between two (nonmated) STPs at the same level in a CCS network hierarchy
C-LINK. Cross link between two members of a mated pair of STPs
CCS (*common channel signaling*)
CCIS (*Common Channel Interoffice Signaling*)
D-LINK. Diagonal link between two STPs at different levels in a CCS network hierarchy
DPC (*destination point code*)
DUP (*data user part*)
E-LINKS. Extended links from an end node in a CCS network to a different STP pair (*see* A-links)
F-LINK. Fully associated link between two end nodes in a CCS network
GLOBAL TITLE. A logical, or alias, address for an application in a CCS network
ISDN-UP (*integrated services digital network user part*)
MTP (*message transfer part*)
OMAP (*operations, maintenance, and administration part*)
OPC (*origination point code*)
OSI (*open systems interconnection*)
OSS (*operator services system*)
REL (*release message*). Indicates the release of a trunk
RLC (*Release-complete message*). Confirms the release of a trunk
ROUTING LABEL. The OPC plus the DPC plus the SLS
SLS CODE (*signaling link selection code*)
SS7 (*signaling system 7*). The seventh recommendation of the CCITT concerning interoffice signaling
SCCP (*signaling connection control part*)
SCP (*service control point*)
SSN (*subsystem number*)
STP (*signaling transfer point*)
TCAP (*transaction capabilities application part*)

Bibliography

Bellcore, Bell Communications Research Specification of Signalling System Number 7, TR-NPL-000246, Issue 1, June 1985; plus Revisions.

Bellcore, Database Services, TR-TSY-000533, Issue 2, July 1987.

Bellcore, ISDN Hold Capability for Managing Multiple Independent Calls: Signaling System No. 7 to Support Hold Notification, TR-TSY-000856, Issue 1, Supplement 1, September 1990.

Bellcore, LSSGR: Common Channel Signaling, Section 6.5, TR-NWT-000606, Issue 1, November 1990.

Bellcore, Service Control Point Node Generic Requirements for IN1, TR-TSY-000029, Issue 1, September 1990.

Bellcore, Signaling Transfer Point Generic Requirements, TR-TSY-000082, Issue 2, Revision 2, June 1990.

Bellcore, Switching System Requirements for Interoffice ISDN Automatic Callback Using Signaling System No. 7, TR-TSY-000855, Issue 1, Supplement 1, January 1991.

Boyles, S., Corn, R., and Moseley, L., Common Channel Signaling: The Nexus of an Advanced Communications Network, *IEEE Communications*, 28(7): 57–63 (July 1990).

Cazzaniga, M., Garavelli, A., and Robrock, A., Implementation of SS7: Italtel's Experience, *IEEE Communications*, 28(7): 84–88 (July 1990).

Goldberg, R., and Shrader, D., Common Channel Signaling Interface for Local Exchange Carrier to Interexchange Carrier Interconnection, *IEEE Communications*, 28(7): 64–71 (July 1990).

Kearns, T. J., and Mellon, M. C., The Role of ISDN Signaling in Global Networks, *IEEE Communications*, 28(7): 36–43 (July 1990).

Kitami, K., and Ogawa, K., Current Role and Future Evolution of the ISDN Signaling System in NTT's Network, *IEEE Communications*, 28(7): 78–83 (July 1990).

Lawser, J., Matsumoto, J., and Pigott, J., Common Channel Signaling for International Service Applications, *IEEE Communications*, 28(7): 89–92 (July 1990).

Marr, F., Signaling System No. 7 in Corporate Networks, *IEEE Communications*, 28(7): 72–77 (July 1990).

Modaressi, A., and Skoog, R., Signaling System No. 7: A Tutorial, *IEEE Communications*, 28(7): 19–35 (July 1990).

Willmann, G., and Kuhn, P. J., Performance Modeling of Signaling System No. 7, *IEEE Communications*, 28(7): 44–56 (July 1990).

References

1. Bellcore, ISDN Access Call Control Switching and Signaling Requirements, TR-TSY-000268, Issue 3, May 1989.
2. Bellcore, Advanced Intelligent Network Release 1 Baseline Architecture, SR-NPL-001555, Issue 1, March, 1990.
3. CCITT, Specifications of Signalling System No. 6. In: *Blue Book*, Fascicle VI.6, International Telecommunication Union, Geneva, 1981.
4. CCITT, Specifications of Signalling System No. 7. In: *Yellow Book*, Fascicle VI.3, International Telecommunication Union, Geneva, 1989.
5. CCITT, Specifications of Signalling System No. 7. In: *Red Book*, Fascicles VI.7, VI.8, International Telecommunication Union, Geneva, 1985.
6. ANSI, Signalling System Number 7 (SS7)—General Information, ANSI T1.110-1987, American National Standards Institute, New York, 1987.

7. ANSI, Signalling System Number 7 (SS7) — Message Transfer Part, ANSI T1.111-1988, American National Standards Institute, New York, 1988.

8. ANSI, Signalling System Number 7 (SS7) — Signalling Connection Control Part, ANSI T1.112-1988, American National Standards Institute, New York, 1988.

9. ANSI, Signalling System Number 7 (SS7) — Integrated Services Digital Network User Part, ANSI T1.113-1988, American National Standards Institute, New York, 1988.

10. ANSI, Signalling System Number 7 (SS7) — Transaction Capability Application Part, ANSI T1.114-1988, American National Standards Institute, New York, 1988.

11. ANSI, Signalling System Number 7 (SS7) — Operations, Maintenance, and Administration Part (OMAP), ANSI T1.116-1990, American National Standards Institute, New York, 1990.

12. CCITT, Specifications of Signalling System No. 7, In: *Blue Book*, Fascicles VI.7, VI.8, VI.9, International Telecommunication Union, Geneva, 1989.

13. Bellcore, Switching System Requirements for Call Control Using the ISDN User Part, TR-TSY-000317, Issue 2, July 1989.

14. Bellcore, Switching System Requirements for Interexchange Carrier Interconnection Using the ISDN User Part, TR-TSY-000394, Issue 2, July 1989.

15. Bellcore, Switching System Requirements Supporting ISDN Access Using the ISDN User Part, TR-NWT-000444, Issue 2, November 1990.

16. Bellcore, CLASS Feature: Automatic Callback, TR-TSY-000215, Issue 1, September 1988.

17. Bellcore, CLASS Feature: Screening List Feature, TR-TSY-000220, Issue 1, April 1989.

18. Bellcore, CLASS Feature: Automatic Recall, TR-TSY-000227, Issue 1, September 1988.

19. Bellcore, Call Forwarding Subfeatures: Switching System Requirements Using Signaling System No. 7, TR-NWT-000972, Issue 1, October 1990.

20. Bellcore, ISDN Call Forwarding: Switching System Requirements Using Signaling System No. 7, TR-TSY-000853, Issue 1, Supplement 2, December 1990.

21. Bellcore, ISDN Message Service Generic Requirements, TR-NWT-000866, Issue 1, January 1991.

RICHARD R. GOLDBERG

Common Channel Interoffice Signaling (CCIS)   (see Common Channel Signaling)

Communication Aids for People with Special Needs

Introduction

It is estimated that of the almost 10% of the world's population that is disabled a significant proportion suffer from some loss in the ability to communicate. Severe sensory or neuromuscular disabilities that inhibit or prevent communication are most destructive when they isolate a person from the outside world. Within the last two decades, a variety of new assistive technologies have emerged that can enable persons with such disabilities to communicate well enough so that they can obtain an education, achieve more independent living skills, obtain employment, and make major contributions to society. This article presents an overview of some the major technologies that are available to aid communication for persons with special needs.

The increased availability of assistive technology over the last 20 years has resulted from a number of reasons. Primarily it has been driven by the extraordinary growth in the semiconductor and computer industries, the allied infrastructures associated with these industries, and the information-processing equipment that they have yielded. Further, these same factors have had a profound effect on the capabilities, cost, and types of assistive technology available for a person with disabilities.

Unfortunately, the use of much of this assistive technology by persons with disabilities is still not widespread. This is due in part due to the lack of awareness of the existence of the equipment and, perhaps more importantly, the lack of necessary funding to purchase equipment.

Devices that help to restore failing powers are called *orthotic aids*, while those that attempt to replace a lost capability are called *prosthetic aids*. This article considers both types and gives examples of their application to communication.

The sections that follow review the state of the art in orthotic and prosthetic devices available for each of the human abilities involved in communication and control: the eyes, ears, voice, and motor control. In addition, a brief discussion of learning disabilities focuses primarily on computer aids for training that have contributed to making communication possible for persons with diminished learning capabilities.

The section that follows comments on the causes and demographics of disabilities. It goes on to present a model of a generic communication aid and outlines issues related to the application of these types of devices to clients.

The next section considers several examples of communication aids for persons who are visually impaired or blind. This is followed by similar treatments of assistive technology for persons who are hearing impaired or deaf and deaf and blind.

Approaches being taken to provide communication aids for persons who have deficits of the central nervous system that leave them motor impaired and sometimes nonvocal are also discussed. Finally, approaches being taken in the use of assistive technology for persons who are nonvocal and learning disabled are discussed in the final two sections, respectively.

Of necessity, the descriptions provided of the many devices that have been produced are schematic, considering only the basic principles and the simplest and most representative models. References that provide more detail are, however, provided in the bibliography.

Fundamental Considerations

In this section some of the basic facts about disability and the application of assistive technology are outlined.

Causes of Disability

The improper functioning of one or more elements of the nervous or muscular systems or the existence of a learning disability can significantly affect the ability to communicate. The causes of these deficits are widespread and may result from a genetic defect, illness, aging, or an accident or trauma suffered before birth or at any time during life. In this last category, an injury to the head can often have an impact on cognitive capabilities. Any of these phenomena may lead to temporary or permanent loss of skills that are essential to communication.

Demographics

In a study sponsored by the United Nations (1), it is estimated that today about 10% of the world's population is disabled. Statistics pertinent to persons with disability are difficult to obtain, inasmuch as they are dependent both on the definition and the degree of the disability. Nevertheless, in the January 1987 *Technology Review*, Frank G. Bowe reported results obtained from a variety of sources (2):

> In the US about 36 million of some 240 million have disabilities. . . . [Of these], 8 million have learning disorders including dyslexia, an impairment that interferes with reading the printed word. . . . [Six] million people are seriously hearing-impaired, including 2 million who are deaf. . . . [J]ust under 1.7 million have poor vision or are legally blind . . . [and] another 3.5 million individuals have disorders of the central nervous system including cerebral palsy, muscular dystrophy, multiple sclerosis, paraplegia, and quadriplegia. (p. 54)

Thus, more than 21 million, or well over half of those who are disabled, suffer from an affliction that interferes in one way or another with effective communication.

Human Communication Fundamentals

Human communication requires that persons be able both to send and receive messages effectively through their external environment. Consider how messages are sent and received. To transmit a message, a person must be capable of manipulating the environment in such a way that signals will be recognized by those with whom they wish to communicate. This requires transmission of signals from the brain to the nerves to control the muscles. Generally, in human communication the control of vocal cords, the mouth, and the throat play primary roles in speech. Control of the hands and fingers is essential to writing. Other parts of the body may be involved in sending messages in secondary ways; these can play a primary role when there is failure in the main system.

To receive a message successfully, a person must be capable of sensing changes that are produced in the environment by the message. The primary senses in this regard are vision and hearing. Other senses, such as touch and smell, do play roles in some communication but they are generally secondary. Here again, these secondary capabilities can be called on by the body to provide the primary communication channels.

The messages causing the changes in the environment are sensed and transmitted by the nervous system to the brain where they are interpreted.

When the ability to send or receive messages is impaired due to a failure in the neuromuscular system, assistive technology can play an important role in providing means for supplementing diminished capabilities. Moreover, technology can also enable the body's natural tendency to substitute one portion of the nervous system, brain, or muscles for another.

A Generic Model

An understanding of the process by which various types of assistive technology work to aid communication can be gained by representing the systems generically as functional elements (see Fig. 1). In our treatment of various communication aids, we discuss how assistive devices achieve these functions.

To aid the receipt of messages, an assistive device may be thought of as having three elements:

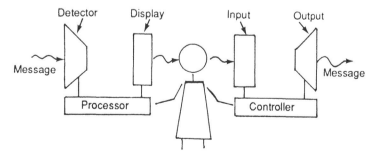

FIG. 1 Generic model of a communication aid.

1. A detector to discern the message being transmitted through the environment
2. A processor to interpret the detected signal so that it can be used
3. A display capable of manifesting the received and processed signal in an amplified or altered form so that it can be interpreted by the person using the assistive device

To aid the transmission of messages, the process can be partitioned and represented by three additional elements:

1. An input to accept commands
2. A controller to take the input signals and convert them in such a way that they can control an output
3. An output or a means of sending the message into the environment more effectively or in an alternate way from that achieved without the aid

Considerations in the Application of Assistive Technology

In the selection and use of assistive devices, it should be noted that two important phases, analysis and proper fitting, are critical to successful use and are vital if any given device is to meet the needs of the user. They are mentioned only briefly here; however, they are discussed extensively in the literature on rehabilitation.

The process must begin with a careful analysis of the capabilities of the potential user of the device. The analysis must, among other things, establish the client's objectives in requesting a device. It must establish the range and nature of the resources that the client can use. It must also determine whether an orthotic or prosthetic device is indicated.

With this type of information, a suitable device can be proposed. Of primary importance in the use of selected devices is that they must be carefully fitted to the user. The human is remarkably adaptive; nevertheless, every effort must be made to minimize the extent of user adaptation required. For the client, this process is slow and demanding of both physical and emotional energy. It can also be a major factor in the use or lack of use of potentially helpful devices.

Finally, it should be noted that the ideal setting in which to introduce a client to a new assistive device is in a so called "team setting," involving both therapeutic and technical resources. In such a setting, one can expect both careful analysis and proper attention to adapting the device to the client's needs.

Communication Aids for Visually Impaired Persons

Probably the most widely used communication aid for the partially sighted person is eyeglasses; for the blind person it is the use of Braille. As electronic technology has been applied to assist persons with special visual needs, the principles used in both of these aids have been adopted. In this section, some approaches are considered that have been used to aid persons who retain partial vision and that work by appealing to retained visual capability. In a subsequent section, consideration is given to the techniques that appeal to senses other than sight to transmit information normally presented visually.

Electronic Systems to Aid Vision

While purely optical techniques are the most widely used and most successful way of providing assistance to the visually impaired, some electronic means have been developed. These are of particular value in special cases, as, for example, in obtaining the output from computers, or devices incorporating computers, that present the output on electronic terminals. Other approaches, such as the translation of information into sound, are helpful to both partially sighted persons and persons without sight and are also considered in this section.

One approach that is helpful for enlargement of an image beyond that achieved with eyeglasses is the use of closed-circuit television techniques. An arrangement of this type is built by Humanware Inc. (Loomis, CA). The system permits the user to scan an object or text with an optical scanner as its input device. In some cases, the light picked up by the device may then be enlarged optically but, in every case, it is converted to electronic signals through the use of the television camera. The electrical signals produced are further processed and then projected onto a television screen or monitor. The user is usually provided with a means of controlling the selection of the segment observed and the degree of enlargement of the image.

Electronically enlarged displays can be used both in standalone configurations and in special displays of text or graphics output from a computer. In the latter case, special enlarging arrangements can make it possible for persons with considerable visual impairment to use the computer for work and recreation.

Two approaches have been taken to provide selected enlargement of computer displays. In the first approach, the program producing the output may have special routines within the application program that can be invoked by the user. These are devised by the application designer. A typical example would be the use of enlarging routines in a word-processor program such as the BEX™ system designed by Raised Dot Computing (Madison, WI). A second approach uses special hardware and software that can be placed between the application software and the screen-driver program. This latter approach has the appeal of working with a number of different application programs (see Fig. 2).

Nonvisual Communication Aids for Visually Impaired Persons

In communication aids for persons with severe visual impairment, there are two techniques that are widely used that do not depend on sight; one technique uses touch and the other hearing. The first, Braille, requires encoding characters and

FIG. 2 Telesensory/VTEK's large-print display processor. The device makes both printed material and computer output available to persons with impaired vision. Photo courtesy of Telesensory/VTEK, Moutain View, CA.

words into a form that can be read by touch and the second translates textual information into speech.

Braille

One of the oldest methods of encoding textual information was invented by Louis Braille in 1824. Many of today's electronic devices still use Braille as their output medium. In Braille, information is represented by an array of up to 6 raised points arranged in 3 rows of 2 columns. The points are arranged with 2.3 mm between rows and columns in a cell and by 4.1 mm between columns of adjacent cells (see Fig. 3). Depending on the character being represented, combinations of the 6 possible points are raised. There are 64 possible arrangements. This is more than enough to cover the letters of the alphabet, digits, and some of the punctuation marks. There are, however, three forms of the Braille code. The first, Type I, uses essentially a configuration of points for each character or item of punctuation. In Braille Type II and Type III, some of the point arrangements are used to represent certain frequently used nouns or verbs. The more complex encoding used in Types II and III permits information to be encoded into fewer symbols and, therefore, allows information to be conveyed at a higher rate.

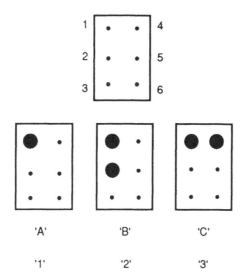

FIG. 3 Standard Braille cells. The dark circles indicate raised areas.

The use of Braille characters preserves for the visually impaired person some of the appeal that printed material has for the sighted. Nevertheless, Braille does have certain limitations. On the positive side, Braille characters arranged on a page allow readers to read selectively at their own pace and to reread. Moreover, information can be arranged in tables and titles and subtitles can be employed.

There are, however, certain disadvantages to the use of Braille. Blindness often occurs late in life (60% of the blind population becomes impaired after age 65). For this segment of the population, Braille often is difficult to learn, especially Type II and Type III encoding. Frequently, the deterioration causing blindness also reduces the sensitivity to touch in the fingertips. In addition, documents presented in Braille are sometimes very bulky. Fortunately, some of the electronic equipment that uses Braille as its output medium partially addresses this last problem by storing information in magnetic form.

The use of the microcomputer has greatly reduced the cost and difficulty of producing Braille and also has eased the problem of information storage.

Consider a typical Braille system on a computer. It permits entering information into the system from other electronic sources or directly from a standard or Braille keyboard, which would enter the Braille characters directly. This keyboard is arranged with a key corresponding to each position in a Braille cell and a seventh key is used to advance the input.

The computer system generally provides some kind of magnetic or electronic storage. The principal output from the system goes to an embosser that places the Braille in the paper. Some systems also provide an electromechanical display arrangement that raises and lowers a set of pins to represent the Braille characters being output. The pin field display can be read with the fingertips. The display generally handles only a limited number of characters at a time.

In computer-driven Braille, system text generally is not sent directly to the embosser. If this were done, a great deal of space would be wasted. A single Braille page measuring $8\frac{1}{2}$ in by 11 in can hold no more than 1000 characters. For this reason, various standards have evolved to represent the maximum information in the minimum space (3). Major space saving is accomplished with contracted (Type II) Braille, where entire words and many letter combinations are represented by one or two Braille characters. Because of the complexity of the encoding, the computer is of considerable assistance in its preparation.

It should also be noted that devices are also available for reading (scanning) Braille documents and converting them to printed form (e.g., MPRINT™ manufactured by Telesensory/VTEK [Mountain View, CA]). The devices are also capable of expanding Type II Braille encoding.

Tactile Converters

A small number of devices has been developed that translate a printed character directly to a tactile form. The most successful of these is the Optacon™ (OPtical to TActile CONverter) developed at Telesensory/VTEK and introduced in 1971. It is a portable electronic device that converts printed material into a tactile form, enabling a blind person to read the printed material. The device does not depend on Braille but instead produces a tactile configuration directly correlated to the character, symbol, or figure it observes.

The device may be considered to consist of the three elements of our generic model (Fig. 1): an input element, a processing element, and an output element. The input element is an optical device that contains a metal-oxide semiconductor component similar to that used in television cameras. The semiconductor is divided into a matrix of elements that can be scanned electrically, and is designed to produce an electrical signal proportional to the light that falls upon each element as it is scanned. The output of the system uses a small matrix of pins, called the *tactile array*, that can be raised and vibrated. The input and the output are scanned in synchronism under the control of the processing element.

The input element, called an *optical probe*, is like a small wand and is held in one hand by the user and moved along a page of printed material. The index finger of the other hand is placed over the tactile array. The pin field covers an area about equal to the fingertip (2.5 cm × 1.25 cm). In operation, the input signal results from scanning with the input light-sensitive matrix in the optical probe. When ambient light reflected from the page, falling upon the matrix element exceeds a certain threshold, a corresponding pin in the output matrix is raised and vibrated. As the optical probe is moved along the text, the user senses with the fingertip the output to the tactile array of the characters observed (see Fig. 4).

In a sense, the Optacon is like a closed-circuit television system except that, instead of causing a television screen to give off light in relationship to light sensed, a matrix of pins is manipulated so that it can be used to transmit tactilely what is being scanned by the input. Additional features included in current models of the Optacon are that it can be used with computer monitors to access the print on the screen and the addition of a set of lenses that read very small fonts.

FIG. 4 Telesensory/VTEK's Optacon print-reading aid, which is used by people who are blind to read both printed and graphical material.

The Optacon definitely meets a need, as it permits the visually impaired person to read the same characters and some of the graphics that are read by the sighted person. Unfortunately, experience has shown that reading with the device is generally slower than with Braille, and that it is difficult to learn to use (4–6).

Voice Output Terminals

The techniques for generating synthetic speech from text have found wide applicability. For many who are blind, the use of speech generation technology has a variety of applications. Several techniques are used to reproduce human speech. These techniques vary in their demand on computer resources and in their realism. Greater realism makes greater demand on computer resources, especially memory. Speech synthesizers are discussed in greater detail in a section below.

Today, visually impaired people are served by over 20 companies that make products aimed at converting computer-generated textual material to speech. A combination of sophisticated software and a speech synthesizer work together to provide computer screen reading capability. This makes it possible for a blind or visually impaired person to hear the information displayed on the computer screen. Moreover, in most designs the user is able to control scanning and reading portions of the screen. In many designs, an effort is made to make the program associated with the speech synthesizer transparent to the application program, thus making it applicable to a variety of application programs. This is often difficult to do, particularly when many graphics features are incorporated in the design. For this reason, in some cases the application and the speech synthesizer are designed together to optimize performance for the visually impaired user.

Reading Machines

With the availability of speech synthesizers and the ability to translate arbitrary text to speech, the logical next step was to produce machines capable of reading printed documents. Indeed, early in 1970 Kurzweil (Cambridge, MA) made headlines in the popular press by developing a machine that could read text for use by blind and severely visually impaired persons (Fig. 5). The key to this effort was the development of optical character recognizers. These devices can scan printed material and, through the use of pattern-recognizing logic in the computer, identify the characters printed in the document.

Today, several companies produce such reading machines. They differ primarily in the range of input text fonts that they can recognize. Beyond the character-recognizing capability, the remaining portion of the design is similar to that used in the screen readers discussed in the section above.

Communication Aids for Hearing-Impaired Persons

Hearing Aids

Perhaps the best known and most widely used communication aid for persons who are hearing impaired is the hearing aid. It is fundamentally a prosthetic device that facilitates impaired hearing by increasing the sound pressure level at the eardrum. The device consists basically of four elements: a microphone, an amplifier, an earphone, and a power supply, usually a battery. The hearing aid amplifier is provided with some filtering to help with noise reduction and to compensate for distortions produced by the other components in the system. Recently, work has also been undertaken to use digital signal-processing techniques to improve hearing aid performance (7).

Hearing aid design has benefited immensely from the achievements of the semiconductor industry, enabling the device to be made quite small and to run on very low power. Conventional hearing aids are designed to be placed at several locations. The most popular design is placed behind the ear; however, in-the-ear designs (i.e., those designed to be placed at the entrance to the ear canal) are almost as popular.

FIG. 5 The Xerox/Kurzweil Personal Reader™, which converts printed text to voice.

When hearing aids cannot be used, some work has been done in placing implants in the cochlea to provide a sensation of hearing through electrical stimulation of the auditory nerve. The cochlea is a small, coil-shaped cavity at the interface with the nerve fibers of the auditory system. The transplant is designed to restore a sense of hearing to individuals who are deaf because of nonfunctioning cochlea. An external processor transforms the acoustical signal into a pattern of electrical signals. It has been found that the technique has been moderately successful in facilitating communication for some persons when the implant sensations are coordinated with lip reading (8).

Aids for the Deaf

Deafness is severe or complete loss of hearing. In the presence of this type of impairment, great emphasis is placed on visual modes of communication. When people are in direct contact, signing and lip reading are generally used. Lip reading or speech reading is a complex task because many words involve similar positions of the face and lips. Successful lip reading depends on several talents in addition to the visual perception of the facial movements. These include understanding learned patterns of movements often associated with speech and careful attention to context (9).

Telecommunications Device for the Deaf (TDD)

A number of items of equipment have been developed to make the telephone system accessible to hearing-impaired persons. These include amplifiers that are placed in a telephone set or that can be attached to a telephone receiver. However, for the person suffering severe hearing impairment or deafness, the Telecommunications Device for the Deaf (TDD) is most widely used (Fig. 6).

The TDD is a simple terminal with a keyboard that allows for the visual display and signaling of messages from the telephone. The device was modeled after the teletypewriter (TTY) although it sends its message as tones over the telephone line. Most of today's TDDs are largely electronic. They also permit dialing to take place directly from the keyboard. Because it requires both parties to have a similar device, the TDD is primarily in use by hearing-impaired persons and institutions wishing to serve them.

History of the Telecommunications Device for the Deaf and the Use of the Baudot Code. The first TDD was developed in 1964 by Robert H. Weitbrecht. At the time of its development, the TTY was in wide use. The TTY had its own network of lines but could not be used on telephone lines. On TTY lines, information was transmitted by various methods of direct-current signaling. This amounted to opening or closing the line or, in some cases, reversing the direction of current on the line to indicate the states of the message code. The code used was the Baudot code in which each character is represented by up to five changes in state.

Weitbrecht wanted a way to use the TTY over telephone lines so that a person who was hearing impaired could use it in place of the telephone. At the

FIG. 6   The Telecommunications Device for the Deaf (TDD), which makes it possible for deaf persons to use the telephone.

time of his work, calls could be made anywhere in the country and direct national dialing was just being introduced in the telephone system.

Weitbrecht hit on the idea of transmitting tones that could be sent over telephone lines as his signaling code elements. In his design, one tone (frequency) was used to represent a closed line and a second frequency an open line. Each party used the same pair of tones. One party transmitted while the other waited to be given the line. This is called half-duplex signaling.

This proposal by Weitbrecht was a genuine breakthrough and was widely adopted. Its use in the deaf community was reenforced when, in 1968, AT&T began offering surplus TTYs to hearing-impaired persons for free. These TTYs could be fitted with tone-generating equipment (modems) and used over the telephone.

TTYs were distributed by Telecommunications for the Deaf Inc. (TDI), which was incorporated in 1969, and initially called Teletypewriters for the Deaf Distribution Committee of Indianapolis, Indiana. The organization exists today as a nonprofit group concerned with telecommunications issues for the hearing impaired.

After many TTYs or TDDs were in use by hearing-impaired persons, other improved methods of transmitting data over telephone lines were developed. These methods had the virtue of permitting both communicating parties to send at the same time (full duplex). Under this full-duplex method, both the originating and the terminating party used different pairs of frequencies. At about the same time as the development of the full-duplex method, 7- and 8-level codes also came into use. The most widely used code today was established as the American Standard Code for Information Interchange (ASCII) (10). This code, generally adopted by the computer industry, has the advantage of permitting a considerably larger set of code elements than the Baudot code.

Unfortunately, for a considerable period of time after improved data-transmission systems and the teleprinter designs were introduced, teleprinters using ASCII were more expensive than TDDs or surplus TTYs. For that reason, the number of the older type devices in use today is well over 100,000. The situation is changing slowly. Some of today's TDDs are being designed to work with either Baudot or ASCII codes. Generally, however, telecommunications for the hearing impaired must be prepared to work with the TDD using the Baudot code and operating in the half-duplex mode.

Dual-Party Relay Service.   The Dual-Party Relay Service makes it possible for telecommunications to take place between persons who use the TDD and regular phone users without the device. In most cases, the service is provided by volunteer operators who act as intermediaries in transmitting calls.

The relay operator is reached by using a set of telephone numbers dedicated to the service. The operator works with a TDD and a telephone receiver and transmitter. Calls originated by a TDD user to the operator are completed on a second line to a person without a TDD (see Fig. 7). Messages that are received over the TDD are spoken to the called party, whose responses are converted to text that is typed on the TDD. Either a party without a TDD or one with a TDD can initiate a call using the relay service. Currently, because of the lack of

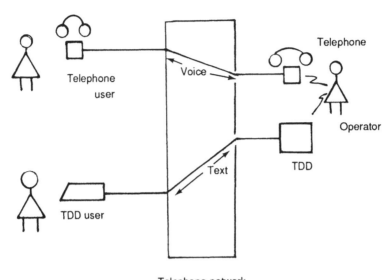

FIG. 7 Dual-Party Relay Service. This service makes telecommunications between the TDD user and persons without the TDD possible.

familiarity with the service among the general public, most calls are initiated by TDD users.

In most volunteer relay services, while the call to the relay operator is paid for by the calling party, the volunteer service pays for the call made to the called party unless the call is a long distance call. The funds to pay these costs must be raised through contributions. On calls involving higher charges, credit-card or third-party billing must be used and the relay operator must obtain the billing information from the calling party and pass it along to the telephone system.

Automated Relay Service.   In the last few years, many states have mandated that the relay service be provided by the telephone companies or special agencies of the government. The largest such service currently available is provided in the state of California. As a result of the increased emphasis on the provision of this service by the operating telephone companies, both AT&T Bell Laboratories (11) and Bellcore (12,13) have undertaken exploratory work to address some of the major problems in the existing relay service, which include the labor-intensive nature of the service, its lack of privacy, and its lack of automatic billing.

In these automated systems, techniques are being planned that permit the TDD user to initiate a call without involving the operator. TDD users in these systems can set up a call themselves, and the operator is attached to the call only when the call is answered. Also, since in many cases the hearing-impaired person does not lack speech capabilities, the systems are being designed to allow the TDD user to deliver a message verbally. In this method of operation, only the response from the hearing party is delivered by the operator through the TDD.

The systems under development also allow relay operators to be multiplexed on the calls. In this method of operation, the operator is attached to a call only when required to perform a task. For example, if the TDD user delivers a message verbally, the operator is removed from the call and is only returned when the person without the TDD wishes to transmit information to the TDD user. The first available operator is then attached to the call to type the message. This is likely to involve several different operators at different points in the call. In addition to using the operators more efficiently, this method of operation affords the conversing parties a measure of privacy.

The systems in development also include automatic techniques to capture billing information, which will speed up the handling of a call and lower its cost.

Speech Synthesis and Speech Recognition in Relay Service. Studies are also in progress to explore the use of both speech synthesis and speech recognition, with the objective of providing additional measures of automation and privacy in relay service. These technologies face severe challenges, however. It has been found that producing error-free conversion of conversational text typed by TDD users, who use many and varied abbreviations, is very difficult to achieve with the text-to-speech logic available today in speech synthesizers. Moreover, speaker-independent speech recognition over the limited bandwidth of a telephone line still seems beyond the current state of speech-recognition techniques. Nevertheless, the work taking place to add additional automation to the relay services will make it easier to introduce these features when they have matured.

Transmission of Sign Language over Telephone Lines

The transmission of visual images rather than tones of speech over the standard telephone line is difficult because of the bandwidth requirements of a video image. It is an appealing idea, however, particularly because sign language (discussed in another section of this article) is a more natural language for persons who are deaf. Exploratory work is in progress to examine techniques of compression and coding that would permit the transmission of a highly schematic visual representation of sign language. At the present time, while the work being done is interesting, a great deal more remains to be done before the approach can be considered feasible (14).

Communication Aids for Persons Who Are Deaf and Blind

Multiple handicaps present substantial technical challenges to the design of communication aids. Creative techniques are, however, on the horizon for persons who are both deaf and blind. There are over 15,000 individuals in the United States who are so afflicted.

Hand-on-Hand Finger Spelling

The principal method of communication with individuals who are deaf and blind is through hand-on-hand finger spelling. In this system, the person sending the message uses the finger-spelling alphabet (Fig. 8) to convey the message

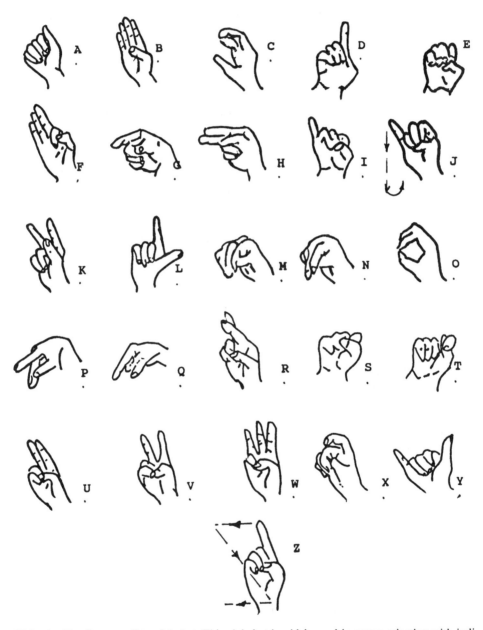

FIG. 8 The finger-spelling alphabet. This alphabet is widely used in communication with individuals who are deaf and blind and between persons who are deaf and those who are not deaf.

while the person receiving the message places his or her hand in contact with the sender's hand while the configurations of each letter in the message is formed.

The Talking Hand

The communication aids being developed for use by persons who are deaf and blind use hand-on-hand finger spelling. Most of the systems under study use a computer-controlled electromechanical hand whose fingers can be formed into the letters of the finger-spelling alphabet (the "talking hand"). The first of these systems was called Dexter and was developed by Deborah Gilden of the Rehabilitation Engineering Center at Smith Kettlewell Eye Research Foundation in San Francisco and David L. Jaffe of the Rehabilitation Research and Development Center of Palo Alto Veterans Administration Medical Center.

With Dexter, letters are entered on the keyboard of a computer. This information is processed by the computer controller and displayed in its associated mechanical hand as the positions of the fingers associated with the finger-spelling alphabet (see Fig. 9). The person who is deaf and blind places a hand in contact with the glove-covered mechanical hand in order to perceive the message. Over the years, several variations in the design have been developed to improve performance and to address difficulties of production, but the basic approach has been essentially the same: to simulate hand-on-hand finger spelling (15,16).

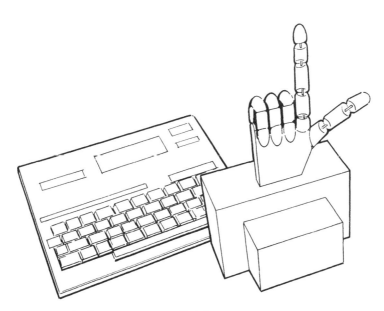

FIG. 9 Dexter, a robotic computer-controlled hand used in communication with deaf-blind persons.

Communication Aids for Motor-Impaired Persons

Neurological and muscular disabilities may impair communication by affecting motor skills used for writing and speaking. The causes of these physical disabilities include Parkinson's disease, multiple sclerosis, amyotrophic lateral sclerosis (ALS, or Lou Gehrig's disease), cerebral palsy, arthritis, stroke, injury, paralysis, and amputation.

Communication needs of many motor-impaired persons may be met by modifying existing tools. If the motor-control impairment is not severe, all that may be needed are special writing tools (such as pen grips) and book holders or page separators (such as a paper clip attached to every page corner) to allow for independent reading.

For many motor-impaired persons who need communication aids, however, more complex devices are necessary. In the generic model of a communication aid discussed above (see Fig. 1), the input element of the device must be adapted to allow for the reduced motor capabilities of a physically disabled person. Neurological diseases may be progressive, so input devices for communication aids should be flexible enough to allow for degrading physical abilities. The range of these input devices is discussed in the section that follows.

For many motor-impaired persons, the disability may affect speech production as well as motor skills. These persons also require special outputs for their adaptive communication aids. Speaking aids and writing aids use similar input techniques but different output formats. Speaking aids output speech, while writing aids output printed material either on paper or on a video monitor. Outputs for adaptive communication aids are also discussed in this section.

Considerations

Although it is common to adapt a device to an individual's specific needs, it is important to consider the appropriateness of a device and investigate alternatives before a choice is made. The abilities, needs, and expectations of the individual should be matched to these considerations.

Communication Rate

An important factor in determining the effectiveness of a communication aid is the rate at which a user can communicate. The average rate of normal human speech is 250 words per minute (wpm), while the rate with an alphabet board (a board that contains the letters of the alphabet at which a person points to spell words) can be as low as 2–26 wpm (17). Three factors can affect the communication rate: the amount of time it takes the user to locate and make the desired selection, the number of selections needed before the desired output is produced, and the amount of information that can be obtained from each output.

Vocabulary

The type of vocabulary used for selection has a great impact on communication rate. A vocabulary can consist of letters or phonemes. Such a vocabulary requires a number of selections to produce an output, but also has virtually an unlimited word selection. A vocabulary can, on the other hand, consist of phrases or sentences. Such a vocabulary requires fewer selections but limits the conversation of the user.

A technique that tries to combine the advantages of both is called *semantic compaction*, in which a series of symbols are used to represent words or phrases. With semantic compaction, the user must memorize a series of symbols for each word or phrase. The symbols are selected to aid that process. *Bliss symbols* use a graphic vocabulary that has a complex structure for combining symbol segments to form many different ideas (18). *Minspeak*™ is another graphic vocabulary that uses symbols with multiple meanings that depend on context for clarity. The Prentke Romich (Wooster, OH) communication aid called Touch Talker™ uses Minspeak (Fig. 10). Different acceleration techniques are also used to reduce the number of inputs; these include encoding, prediction, and scanning order. *Encoding* uses a predefined shorthand version to represent a word or phrase. *Prediction* uses initial letter(s) to cause a menu of possible word choices to be presented. *Scanning order* can be defined to present the most frequently selected items first.

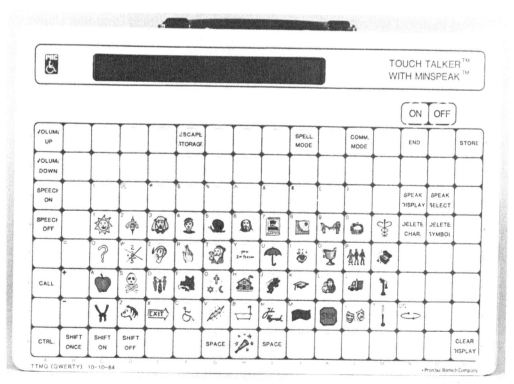

FIG. 10 The Prentke Romich Touch Talker with Minspeak. A voice output communication aid in which symbols are used to represent words and phrases.

Other Factors

There are a number of other factors that should be considered in choosing the most natural and effective communication aid for a nonvocal, motor-impaired person. These include portability; flexibility for custom fit; quality and gender of speech; ability to update vocabulary; programmability; expense, including hidden costs of maintenance and repair; downtime due to updating, maintenance, and repair; ease of instruction, including good documentation; and the manufacturer's service support. Also important are the mobility, positioning, and endurance of the physically disabled person who will use the aid.

Types of Communication Aids

Communication aids can be divided into nonelectronic and electronic aids. Non-electronic aids are generally custom-made for the individual. They are less expensive because they usually have a limited feature set. Electronic aids have an expanded feature set and are more flexible; however, they are expensive and may be complex to use. It is often necessary to have a nonelectronic aid for backup when the electronic aid is being recharged, repaired, or updated.

Electronic aids can be further divided into microcomputer-based aids, adapted aids, and dedicated aids. Microcomputers can be transformed into communication aids with specialized software and hardware. *Adaptive aids* are devices designed for able-bodied individuals (i.e., educational toys) that can be used as communication aids. *Dedicated aids* have been specifically manufactured as communication aids for the disabled.

Many of the electronic input devices used for communication aids may also be used with other control aids, such as environmental controls, feeders, page turners, and wheelchairs.

Input Devices

For a physically disabled person, the input device of the communication aid is the most important part, as it must be adapted specifically to the limitations of that person. Basically, the input device is used to make a selection to achieve the desired output. Methods for selection include use of controlled physical motion (such as pointing or eye gaze), transducers or switches, standard or nonstandard keyboards, and speech recognition.

Selection Techniques.   Selection technique may vary according to the motor limitation and needs of the physically disabled person. The set of choices of selection technique must be presented to the user so that he or she can choose the manner most appropriate to his or her capability (i.e., tactile, auditory, or visual form). Two types of selection are direct selection and scanning. Methods of selection include keyboard, switches, ultrasound, electropotential, transducers, Morse code, and voice recognition.

Direct Selection. With direct selection, the individual points to the selected element in the vocabulary by using a part of the body or some sort of transducer. The different type of pointers include head pointers, mechanical switches, light pointers, and eye-position monitors.

Scanning. When motor-control capabilities restrict movement for direct selection, scanning selection may be used. In scanning selection, vocabulary choices are displayed sequentially until the user makes a selection. Generally, scanning is used for individuals with little motor control because it requires fewer selections. Vocabulary selection with this technique must be limited to minimize scanning time. Encoding or prediction techniques can also be used to limit scanning time. Also, most direct-selection communication aids can be modified to scanning aids by using the receiver or some other device to scan the available selections until the desired choice is indicated by the individual. Methods for making a selection either directly or with scanning are described below.

Communication Boards. One example of a nonelectronic input device is the *communication board*. An individual may point to the desired vocabulary item on a board that has printed words, phrases, or graphical symbols (see Fig. 11). The pointing is accomplished by using any available controlled motion, such as moving a finger, arm, foot, head, or even eye gaze. Head pointers, mouth sticks, and hand sticks also may be used to aid the selection (see Fig. 12).

Communication boards are adapted to individuals for portability and for all selections to be within reach. This type of device has many advantages. It can be quite inexpensive to make, the vocabulary can be updated, and it is easy to use. The disadvantages include a limited vocabulary, which often must be changed by someone other than the disabled individual, and it is also somewhat difficult to correct mistakes. Additionally, it may not get the intended receiver's attention and it may require the receiver to remember previous vocabulary until a thought is completed.

Eye Board. Directed eye gaze may be used to select an element on an eye board (see Fig. 13). An eye board is made of clear plastic with letters or codes placed on it. The eye board is placed between the user and a partner, and the partner follows the eye movements of the user and interprets a selection when the user fixes his or her gaze on an element. An open space is provided in the middle of the plastic eye board so the partner can better follow the user's eye movements and also to provide a space to indicate a break or pause. Although letters and numbers may be placed on the eye board, encoding is often used because of the limited space.

Transducers and Switches. A *transducer* converts input energy into output energy, which is usually a different kind of energy but is in relation to the amount of input energy. A transducer may be used to measure a signal from the body and to generate an electrical signal to control a communication device. A *switch* is a type of transducer that has discrete states and typically is used to open and close electrical circuits. Transducers and switches may measure three

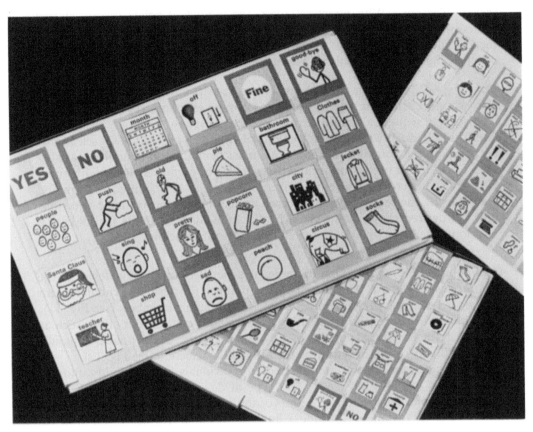

FIG. 11 Simple communication boards. The user communicates by pointing at pictures of objects on the board.

types of energy from the body: mechanical, electromagnetic, and biopotential. Examples of mechanical switches include paddle switch, wobble switch, joystick, wrinkle switch, and puff/sip (pneumatic) switch. Electromagnetic switches include light beam, light-emitting diode (LED), or long-range optical pointer switches. Biopotential switches are activated through measurement of electrical activity of different parts of the body. Such switches include those that measure electric potentials to detect muscular movement or that detect eye position through skin electrodes placed near the eye. Figure 14 shows examples of various types of switches.

Morse Code. When a person with limited physical control is able to send two distinct signals, it is possible through the use of Morse code (17) to achieve a considerable increase in his or her input rate to a communication aid over that achieved by scanning with single-switch selection. Two input states corresponding to the dot and dash of the Morse code can be obtained by changing the duration of the input signal or by using two switches. With Morse code, the

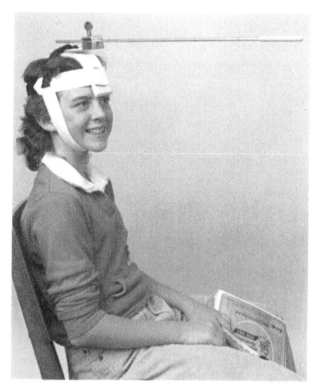

FIG. 12 Fred Sammon's adjustable headpointer (Fred Sammons Inc., Burr Ridge, IL).

user can encode the entire alphabet, numbers, and punctuation. The code has a variety of applications as an input method for communication aids.

Keyboards.   The most common type of input device is a keyboard. Many physically disabled persons may be able to use a standard keyboard by using only one or two fingers or, if a person is more severely motor impaired, the input to the communication aid may originate from other controllable motions using head pointers, hand sticks, or mouth pointers.

Software and hardware may also be modified to allow for keyboard manipulation with one hand, one finger, or reduced fine-motor control. A key guard may be placed over the keyboard to stabilize hand and finger movement so that it is easier to depress only one key at a time. The autorepeat may be turned off for persons who cannot immediately take pressure off the keys and prevents one keystroke from appearing as multiple keystrokes.

For those who cannot depress more than one key at a time, single-finger software may be used to replace simultaneous keystrokes sometimes required with a series of single keystrokes. A keylatch is a technique that may be used to toggle a key on or off, and it is also used for simultaneous keystrokes. Also, keys on a keyboard can be redefined through the use of software for persons with limited access to the keyboard.

FIG. 13 An Imagine Art Eye Communication Board. This device provides a simple, low-cost means of communicating for nonvocal, physically disabled persons, who need only use their eye gaze to operate the board.

A standard keyboard may also be replaced with a modified keyboard for use by people with varying impairments of motor abilities. For example, the amount of pressure needed to depress keys may be adjusted on standard keyboards. A membrane keyboard is used by persons who cannot depress individual keys, but who can apply pressure to flat surfaces.

Many expanded keyboards are available in which the keys are enlarged for persons who are unable to direct motion to a smaller standard-size keyboard. EKEG (Vancouver, BC, Canada) makes a variety of enlarged switches for special purposes. Figure 15 shows the Narwhal™ Expanded Keyboard for computer input and Fig. 16 shows a large switch for use with a telephone. Unicorn Engineering (Richmond, CA) builds the Unicorn™ Expanded Keyboard, which is a programmable membrane keyboard with 128 touch-sensitive keys that can be defined by the user to represent any string of characters up to 40 characters long. For persons with only limited range of motion, miniature switches are available. One built by EKEG is shown in Fig. 17.

Instead of using a keyboard, other types of input devices may be used to emulate a keyboard for use on a computer. In general, these input devices require special interface circuitry. For Apple products, the Adaptive Firmware Card™ produced by Don Johnson (Wauconda, IL) provides this function; for IBM and IBM-compatible computers, the PC Aid™ and the PC Serial Aid™ built by Dada (Toronto, Ontario, Canada) are representative of equipment

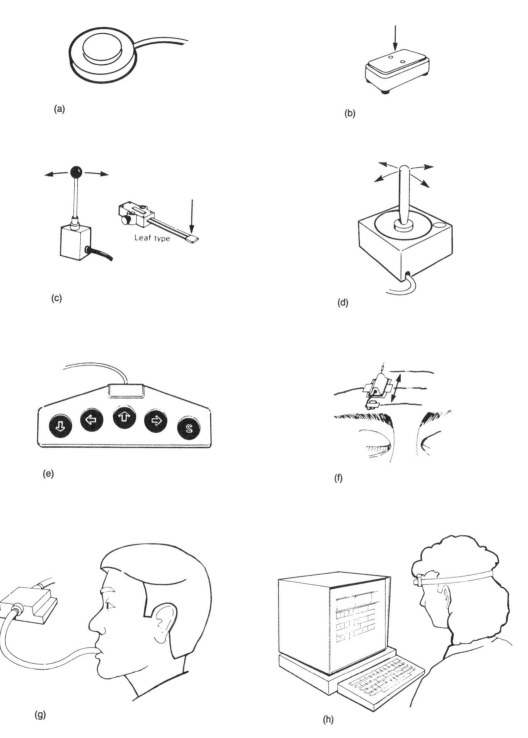

FIG. 14 Various types of switches: *a*, round pad; *b*, plate; *c*, joystick, leaf; *d*, joystick; *e*, wafer; *f*, eyebrow; *g*, pneumatic (puff/sip); *h*, optical pointer.

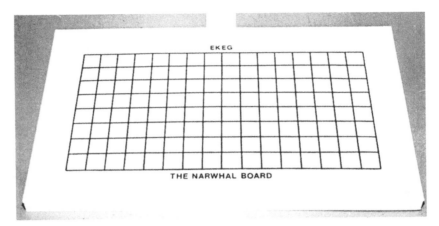

FIG. 15 Narwhal expanded keyboard, which enables computer access for a person with little dexterity.

performing the interface function. These interface circuits permit alternate input devices to enter characters into the computer. The alternate input devices range from specially adapted keyboards to single or dual switches. The type of device is usually indicated to the computer by a selection made from a menu. Of particular interest is a keyboard manufactured by ComputAbility (Novi, MI) called Aid-Me™. This device is able to work with the Apple, Macintosh, and IBM compatible computers. It replaces the keyboards provided by the computer manufacturer and accepts input from a wide variety of alternate input devices. It does this without requiring any change in the application software running on the host computer. The keyboard also provides for speech output.

Speech Recognition. For motor-impaired persons who are not speech impaired, speech recognition may be used as an input technique. Within the last decade, significant progress has been made in speech-recognition technology. In fact, a number of systems are now commercially available and in use as

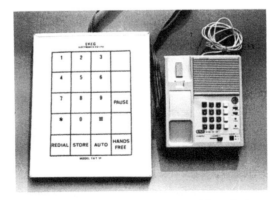

FIG. 16 Expanded keyboard for use with a telephone.

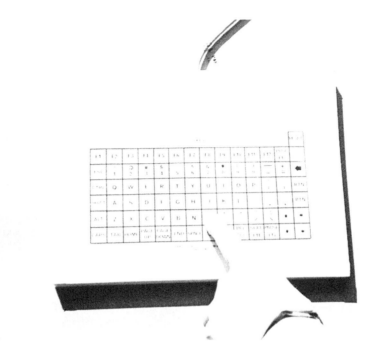

FIG. 17 Miniature keyboard that can be used by a person with restricted movement.

assistive devices (19). The systems function primarily in communications between humans and machines. The technique is used most often to provide commands usually input from the keyboard in environmental control, word-processing, and workstation applications (20).

Most speech-recognition systems work by digitizing selected intervals of speech and then using one or more pattern-recognition strategies to select from stored templates of prerecorded vocabularies. Generally, the stored vocabularies are produced in training sessions carried out by the user working with the system. There are, however, some systems with very limited vocabularies that are relatively speaker independent.

Currently, the applications for speech-recognition systems by persons with physical disabilities are limited due to the expense of the equipment, the difficulty of the training for use, and the lack of reliability in performance. In rehabilitation applications, the sensitivity of the systems to variability in a user's voice quality due to such factors as changes in tone, direction of projection, loudness, temporary illness (i.e., a cold), and progressive physical disabilities is especially troubling. These factors, all of which can affect the signal with which the system has to work, can cause it to fail to recognize a previously recognizable command. Limitations in the applications for speech recognition also arise due to the difficulty in filtering out unwanted background noise.

In spite of these problems, work continues in a number of laboratories throughout the world, and the long-term picture for the use of the application in assistive technology is bright.

Output Devices

Possible output formats for communication aids are visual displays, printers, and speech synthesizers. Considerations for output selection include correctability, ability to obtain a permanent record, allowance for private conversations, ability to get the intended receiver's attention, ability to communicate to a group, and allowance for distance between user and receiver.

Displays.   Visual outputs can take the form of characters, words, phrases, or graphical symbols that are pointed to or displayed. The display may be static, such as that for a communication board, or changeable by the user, such as output to a computer monitor. A visual display, as opposed to voice output, allows for nondisruptive and private conversations and is easily correctable. On the other hand, visual displays require the receiver of the message to watch the output display for the entire length of the conversation.

Printers.   The appeal of printers as output devices is that they provide a means of permanently recording the user's message and providing an alternative to writing if the user is motor disabled. Printers are most often necessary in an educational or vocational setting. They can be attached to a computer, monitor, or speech synthesizer, and the user can select the appropriate output format. The disadvantage is that they are generally not portable.

Voice Output: Speech Synthesizers.   Communication aids that use speech synthesizers come closest to replacing natural speech. Generally, the higher the price of a speech synthesizer, the better the quality of the speech output. Speech synthesizers that provide male and female voices are also available. Speech synthesizers can be connected to computers to change text to speech as an alternate output device, or can be used in specialized voice-output communication aids (see Figs. 18 and 19). Today, there are three technologies used for speech synthesis: digitized waveform, phoneme coding, and predictive encoding.

Digitized Waveform.   In digitized-waveform synthesizers, actual human words are stored digitally in a computer to be converted to analog sounds when needed. This type of speech synthesizer produces the most natural voice; however, it requires a large memory (even when compression techniques are used) so it tends to be an expensive technology. Often, when it is used, the vocabulary is limited to a set of words.

Phoneme Coding.   Phoneme coding stores the phonetic components of speech (*phonemes*), which are concatenated to output words. Any word can be produced by outputting the appropriate string of phonemes. Phonemes require less memory storage than entire words; however, this method of speech synthesis produces a monotonic speech quality.

Predictive Encoding.   In predictive encoding, an electronic model of the hu-

FIG. 18 The Prentke Romich Introtalker, a voice-output communication aid designed for use by nonvocal persons.

man vocal system is stored digitally. Then, language and pronunciation rules are used to generate speech. A computer algorithm is used to analyze input text and then generate the appropriate string of phonemes. Pitch, energy level, and inflections may be changed by adapting the algorithms. Again, the speech produced tends to have an artificial quality.

Communication Aids for Nonvocal Persons

Background

Persons who are nonvocal are unable to use natural speech to meet all of their communication needs. The causes of this disability vary. Some individuals have no disability other than loss of speech production, and still have the ability to use other forms of communication, such as writing. Laryngectomies and neurological damage are the major causes of this type of impairment.

Other groups of nonvocal persons have more severe communication problems. While they do not have speech, they may also not have the manipulative control needed for other forms of communication. In these groups, the amount of physical control determines the mechanism needed for the communication aid. Some have lost their speech-production capabilities later in life due to head injury, stroke, or ALS. Communication aids for this last group are discussed in a section above. Others have disorders that have prevented the development of

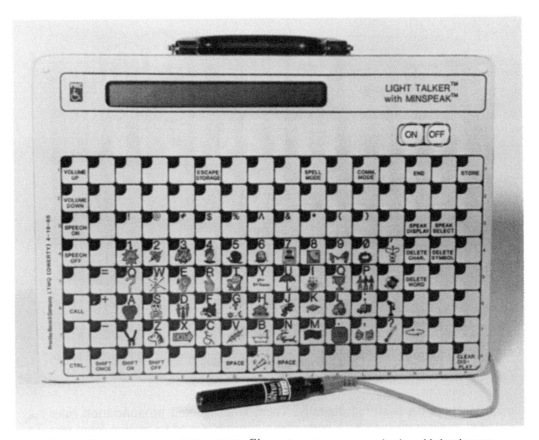

FIG. 19   The Prentke Romich Light Talker™, a voice-output communication aid that is operated by an optical pointer.

speech, such as cerebral palsy or mental retardation. Communication aids for
this group may have to overcome poor language skills, since these disorders
sometimes affect language performance.

Nonelectronic Communication Aids

Nonelectronic communication aids used by persons who are nonvocal use the
individual's existing capabilities to replace speech. These include sign language
and, in some cases, esophageal speech.

Sign Language

Sign language is considered a separate language from spoken English. The term
sign language is a generic term that refers to all forms of manual communication. American Sign Language (ASL) evolved in the early 1800s from American
and French signs. It is a language used by deaf people when they communicate.

It is estimated that some 400,000 deaf persons in North America use the language daily (21). The language is optimized for nonverbal communication, having its own syntax and grammar. It involves the use of the hands, head, and shoulders and is physically very expressive.

When deaf persons communicate with hearing persons, they generally use a combination of ASL and English. The result is a pidgin signing system. Neither of these languages, ASL or pidgin signing, use strictly English grammar. For this reason, most educators of deaf children use and teach a signing system that follows English grammar closely and is referred to as manually coded English.

Finger spelling plays a significant role in both pidgin signing and speech based on manually coded English. It is used in both systems to provide words that do not have signs (see Fig. 8).

Esophageal Speech

To compensate for the loss of a larynx, esophageal speech may be used. A laryngectomy eliminates use of the lungs as the source of air for producing speech, since the lungs are no longer connected to the throat. Esophageal speech uses air from the stomach to create sounds that are accomplished by various forms of burping. In all, 60% of the people who have had laryngectomies can develop esophageal speech.

Electronic Aids

Electronic communication aids currently available for nonvocal persons include the speech amplifier, the artificial larynx, and, more recently, new concepts like the talking glove.

Speech Amplifiers

A number of companies produce personal speech amplifiers for use by persons who have diminished speech-production capability. The amplifiers vary in size, quality, and price but are generally composed of a microphone placed close to the speech-production source of the user, an amplifier to heighten the sound, and a speaker. Using the systems is often difficult because of the need to avoid feedback. Generally, this is managed with careful placement of the microphone and the use of a directional microphone.

Artificial Larynx

The artificial larynx uses electronic devices to produce desired frequencies that are then resonated in an individual's resonant chamber to produce speech. This electrolarynx may be outside the body or surgically implanted inside the body.

If it is used outside the body (an *extraoral electrolarynx*), it requires exact placement next to the mouth or neck and the user mechanically changes the frequency to produce the desired sounds. Although the internally placed electrolarynx (*intraoral electrolarynx*) is more cosmetically appealing and is easier to learn how to use, it is still very controversial.

The Talking Glove

The Talking Glove is a product of research and is being developed to break the barriers to communication that exist for nonverbal deaf persons, persons who are deaf and blind, and persons who do not know sign language. The system was invented by James Kramer at Stanford University (22,23). Through the use of the Talking Glove, finger spelling can be converted to speech.

In the operation of the Talking Glove, a nonverbal person puts his or her hand inside a glove that contains a set of strain gauge sensors that run along the back of the glove over each finger. Wires are attached to the sensors in the glove and run to a computer that is equipped with a speech synthesizer. In operation, the person finger spells with the gloved hand the words he or she wishes spoken.

For each hand position, the multiple sensors in the glove produce a set of output voltages that correspond to the hand position. These voltages are matched against a record stored in the computer of voltage patterns that were produced when the user trained the glove by taking up the specific hand positions of each letter. The hand positions, while being in general similar from one person to another, vary slightly for each person. In eye or hand-on-hand communication, the brain makes the adjustment for individual variation.

When the glove is in use and as the fingers are configured, the computer determines in real time what letter is being formed. The letters are gathered into words that are then passed into the text-to-speech logic of the speech synthesizer. The synthesizer then pronounces each word.

By using a combination of Dexter and the Talking Glove, two-way communication has been successfully achieved between nonimpaired persons and individuals who are both deaf and blind (22). These systems are still largely in the developmental stage but use among people who are deaf and blind seems near.

Communication Aids for Learning-Disabled Persons

Communication aids for persons with learning disabilities fall into two categories. When the disability is limited to cognitive and linguistic impairments, educational tools are used. Computers are invaluable in this role because they allow a person to learn at his or her own pace. If the learning disability is present in conjunction with other disabilities such as those described in the sections above, use of the communication aids for those disabilities is also appropriate.

Communication and educational aids are often used together and, because the learning abilities vary widely, aids frequently must be adapted for individual abilities. In some cases, the absence of some abilities in and of themselves result in extremely challenging learning situations, as for example with a child born deaf. Here again, computer-based educational tools in addition to communication aids can sometimes be used to considerable advantage.

Conclusion

This article considers a large variety of assistive technology for persons with disability and focuses primarily on technology directed at enhancing and restoring lost or impaired communication capabilities. The devices considered are in some cases very simple and widely used and in other cases very complex, used by few, and still largely in the laboratory. All of the devices have played or are destined to play important roles in improving the quality of life for many with communication disabilities.

A number of organizations in this country and in other parts of the world are producing an increasing variety of new assistive technology. For the seeker of additional information concerning this field, two organizations are singled out as resources. The first is RESNA, an international, interdisciplinary association for the advancement of rehabilitation and assistive technology whose headquarters is located in Washington, D.C. The second is the Trace Research and Development Center on Communication, Control, and Computer Access for Disabled Individuals, which is located at the University of Wisconsin–Madison. Both of these organizations play important roles in the advancement of knowledge on assistive technology throughout the world.

Acknowledgments: BEX is a registered trademark of Raised Dot Computing. MPRINT, Vista, and Optacon are registered trademarks of Telesensory/VTEK. Personal Reader is a registered trademark of Xerox/Kurzweil. Narwhal is a registered trademark of EKEG. Unicorn is a registered trademark of Unicorn Engineering. Touch Talker, Light Talker, Minspeak, and Introtalker are all registered trademarks of Prentke Romich. Aid-Me is a registered trademark of ComputAbility.

Bibliography

Apple Computer, Inc., Office of Special Education Programs, *Apple Computer Resources in Special Education,* DLM Teaching Resources, Allen, TX, 1988.

Bergan, A. F., Presperin, J., and Tallman, T., *Positioning for Function: Wheel Chairs and Other Assistive Technology*, Valhalla Rehabilitation Publishing, Valhalla, NY, 1990.

Berliss, J. R., Borden, P. A., and Vanderheiden, G. C., *Trace Resource Book,* Assistive Technologies For Communication, Control, and Computer Access, 1989–90 ed., Trace R&D Center, Madison, WI. 1989.

Boone, D. R., *Human Communication and Its Disorders*, Prentice-Hall, Englewood Cliffs, NJ, 1987.

Bowe, F., *Personal Computers and Special Needs*, Sybex Computer Books, Alameda, CA, 1984.

Breuer, J., *Handbook of Assistive Devices for the Handicapped Elderly*, Haworth Press, New York, 1982.

Carter, J. P., *Electronically Hearing: Computer Speech Recognition*, Howard Sams & Co., Indianapolis, IN, 1984.

Cattoche, R. J., *Computers for the Disabled—A Computer Awareness First Book*, Franklin Watts, New York, 1986.

Crewe, N. M., et al., *Independent Living for Physically Disabled People*, Jossey-Bass, San Francisco, 1983.

Davis, W., *Aids to Make You Able: Self-Help Devices and Ideas for the Disabled*, Beaufort Books, New York, 1981.

Eisenson, J., *Language and Speech Disorders in Children,* Pergamon Press, Elmsford, NY, 1986.

Green, P., and Brightman, A. J., *Independence Day: Designing Computer Solutions for Individuals with Disability,* DLM Teaching Resources, Allen, TX, 1990.

Helfman, E., *Blissymbolics*, Elsevier/Nelson Books, New York, 1981.

Horstmann, H. M., Levine, S. P., and Lincoln, A. J., Keyboard Emulation for Access to IBM-PC-Compatible Computers by People with Motor Impairments, *Assistive Technology*, 1(3): 63–69 (1989).

McWilliams, P. A., *Personal Computers and the Disabled*, Quantum Press, Doubleday & Co., Garden City, NY, 1984.

National Support Center for Persons with Disabilities, *Resource Guide for Persons with Vision Impairments*, IBM, Atlanta, GA, 1989.

1990 Resource Directory, *Closing the Gap*, 8 (6) (February–March 1990).

Rehabilitation Research and Development 1988 Progress Report, Veterans Administration Medical Center, Palo Alto, CA.

RESNA, *Proceedings of the 12th Annual Conference of RESNA*, RESNA Press, Washington, DC, 1989.

Smith, R. O., *Technological Applications for Enhancing Performance: Human Performance Deficits*, Slack, Thorofare, NJ, 1990.

References

1. 10% of the World's Population is Disabled, *New International*, (Special publication, DESI/DEPI), United Nations, New York.

2. Bowe, F., *Technology Review*, 90:52–59 (1987).

3. *Code of Braille Textbook Formats and Techniques*, 1977.

4. Webster, J. G., et al., *Electronic Devices for Rehabilitation*, John Wiley & Sons, New York, 1985, pp. 53–59.

5. Goldish, L. H., and Taylor, H. E., The Optacon: A Valuable Device for Blind Persons, *New Outlook for the Blind*, 68:49–56 (1974).

6. Gadbaw, P. D., Dolan, M. T., and De L'Aune, W. R., Optacon Skill Acquisition by Blinded Veterans, *Journal Visual Impairment and Blindness*, 71:23–28 (1977).

7. Levitt, H., and Neuman, A., Digital Hearing Aids, *RESNA '87 Conference Proceedings,* 389–391 (1987).

8. Boothroyd, A., Development of Speech Perception Skills in Cochlear Implantees, *RESNA '87 Conference Proceedings,* 428–430 (1987).

9. Webster, J. G., et al., *Electronics Devices for Rehabilitation*, Wiley Medical, New York, 1985, p. 148.

10. Taddonio, L. C., ASCII Code. In: *The Froehlich/Kent Encyclopedia of Telecommunications*, Vol. 1 (F. E. Froehlich and A. Kent, eds.), Marcel Dekker, New York, 1990, pp. 389–395.

11. Jackson, J., AT&T Bell Laboratories. In: *The Froehlich/Kent Encyclopedia of Telecommunications*, Vol. 1, (F. E. Froehlich and A. Kent, eds.), Marcel Dekker, New York, 1990, pp. 397–406.

12. Holder, A., Bell Communications Research Inc. (Bellcore). In: *The Froehlich/Kent Encyclopedia of Telecommunications*, Vol. 2 (F. E. Froehlich and A. Kent, eds.), Marcel Dekker, New York, 1991, pp. 57–64.

13. Terhune, W. R., Telecommunications Network for the Deaf, *Bellcore Digest of Technical Information*, 6(11):1–7 (February 1990).

14. Boubekker, M., Automatic Feature Extraction for the Transmission of American Sign Language over Telephone Lines, *RESNA '87 Conference Proceedings,* 434–436 (1987).

15. Smallridge, B., and Gilden, D., DEXTER III: A Product in Hand, *RESNA '89 Conference Proceedings,* 137 (1989).

16. Jafee, D., DEXTER II – The Next Generation Mechanical Finger Spelling Hand for Deaf-Blind Persons, *RESNA '89 Conference Proceedings*, 349–350 (1989).

17. Fishman, I., *Electronic Communication Aids*, College-Hill Publication, Little, Brown and Company, Boston, MA, 1987.

18. Helfman, E., *Blissymbolics*, Elsevier/Nelson Books, New York, 1981.

19. Berliss, J. R., Borden, P. A., and Vanderheiden, G. C., *Trace Resource Book*, University of Wisconsin–Madison, 1990, p. 245.

20. Snell, E., Neumann, S., and Atkinson, C., Comparison of Three PC Based Voice Recognition Systems, *RESNA '89 Conference Proceedings,* 717–719 (1989).

21. Shroyer, E. H., *Signs of the Times*, Gallaudet University Press, Washington, DC, 1982.

22. Goldsmith, M. F., Computers Star in New Communication Concepts for Physically Disabled People, *Journal of American Medical Association,* 261(9):1257.

23. Kramer, J., and Leifer, L., The Talking Glove: A Speaking Aid for Non Vocal Deaf and Deaf Blind Individuals, *RESNA '89 Conference Proceedings*, 471–472 (1989).

VICTOR L. RANSOM
LAURA S. REDMANN

Communication Gateway Services for Videotex

Introduction

Definitions and Roles

Videotex is an interactive electronic medium that allows communications, information storage and retrieval, and transaction capability. There are generally a variety of services included on any given videotex system that may be accessed through a menu or list of services. Examples and types of these services include home banking (transaction), home shopping (transaction), White Pages directory (retrieval), database searches (retrieval), electronic mail (communications), and "chat" services (communications). "Videotex" may be packaged or trademarked in a fashion similar to any other brand-name product. Services may be international, national, or local. Depending on the type of service, customers may be billed by subscription, transaction, minutes of use, or services may be free to the end users.

Access to these services is typically via a personal computer equipped with a modem to allow interconnection with a telephone line, although services can also be distributed through coaxial cable or satellite broadcast. There are a number of specialized terminals used with specific services, as well as a "simple" terminal (keyboard and screen) that can be used in lieu of a personal computer and modem for connection via telephone line.

Several terms typically are used in describing the roles various videotex suppliers take, including gateway, network operator, information provider, service bureau, and systems packager. Figure 1 illustrates these videotex industry roles. The terms are defined below, but the main focus of this article is on gateway services. The article explains what a gateway is, the various roles industry players may assume in connection with a gateway, and what issues are involved in implementation.

A *gateway* is defined by the Gateway 2000 report of the Videotex Industry Association (VIA) as "consisting of a set of functions intended to facilitate electronic access to users to remote services and vice versa. Gateways are intended to provide a single source through which users can locate and gain access to a wide variety of services" (1, p. 64). In other words, a gateway is a meeting point of a group of users and a group of information services.

This meeting point is often facilitated by a network operator. A *network operator* typically is a common-carrier-based company that offers basic transport services and billing. The network operator provides the technical interconnection to the distribution network, typically the telephone or cable network.

An *information provider* is the person or organization that provides the service. Historically, information providers were only content providers (owners of the original data), but the term often is applied in a broader sense to those who package and market the information.

An evolving industry role has been that of the service bureau. A *service*

219

Conduit   **Content**

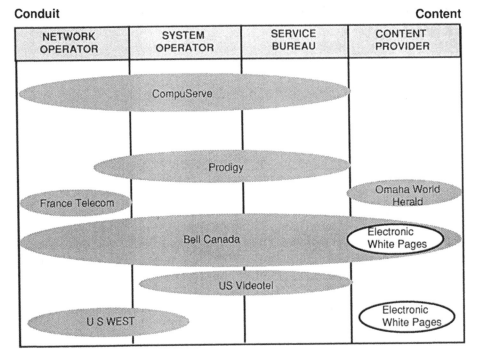

FIG. 1   Videotex services industry roles. Industry roles vary on a continuum of conduit versus content. Most players play more than one role, with most seeking to gain better leverage by participating either individually or with partners in multiple roles.

bureau is an entity that can assist less-sophisticated information providers in entering the videotex market by providing a whole range of services such as hardware, software, and service-development expertise. They often market the services of their clients as well. Additionally, service bureaus offer a data-storage platform for information providers.

An even more expansive role in the videotex industry is taken on by those who function as system packagers. A *system packager* is a total packager of videotex products that include network operation, information-service development and storage, as well as end-user market development. Examples of system packagers today are firms like PRODIGY and CompuServe®.

Gateway 2000 Study

In October of 1988, the VIA issued the Gateway 2000 report, a study of North American gateways (1). The industry began this work because it recognized a need to bring the industry together in order to facilitate the evolution and success of mass-market videotex. Over 24 organizations participated in this effort.

The VIA's premise for the Gateway 2000 study was that "gateways can play a critical role in stimulating growth of the market for electronic information, transaction, and communication services by facilitating access of users to services and vice versa" (1, p. 1). They believed that this premise was affected by the fact that North America neither shared a common approach nor ensured a common set of practices, which are believed necessary to stimulate a mass market for gateways. This belief has been supported in the development of other mass markets such as telephone, television, and automobiles. A car is a car, a phone is a phone, a television is a television, and across each of these markets there is common mode of operation. The VIA has defined as the key principles of mass-market gateways that they must be easy to use, affordable, ubiquitous, and uniform. These key principles share a commonality with the telephone and the key principles for its mass market. Therefore, the study focused on defining the functional characteristics required by a broad set of users of mass-market gateways.

The Gateway 2000 report defines a core set of characteristics as user interface, connectivity, and administrative support functions. In addition, it has specifically defined the items below as necessary for a common approach by operators of mass-market gateways:

Access
Directory
User-interface activities and functions
Connectivity specifications
Gateway-to-gateway interconnectivity
User authentication
System messages
Transaction formats for services
Billing formats
Notices such as trademarks, copyrights, and disclosures

These characteristics are believed to foster a successful mass-market gateway operation inasmuch as they would develop a commonality that users could expect from all mass-market gateways. The list above, while not inclusive, does highlight the critical components to a common gateway approach. A full reading of the Gateway 2000 report is recommended for those who require an in-depth understanding of the core set of gateway characteristics.

Schematics of Major Gateways

Gateways typically use the telephone network, coaxial cable, or some hybrid to distribute electronic services. Access devices include the television set, a personal computer, or a dedicated terminal to allow the user entry to the service. Examples of three typical gateway schematics are shown in Figs. 2, 3, and 4.

A typical videotex gateway on the telephone network (such as Community Link[SM] — see Fig. 2) consists of the public telephone network, a public or private data network, and a processor. These components are used to log end users on and off, provide a directory of services, connect and disconnect information providers, and track and verify billing. Usually a standard X.25 data connection is used by local information providers to connect to a gateway. An X.75 connection may be used by national and international information providers through a long-distance carrier.

This type of gateway is accessed through either a personal computer or a

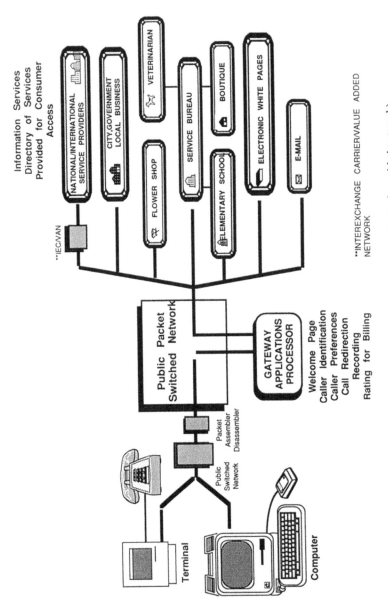

FIG. 2 Community Link^{SM} service (*interexchange-carrier/value-added network).

simple terminal, and generally supports either American Standard Code for Information Interchange (ASCII), Teletel, North American Presentation Level Protocol Syntax (NAPLPS), or some combination of these protocols.

A typical videotex application on coaxial cable is X-PRESS Information Service. This service is provided using a host computer, satellite, cable television company, and a personal computer. Through continual feed from the cable television company to the personal computer, the data requested is captured and stored for display later. X-PRESS can be used to request data on mutual funds, interest and money rates, business and financial news, and the like. X-PRESS supports a variety of personal computers, including IBM, Apple, Macintosh, Atari, and Amiga. Figure 3 illustrates how this service is deployed.

An example of a hybrid technology is being offered by Singapore Telecom through a product called Teleview. The system utilizes both the telephone and the television to convey text and image data and uses a UHF television transmitter to relay coded information to users phoning in a request. The bandwidth of the UHF signal enables high-quality color pictures to be sent with the information. Teleview is one of only two videotex systems in the world that can transmit photographic quality images. A consumer is required to have an adapter and keyboard connected to a telephone and television to interact with Teleview. The consumer's request is sent to the Teleview computer, which sends the coded response back to the adapter via the UHF transmitter. In addition, data may be sent through the phone line but, due to the narrower bandwidth, the service is then considerably slower. Teleview integrates the technologies of the telephone

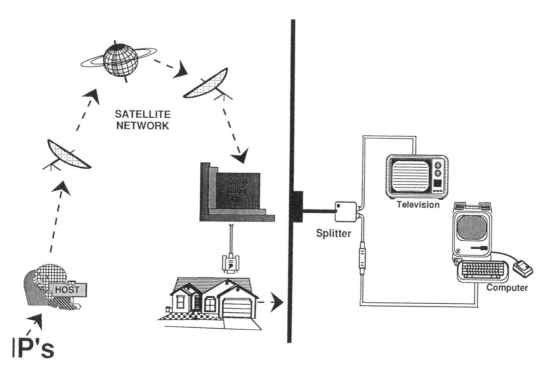

FIG. 3 X-PRESS Information Service.

and television to provide a unique videotex application. Figure 4 illustrates how this service is deployed.

Figure 5 compares Community Link, Teleview, and X-PRESS, highlighting the following areas: access device, network technology, provisioning of service content, and cost to consumers.

Survey of Current Activities

Overview

Commercial videotex gateways were initially offered in the late 1970s, with activities increasing in number and in marketing intensity on a worldwide basis in recent years. Early entrants such as the Times-Mirror Gateway and Knight-Ridder's Viewtron were discontinued in the United States after a few years of operation. The first large-scale, consumer-oriented service was France Telecom's Minitel service. Services are now provided in numerous countries in

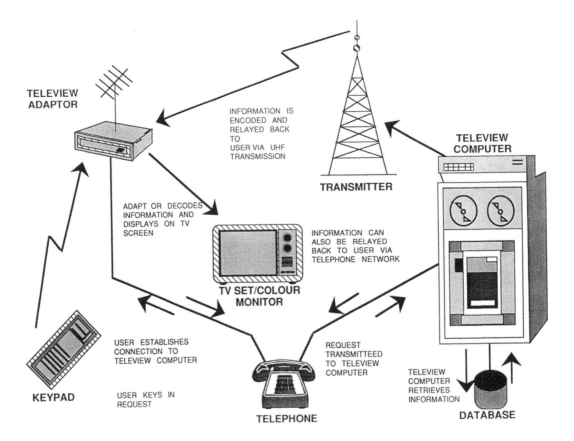

FIG. 4 Teleview network.

	Community Link^sm	Teleview	X-Press Info Svcs
Access Device	Personal Computer with modem or Terminal	Teleview adaptor connected to television set or color monitor	Personal Computer connected to cable system
Network Technology	Public Packet Switch Network	UHF Transmission	Cable or Satellite System
To Provide Service Content	Information Service Provider connects to the gateway system to provide content via a direct computer connection or through a service bureau is displayed on a screen in Teletel or ASCII format	Studio edited information is provided to the Teleview computer which is displayed on the screen in photographic quality.	Information providers generate information for the host company that transmit the information in a continued feed format to be customized by the subscriber format as what information will be displayed on the screen.
Cost of Service	Directory access at no charge. Charge occur when accessing the various information service providers' service	Free service for 1989	X.Change -No additional charge to basic cable service Executive -$19.95 monthly charge in addition to monthly cable service

FIG. 5 Comparison of videotex systems.

Europe, North America, South America, the Far East, the South Pacific, and Africa. Tables 1 and 2 list current activity worldwide and in North America, respectively. In addition to these countries, other countries are investigating and planning for the introduction of commercial videotex gateway services, including Korea and Ireland.

The characteristics of these gateways vary considerably in terms of marketing and technical standard variables. Marketing variables include how the service is priced (flat subscription vs. usage based), what market it addresses (generally business vs. consumer), and whether the services offered are primarily local or national in scope.

Technical standards can vary considerably as well. Some of these variables include information provider access arrangements with the gateway provider (access through a VAN, or value-added network, direct access, etc.), network configurations (single-applications processor, multiple-applications processors in a distributed network, etc.), information-presentation protocols (ASCII, Teletel, NAPLPS, Prestel, etc.), and terminal-display standards. Figure 6 compares the three most common gateway information-presentation protocols.

Marketing Trends

Many international gateway providers that have focused only on the business community in the past are now developing mass-market approaches that address consumers. Some are emulating the French model by offering low-cost terminals to subscribers. For example, the Italian PTT (telephone company) plans to

TABLE 1   Worldwide Videotex Service Gateways: Technical Standards

Country	Videotex Service Name	Year Commercially Introduced	Gateway Protocol	Terminal Display Standard
Europe (EC)				
Belgium	Videotex	1986	Prestel	Prestel, CEPT, Teletel, ASCII
Denmark	Teledata	1984	Prestel	CEPT
France	Teletel	1981	Teletel	Antiope
Italy	VIDEOTEL	1985	Prestel, Teletel	Prestel, Teletel
Luxembourg	Videotex	1986	Prestel	CEPT
The Netherlands	Viditel	1981	Prestel	Prestel
Spain	IBERTEX	1986	IBTX	CEPT
United Kingdom	Prestel	1979	Prestel	Prestel
West Germany	Bildschirmtext	1984	EHKP	CEPT
Europe (EFTA)				
Austria	Bildschirmtext	1984	Prestel	CEPT
Finland	Telset	1984	N/A	N/A
Norway	Teledata	1986	Prestel	Prestel, CEPT, Antiope
Sweden	Videotex	1982	Prestel	CEPT
Switzerland	Videotex	1987	Prestel	CEPT
Far East				
Hong Kong	Infokey, Viewdata	1988	Prestel	N/A
Republic of China	N/A	1987	NAPLPS	N/A
Japan	CAPTAIN	1984	CAPTAIN	NAPLPS
South Korea	Chollian	1986	NAPLPS	NAPLPS
South America				
Brazil	VIDEOTEXTO	N/A	N/A	N/A
South Pacific				
Australia	Viatel	1985	N/A	N/A
Malaysia	Telita	1983	Prestel	N/A
New Zealand	Vapnet	1986	N/A	N/A
Africa				
South Africa	Beltel	N/A	Prestel	Prestel, ASCII

TABLE 2 North American Videotex Service Gateways and Other Selected Interactive Electronic Services: Technical Standards

Country and Company	Videotex Service Name	Year Commercially Introduced	Gateway Protocol	Terminal Display Standard
Canada				
Bell Canada	ALEX	1988	NAPLPS	NAPLPS
United States				
RBOCs				
Bell Atlantic	IntelliGate	1989	ASCII	ASCII
BellSouth	Transtext Universal Gateway (TUG)	1988	ASCII	ASCII
NYNEX	InfoLook	1988	NAPLPS, ASCII	NAPLPS, Teletel,
US WEST	Community Link	1989	Teletel, ASCII	Teletel, ASCII
Other Selected Interactive Electronic Services:				
H&R Block	CompuServe	1979	ASCII	ASCII
General Electric	GEnie	1985	ASCII	ASCII
Quantum Computer Services	Q-Link, PC-Link America Online	1985	ASCII	ASCII
IBM/Sears	PRODIGY	1988	NAPLPS	NAPLPS
US Videotel	US Videotel Network	1989	Teletel	Teletel
General Videotex Corporation	Delphi	1980	ASCII	ASCII
Dow Jones & Co. Inc.	Dow Jones News/ Retrieval	1978	ASCII	ASCII
Mead Corp.	Lexis/Nexis/ Medis	N/A	ASCII	ASCII
Knight-Ridder Inc.	Dialog	1972	ASCII	ASCII

distribute over 600,000 terminals by the end of 1993. Tables 3 and 4 provide an overview of marketing variables worldwide and in North America, respectively.

Efforts are also underway to interconnect gateways around the world. This includes not only interconnection among European gateways, but interconnection between gateways in North America and those in Europe as well. The Italian gateway already has been connected to the French gateway; and an interconnection among Spain, France, Portugal, and Italy is planned. West Germany's gateway should be interconnected with Teletel in France and with the United States in 1990.

	DISPLAY FEATURE	NAVIGATIONAL FEATURES	ADVANTAGES
ASCII (AmericanStandard Code for Information Interchange)	- 128 character set - Scrolling - 24 rows of 80 characters	- Specific characters followed by carriage return	- More information can be presented on a screen - Scroll oriented - continuous flow of text - Ubiquity amongst Computer Community
TELETEL	- 25 rows of 40 characters - Alphamosiac characters (graphics) - 8 Foreground and 8 background colors - Page oriented display	- Specific characters followed by function key - Some special function keys like *Guide* for help	- Standard function keys - enhance user friendliness - Graphic capability makes services more dynamic
NAPLPS (North American Presentation Level Protocol Syntax)	- Extensive character set, higher resolution - Bitmapping - 16 colors	- Function keys - No official standards - Does not have definitions for function keys or navigation	- More complex graphic language allows higher definition of images - More efficient with information

FIG. 6 Comparison of gateway information-presentation protocols.

Technical Standard Trends

The major technical standard trend is a growing acceptance among gateway providers, especially those in Europe, that there will not be one protocol that is standard for all videotex gateways. In lieu of trying to create a universal standard, many of the gateway providers are concentrating on multistandard terminals to hurdle the standard-incompatibility barrier. The multistandard terminal that appears to be gaining widespread support in Europe is the terminal that allows access to services in Prestel, Teletel, and ASCII protocols. The most common link between countries at this time appears to be the Teletel protocol. Tables 1 and 2 provide an overview of technical standards by country.

Legal Differences

Most telephone subscribers in the United States are served by one of several corporations that were formerly part of AT&T; these companies are subject to the restrictions of the Modified Final Judgment (MFJ) that forced the divestiture of AT&T's local telephone exchanges in 1984. The MFJ prohibits these local companies, generally referred to as the Regional Holding Companies (RHCs) or Regional Bell Operating Companies (RBOCs), from generating, acquiring, transforming, or processing information, but does allow gateway functions. This applies to a number of gateway functions that have nothing to do with being a service provider and means most American telephone companies cannot play the role of a gateway provider without severe limitations (see Fig. 7).

Market Development Issues

Developing Information Providers

The electronic delivery of information/services to users is done today to specialized niche markets. The videotex applications required to attract a larger base of end users do not exist today and must be developed. This requires first convincing potential service providers that they have a viable application and then developing software to deliver the service.

The development of the videotex industry is just beginning, making participation at this time a risk. To date, there is no positive proof that the end-user market will develop in the United States. There is still no common agreement on how fast it will develop. Therefore, convincing nontraditional potential information providers that videotex is an opportunity is often a challenge.

The lead time required to develop videotex services varies according to the degree of functionality or interactivity between the user and the service provider and whether the provider is currently in the on-line or traditional publishing business. Services that allow users to communicate with others and to conduct transactions or manipulate information are generally of more value than services that provide information only. The software development for communication, transaction, and manipulation services is much more complex and therefore takes more resources to develop.

An existing on-line provider or publisher already has many of the necessary components in place to offer a new videotex application. A publisher may already have the information residing in a database and only needs to make modifications and establish a delivery channel. An existing on-line provider already has the delivery channel in place and needs to acquire and package the information. However, as with most media changes, somewhat different skills are required in packaging electronic services than those needed in traditional publishing businesses.

Many potential applications can be offered by the retail and service businesses and by government and community organizations. Once the decision is made to participate, the information must be gathered and organized and the application software must be developed. This increases the time required to bring a new application to market.

Figure 8 illustrates the approximate lead time required to bring an information service on line. Increased interactivity between users and service provider increases the time it takes to develop a service. Typically, current on-line providers require the shortest lead time, while government and community organizations (particularly nonprofit organizations) require the longest lead time.

Developing the Consumer Market

Videotex is fairly common in the United States for business applications such as legal research, stock quotations, and news retrieval services. It is particularly valuable where information is perishable (stock quotations) or frequently updated (legislation and case law). Consumer services are far less prevalent in the

TABLE 3 Worldwide Videotex Service Gateways: Marketing Variables

Country	Videotex Service Name	No. of ISPs or Services	No. of Subscribers	Access Device(s)	Monthly Connect Hours	Targeted Subscriber Segment(s)
Europe (EC)						
Belgium	Videotex	150 ser. ('89)	6,500	Terminal	N/A	N/A
Denmark	Teledata	85 ISPs ('88)	4,100	Terminal	N/A	N/A
France	Teletel	12,000 + ser.	15,000,000+	PC, terminal	7,000,000	Business, consumer
Italy	VIDEOTEL	1,500 ser.	100,000	PC, terminal	300,000	Business, consumer
Luxembourg	Videotex	N/A	4,000	N/A	N/A	N/A
The Netherlands	Viditel	1,000 ISPs ('88)	25,000	N/A	N/A	Business
Spain	IBERTEX	150 ser.	1,000 ('88)	PC, terminal	30,000	Business, consumer
United Kingdom	Prestel	1,300 ISPs ('88)	95,500	Terminal	N/A	Business, consumer
West Germany	Bildschirmtext	6,000 ser.	225,000	PC, terminal, TV Adapter	1,300,000	Business
Europe (EFTA)						
Austria	Bildschirmtext	750 ISPs ('88)	9,500	Terminal	N/A	N/A
Finland	Telset	40 ISPs ('88)	8,000	N/A	N/A	Business
Norway	Teledata	100 ISPs ('88)	3,500	N/A	N/A	Business, consumer
Sweden	Videotex	332 ser.	25,000	PC, terminal	N/A	Business
Switzerland	Videotex	300 ISPs ('88)	40,000	N/A	N/A	N/A

Far East						
Hong Kong	Infokey, Viewdata	N/A	N/A	N/A	N/A	Business
Republic of China	N/A	N/A	10,000 ('89)	N/A	N/A	N/A
Japan	CAPTAIN	580 ser.	10,000 ('89)	Terminal	80,000	Business, consumer
South Korea	Chollian	N/A	N/A	Terminal	N/A	N/A
South America						
Brazil	VIDEOTEXTO	N/A	18,000 ('88)	N/A	N/A	Business
South Pacific						
Australia	Viatel	250 ser. ('86)	16,000 ('86)	N/A	N/A	N/A
Malaysia	Telita	N/A	N/A	N/A	N/A	N/A
New Zealand	Vapnet	N/A	N/A	N/A	N/A	Business
Africa						
South Africa	Beltel	157 ser.	12,000	N/A	30,000	Business

Note. All data is 1990 data unless otherwise noted.

TABLE 4 North American Videotex Service Gateways and Other Selected Interactive Electronic Services: Marketing Variables

Country, Company, and Location	Videotex Service Name	No. of ISPs or Services	No. of Subscribers	Access Device(s)	Monthly Connect Hours	Targeted Subscriber Segment(s)
Canada						
Bell Canada	ALEX	650 ser.		PC, terminal	10,000	Business, consumer
Montreal			16,000			
Toronto			14,400			
United States						
RBOCs:						
Bell Atlantic	IntelliGate	100 ISPs		PC	N/A	Business, consumer
Philadelphia			4,900			
Washington			4,800			
BellSouth	Transtext Universal Gateway (TUG)	57 ISPs		PC	N/A	Business, consumer
Atlanta			3,000			
NYNEX	InfoLook	180 ISPs		PC	N/A	Business, consumer
Burlington			500			
Boston			5,000			
New York			5,000			
US WEST	Community Link	750 ser.		PC, terminal	N/A	Business, consumer
Omaha/Council Bluffs			4,800			

Other Selected Interactive Electronic Services:

H&R Block national	CompuServe	1,400 ser.	625,000	PC	1,200,000	Business, consumer
General Electric national	GEnie	200 ISPs	215,000	PC	N/A	Consumer
Quantum Computer Services national	Q-Link, PC-Link, America Online	200 ISPs	50,000 50,000 45,000	PC	250,000 (89)	Consumer
IBM/Sears national	PRODIGY	800 ser.	304,100	PC	N/A	Consumer
US Videotel Dallas Houston	US Videotel Network	75 ISPs	2,500 10,000	PC, terminal	250,000	Business, consumer
General Videotex Corporation national	Delphi	N/A	80,000	PC, terminal	N/A	Business, consumer
Dow Jones & Co. Inc. national	Dow Jones News/ Retrieval	N/A	315,000	PC	N/A	Business
Mead Corp. national	Lexis/Nexis/Medis	N/A	237,500	PC	N/A	Business
Knight-Ridder Inc. national	Dialog	N/A	120,000	PC	N/A	Business

MFJ Restriction on Content	Impact on Videotex Gateway
Being an information provider.	Key "high value" services are not available to consumers.
Providing advice on designing services or hardware/software selection to potential service providers	Difficult for entrepreneurs or small businesses to participate.
Providing Information Providers feedback that would improve the value of their services.	Improvements in service quality are not made.
Setting common navigation standards within services.	Navigation within services is not consistant, negatively impacting ease of use.
Providing a Service Directory with Keyword search and placing IP services in proper categories.	Services are not easy for users to find
Identifying users segments based on gateway usage patterns.	Advertising/promotion plans targeted towards past or low users to minimize churn cannot be implemented.
Serving markets that are limited geographically	Many information providers customers are in more than one geographic market. Users would incur long distance to call.
Data processers serving more than one market.	Centralized data processers serving multi-geographic markets cannot be used, increasing costs for developing videotex markets.

FIG. 7 Impact of Modified Final Judgment restrictions on Regional Bell Operating Companies.

videotex market. Presently, only about 1% of U.S. households subscribe to a videotex service. Large-scale consumer gateways probably will not develop until these gateways are low cost, easy to use, and offer services that are of more value than the alternative means of acquiring information, doing a transaction, or communicating.

The use of videotex requires a modification of behavior. Instead of looking in the newspaper for the events occurring in the community this weekend, one could use a videotex service that provides a comprehensive listing of all events categorized by location. Users must be convinced that the new way of doing things is better than the old way before a new product or service is adopted. A new service is not really adopted by a consumer until the service is incorporated into his or her everyday routine.

Figure 9 lists the stages a consumer normally must go through in adopting a videotex service. The advertising and promotion of videotex services must facilitate the consumer's movement through these stages. The potential user must first be made aware that videotex services exist. Next, the benefits of using videotex must be communicated so the potential user will become interested and seek additional information from which to evaluate whether or not this new service could benefit him or her. The potential user must be given the opportunity to try the service to see if it really does benefit him or her. It is critical that the user has a positive experience and finds services of value during

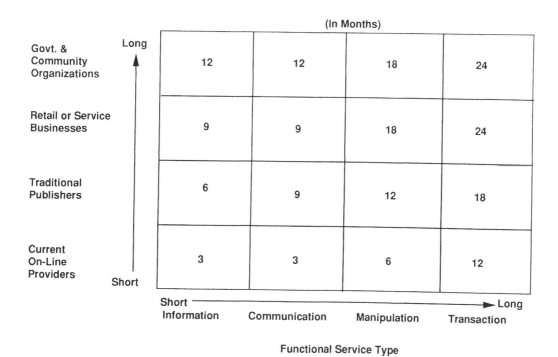

FIG. 8 Videotex application sales and development cycle.

Awareness	Knowledge of the existance of videotex
Interest	Is interested and seeks further information on videotex
Evaluation	Evaluates information gathered and evaluates whether or not to try videotex
Trial	Tries innovation to determine the value of videotex
Adoption	Makes full and regular use of videotex

FIG. 9 Stages of consumer adoption of videotex.

the trial stage or he or she will not adopt videotex. Therefore, the timing of offering high-value services and the consumer trial stage is critical to the development of the end-user market.

Most sources estimate that between 10% and 15% of U.S. households have a personal computer (PC) and modem. Whether or not this comprises a sufficient number of access devices to truly tap a consumer market is the subject of significant debate in the industry. In the short term, it tends to encourage gateway providers to offer services nationally to get a large enough base of users, because there are not enough users in a single geographic market. Notable exceptions to the focus on national services are local bulletin boards that are offered free, such as Cleveland Freenet or Tri-State Online.

A low-cost, easy-to-use alternative is not readily available to consumers in the United States at this point in time. Only two major U.S. companies, U.S. Videotel and US WEST, are currently working with terminal manufacturers to insure the availability of terminals in conjunction with the deployment of their videotex gateways.

Future Implications

The real power of gateways is in bringing together in one place access to an almost limitless array of databases, transactions, and other people using interactive text and graphics rather than just voice. Information can be shared and transactions concluded any time at the convenience of the user. It is possible to do a wide variety of futuristic tasks using technology available today, including full-motion video, touch screens, and voice recognition/response, for example. Such technology is currently not commercially viable because the market is not yet large enough to support the volume required to make it economical.

Impact on Consumer Behavior

The most difficult challenge for any new technology is to change consumer behavior. Gateways start that process by educating the consumer to think first of doing a task electronically, much like people learned to use the telephone 80 or 90 years ago.

There are some important analogies between early telephone use and the introduction of videotex gateways. A behavior change was not easy even for the telephone. For the first 10 or 20 years, telephone service was mostly used by businesses, and mostly by men, much as videotex is used today (about 90% of videotex users are men). Use of the telephone was considered too technical for most women. Telephone use generally was considered a means of doing business only. Telephone use exploded when there gradually got to be enough market penetration that users could interact with each other to communicate with friends and family, not just to conduct business.

The time-independent nature of electronic services makes them applicable to either time-starved or time-rich users. The former includes the many dual-income families with very busy schedules who could do their grocery shopping, communicate with their child's teacher, or apply for a building permit on line at their own convenience. Electronic services also provide entertainment, education, and a way to meet people for those looking for a way to fill time.

Impact on Business Operations

Most businesses use electronic applications either to distribute products and services, acquire new customers and markets, or service existing customers.

The tremendous growth in selling through mail order has a logical extension to electronic services, where product information can be delivered more cheaply, targeted, and combined with order entry. The availability of high-quality graphics and full-motion video can be expected to accelerate this process. This is an especially valuable application for "time-sensitive" transactions such as concert tickets, airline seats and travel tours, or classified advertising. The reader can see exactly what is available at the time it is available; once the item is purchased it is removed from the system.

Retail stores find electronic selling a way of extending their geographical markets and/or more effectively targeting their efforts. As mentioned above, a variety of demographic data can be collected about the particular user, and only certain messages or advertising then targeted to that user. In this manner, a small boutique can be featured on a gateway in another part of the country.

Customers can be expected to grow more comfortable over time with interfacing electronically, allowing many options for service. Even now, banks find that electronic-banking customers tend to turn over more slowly and to consolidate their accounts—both of which make the account more profitable for the bank. Store-account customers could get a bill and pay it electronically, as well as order merchandise, be informed of a sale, get a note of appreciation from their salesperson, or be notified that the item they ordered was in. What appears to be an impersonal form of communication can become very personal, as databases can be used to keep track of items that enable humans to communicate specific items that are of interest to the receiver.

Impact on Society

This behavior change can have some interesting implications for its users. The French have already discovered that one of the attractive attributes of the Minitel service is its anonymity. Even in the United States, electronic text communication can eliminate the barriers of gender, color, class, or ethnic background because text communication is generally independent of these factors. At the same time, these demographic characteristics can be well known to the electronic service deliverers, either allowing sophisticated target marketing or endangering individuals' privacy, depending on one's point of view.

Electronic services allow equal access for most of the physically disabled. For example, the concept of a mainstream application that can be used by the hearing impaired for shopping, making reservations, or just communicating is very powerful. As those in society age, people with limited mobility but clear minds may find this kind of access a way of interfacing with the rest of the world. The concept of "equal access" has also raised interesting issues around whether there will be an "information elite." Telephone service is considered a basic necessity by most, with government subsidies or "lifeline" services relatively available, resulting in household penetration for telephone service of over 90%. Will electronic access to information only be by those who can afford a computer? Is access to information services a luxury or a fundamental right? What role, if any, does the U.S. government play in providing access, as most international government-owned PTTs have? Who is responsible for insuring the privacy of those who access these services? Is this an "alternative technology" that can integrate the needs of the physically impaired with those in the "mainstream"?

The answers to these questions will likely be the subject of much debate over the next few years.

Acknowledgment: Community Link is a service mark of US WEST.

Bibliography

Cutler, B., The Fifth Medium, *American Demographics*, 24–29 (June 1990).

Datapro International, *Datapro Reports on International Telecommunications*, McGraw-Hill, New York.

Diebold, J., Videotex in the U.S.: An Assessment, *Telecommunications,* North American ed., 22(7): 78–83.

Fleming, M., et al., *Information Industry Factbook*, 1989–90 ed., Digital Information Group, Stamford, CT, 1989.

IDP Report, SIMBA Information, Wilton, CT, 1990.

Information Industry Bulletin, Digital Information Group, Stamford, CT, 1990.

Interactivity Report, Arlen Communications, Bethesda, MD, 1990.

Malone, T. W., Yates, J., and Benjamin, R. I., *Electronic Markets,* Sloan School of Management, Massachusetts Institute of Technology, Cambridge, MA, 1988.

National Association of Broadcasters, *Videotex Success: Some Basic Guidelines*, 3 (March 1989).

National Telecommunications and Information Administration, *NTIA Information Services Report,* U.S. Department of Commerce, Washington, DC, August 1988.

Nielson, K. O., The Winning Formula at Quantum Computer Services, *Electronic Services Update*, 3 (February 1990).

Purton, P., Videotex Struggles to Leave Its Stamp on Europe, *Telephony*, 214(26): 44–46.

Roberts, S., and Hay, T. (eds.), *International Directory of Telecommunications*, Longman Group Limited, Essex, England, 1986.

Thomas, H., Requirements for Success in Videotex in the Nineties, address delivered at VIA annual conference, Toronto, Canada, June 1, 1990.

U S WEST Communications' Community Link, *Worklife*, 3:2 (Summer 1990).

VIEWTEXT, Phillips Publishing, Potomac, MD, 1990.

Yankeevision, The Yankee Group, Boston, MA, 1990.

Additional Sources of Information

Connect Times, Jupiter Communications Company, New York.
Enhanced Services Outlook, Telecom Publishing Group, Capitol Publications, Alexandria, VA.
European Telecommunications, Probe Research, Cedar Knolls, NJ.
LINK Resources Corporation, 79 Fifth Avenue, New York, NY 10013 (212) 473-5600.
Singapore Tests the Teleview Market, *Asian Computer Month* (November, 1988).
Videotex Industry Association, 8403 Colesville Rd., Suite 865, Silver Spring, Rosslyn, VA (301) 495-4955.

Reference

1. Videotex Industry Association, *Gateway 2000: Report of the Videotex Industry Association Study of North American Gateways,* Videotex Industry Association, Rosslyn, VA, October 1988.

LINDA J. LASKOWSKI
NANCY K. METZLER
HEATHER S. TOOKER
KEITH D. WALLIN

Communication in Education

Introduction

Education can be defined as the outcome of the learning process. In our time, the highly structured, hierarchical, credentialing mammoth that is the sum total of educational institutions is modified and, in some cases, supplanted by a flexible and freeform collection of learners, learning objectives, and information resources loosely connected in public, private, and alternative education systems (Fig. 1). Increasingly, the connection between the learner and his or her object are provided by telecommunications. For the purposes of this article, the term *education* encompasses learning received from traditional institutions of learning (schools, colleges, universities, and libraries); vocational, employer-provided and employment-related training; and self-help and self-improvement activities. If we accept the definition of education as being the product of learning, we can begin to evaluate the central and emerging role of telecommunications now and in the future.

Individuals in our society are increasingly engaged in multiple, sometimes simultaneous self-directed learning enterprises that continue throughout their lives. Telecommunications is a facilitator of these learning enterprises — whether our objective is to pursue a high-school diploma, to develop proficiency in household maintenance, to obtain professional development and recertification, or to pursue advanced Bible study.

The role of telecommunications in education is being redefined by the change in the very nature of education itself. We examine some of the ways telecommunications are used to enable education, with particular emphasis on telecommunications' support for the learning enterprise.

The use of telecommunications in education parallels yet lags behind telecommunications uses in business and industry. The traditional education establishments (schools, colleges, universities, and libraries) use telecommunications to maintain contact with their respective constituencies, to facilitate interaction between collaborators, to create linkages between human and electronic resources, and to achieve operational efficiency by substituting telecommunications expenditures for other costs of doing business.

The use of telecommunications varies by the size and mission of the institution. Generally speaking, colleges and universities are more experienced in the use of telecommunications for administrative and academic computing purposes. Schools and libraries, however, were among the first institutions to explore the potential of telecommunications for distance learning and alternative curriculum delivery.

Within the formalized educational structure, telecommunications applications underpin an institution's administrative operations. Increasingly, applications that have an impact on the delivery of curriculum and that support an institution's teaching/learning mission are growing in acceptance and use.

Educators, policy makers, learners, and the telecommunications industry are increasingly applying creative energy toward harnessing the potential of an

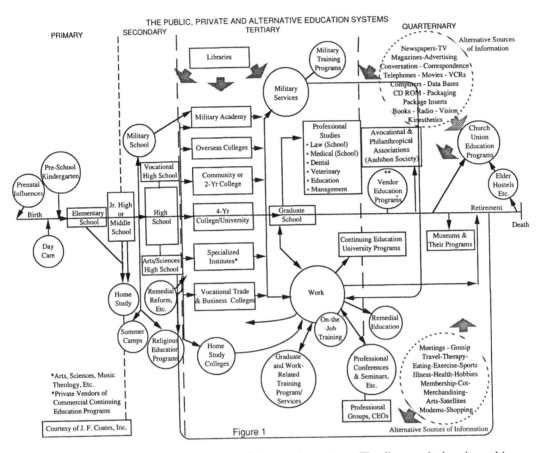

FIG. 1 The public, private, and alternative education systems. The diagram depicts the multi-plicity of sources from which society obtains education. The explosion in our need to learn continuously and specifically throughout our lifetimes and the availability of education/information sources place telecommunications in a central, enabling role.

evolving telecommunications infrastructure in new and sophisticated ways. As digital signaling permeates the network from end to end, as available bandwidth grows exponentially, and as the cost/performance ratio of optoelectronic and computing devices steadily decreases, telecommunications is poised to touch every aspect of education.

The telecommunications infrastructure, both network and customer prem-ises components, have become critical to the education/learning enterprise. So-ciety has reacted to the explosion in information with a commensurate explosion in learning. A fundamental paradigm shift is in the offing.

In Iowa, a leased wire hook-up between schools and the homes of shut-in pupils permits them to "go to school" despite physical afflictions, such as heart disease, limbs withered by infantile paralysis, or bones broken by accidents. The Iowa Super-intendent of Public Instruction has installed the telephone intercommunicating sys-tem in about a dozen districts; the state provides the equipment and the school district pays the telephone tolls. The box-like instrument at school is carried from room to room as classes change, a special plug providing the connection in each room. In the child's room at home is a similar device. The shut-in child not only hears everything that goes on in the classroom; he also "recites" when called upon. One such shut-in was elected president of his class. (1, p. 13)

The use of telecommunications to enhance and expand the reach of educational opportunities is not new. Indeed, because education is, by its nature, communications intensive, any advances in our fundamental technical ability to communicate have found their way into our systems of education.

Telecommunications systems in education range from the most basic—a single, nominal 4-KHz telephone line and a single line telephone set in an administrator's office—to a broadband pipe supplying integrated digital audio, data, and video to an electronic classroom. These services are derived from the public telephone networks, customer premises equipment (CPE), and, increasingly, a combination of public and private elements or a hybrid system. Almost all commercially available telecommunications systems have found a practical application in education and learning.

This article examines a range of telecommunications applications and the underlying systems that make them possible.

Telecommunications Applications in Education

The increasing development and use of telecommunications applications by schools, colleges, universities, and libraries has several driving forces. Applications are driven by:

- A geometric increase in information and knowledge available from a wide variety of sources
- Advances in telecommunications technologies that enable heretofore unimaginable possibilities
- An exponential growth in computers and computing (especially microcomputers), which stimulates a requirement for connectivity
- The growing use of multimedia formats of communication that incorporate voice, data, images, and video into curriculum
- Student, faculty, and administrator expectations of access to advanced telecommunications resources
- The relative inexpensiveness of telecommunications services versus alternatives (e.g., physical transportation of resources)
- The very presence and availability of ubiquitous telephone service on the public network

Voice Applications

Voice telephony is universally available in educational institutions. In schools, colleges, and universities, basic voice services are as essential to the efficient conduct of education's business as they are at home and in industry.

A notable development in the university community is the resale of voice services to students and faculty. Resale gives a university a profit center that offers a wide range of communications services. These services can be offered economically by university "telephone companies" in cooperation with selected local and long-distance providers that offer volume discounts. The local telephone company can provide the university itemized billing for features and services against not only individual extensions, but also against authorization codes used per extension.

Revenue generated from retailing these services are increasingly invested in the modernization of the university's telecommunications plant and equipment. Modern telecommunications then becomes a competitive advantage in the competition for students, faculty, and research grants.

Educational institutions are exploring new ways to use basic voice telephony to improve services and efficiency. Many schools offer student help and information lines that improve access to on-line support and timely information. Outbound wide-area telephone services (WATS) and 800 service are used increasingly in recruitment and research activities.

Audioconferencing

Audioconferencing is an electronic meeting in which participants in different locations use telephones to communicate with each other simultaneously. In education, groups of researchers, collaborating teachers, or groups of students might use this basic, "tried and true" telecommunications service to exchange ideas. Advances in network-based audioconference bridging products and services now make it possible to conduct audioconferences on an ad hoc basis with a seemingly unlimited number of participants. All that is required is a telephone set, a telephone line, and a directory number that reaches the conference bridge.

Voice Processing

Abreast of the general popularity of answering machines, voice-mail applications are playing an expanding role in education. Beyond the obvious application as a communications efficiency tool for administrators, teachers, and learners, voice mail provides a facility for asynchronous communications, for example, conducting dialogues in non–real time. For example, a parent might access a given teacher's voice-mail box to learn the day's homework assignment or to hear, in the teacher's own voice, a characterization of the day's classwork. The same parent might also choose to leave a message in the teacher's voice-mail box regarding a student's performance or impending absence.

Similarly, voice messaging facilitates access to recorded information via a touch-tone pad. The touch-tone pad allows a caller to navigate through a topical voice index to reach a prerecorded message of interest.

When combined, voice mail and voice messaging capabilities are a powerful and user-friendly vehicle for sharing information from one to one, one to many, or many to one (see Fig. 2).

CLASSNOTES℠ Voice Messaging Service for Schools

A Community Service Project of BellSouth, Memphis City Schools and U.S. Sprint

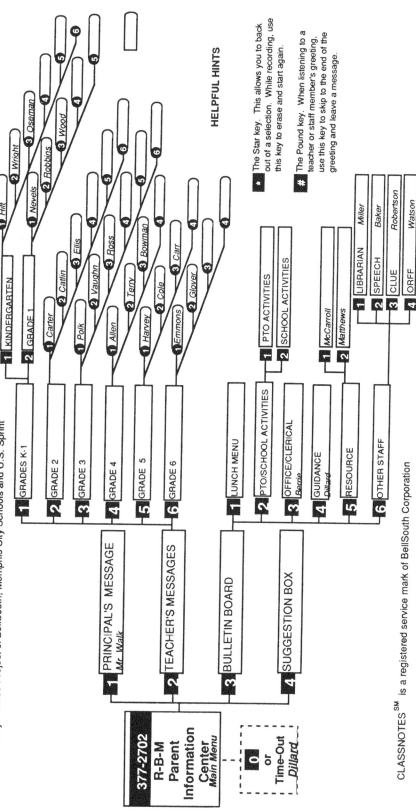

HELPFUL HINTS

[*] The Star key. This allows you to back out of a selection. While recording, use this key to erase and start again.

[#] The Pound key. When listening to a teacher or staff member's greeting, use this key to skip to the end of the greeting and leave a message.

CLASSNOTES℠ is a registered service mark of BellSouth Corporation

FIG. 2 Voice Messaging Service for Schools. Voice messaging makes it possible for timely and relevant information to be disseminated throughout an educational institution's community of interest. Voice mail makes it possible to conduct an asynchronous dialogue.

Data Applications

The proliferation of computers and computing in education has necessitated a corresponding improvement in telecommunications capabilities. While some data applications are easily supported by the existing public telephone network, most others require higher transmission rates and greater network bandwidth. This is especially true when the data application must integrate simultaneous voice and data over the same telephone line. Table 1 identifies certain applications and their corresponding data-transmission requirements.

The simplest data applications use a telephone keypad as the basic input device. Some applications involve terminal-to-mainframe connections and others require computer-to-computer links via a network. The speed at which data are interchanged and/or the amount of data to be transmitted determines the required transmission speeds and, hence, the network bandwidth requirement.

Generally available voiceband data services operate at transmission speeds of up to 9.6 kilobits/second (kbps) and permit the combination of both voice and data on a single telephone line through the use of a CPE device called an *integrated voice/data multiplexer* (IVDM). Basic rate Integrated Services Digital Network (ISDN), with two B channels operating at speeds of 64 kbps and a D channel for signaling, dramatically boost the speeds at which data applications take place. Colleges and universities were among the first to identify uses for ISDN capabilities.

Education data applications parallel applications in business and industry. Learners and educators seek to access the store of electronic information in remote databases. Libraries increasingly store their journals, reference works, and entire card catalogues in electronic media accessible via telecommunications. Other data applications prevalent in education are discussed below.

Electronic Mail

Educators and learners are discovering the benefits of electronic mail (E mail) for asynchronous communications and on-line conferencing. In the university environment, access to mainframe-based electronic-mail systems is provided by the institution's private branch exchange (PBX) or telephone company central office. Students and their academic counselors, collaborating professors, ad-

TABLE 1 Applications and Data-Transmission Rates

Application	Transmission Speed Required
Electronic mail	1.2 or 2.4 kbps
Word processing	4.8 or 9.6 kbps
Still video with teleconferencing	19.2 kb–1.5 Mbps
Compressed video	56 kb–1.5 Mbps
Full motion video; interactive TV	44.7 Mbps
Computer-to-computer communications	19.2 kb–4.8 gigabits (Gb)

ministrators, and their staffs use E mail to augment or supplant their face-to-face or written communications.

Teleregistration

Electronic registration (or teleregistration) eliminates most of the manual labor and frustration from the registration process. Telecommunications, when linked to an institution's computers, provides the answer to the tasks of transporting requests, record matching, and data processing, right down to issuing a bill—all without the appearance of the student on campus (Fig. 3).

Each student enters data into the university computer using a touch-tone telephone keypad. A voice-response system prompts the data entry. The student enters class choices. The administrative computing system then matches the requests with class availability. Confirmation of class assignments is completed while the student is on line.

In addition, most teleregistration systems incorporate features for access security.

Computer Conferencing

Computer conferencing allows individuals at different locations to communicate directly with each other through computers. Communication may be in

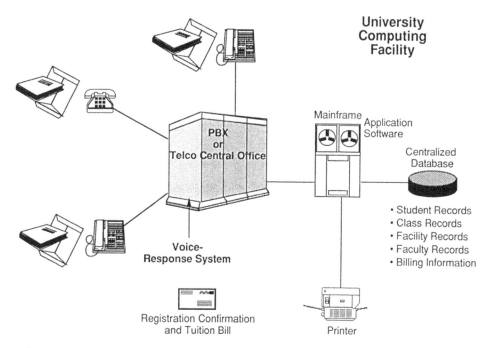

FIG. 3 A typical teleregistration system. Teleregistration is but one of the many ways that educational institutions are using their telecommunications infrastructure to improve service and achieve a competitive advantage.

real time or delayed. Educationally focused bulletin boards are one example of computer conferencing, yet many universities conduct entire course sections using computer-to-computer conferences. The key enabler is the telecommunications transmission path that links the conferees.

Computer conferencing is an important enabler for the alternative systems of educational delivery mentioned above. Individuals or ad hoc groups of learners use computer-conferencing facilities to interact with remote information resources in a self-defined learning enterprise. Increasing numbers of self-directed learners and increasing demands for information resources in voice, data, image, and video formats will drive demand for increased transmission speeds and network bandwidth to homes, businesses, and traditional educational institutions.

Image and Video Applications

It is difficult for those educated in a bygone era to appreciate the extent to which education has become a multimedia experience. A generation weaned on television and videogames has a predisposition toward learning visually. Consequently, educators and learners see a growing need for visual media in the learning enterprise.

Our existing telecommunications infrastructure is just beginning to provide the functionality required to deliver visual resources to the learner whenever and wherever they are needed. As speed and bandwidth capabilities grow, so do applications that integrate visual formats into curriculum delivery.

While past efforts have focused on integrating visual media into the classroom environment through the use of such on-premises devices as slide projectors, fax machines, and videocassette recorders (VCRs), a growing movement toward the delivery of all required teaching resources via telecommunications is developing. This effort is called generically *distance learning* and constitutes the primary application of telecommunications' image and video capability in education.

Distance Learning

Distance learning has a long and effective history. As settlers pushed westward, circuit teachers traveled from settlement to settlement offering a rudimentary educational experience to pioneer children. Correspondence courses have been around for as long as the mails and have provided a means for remote learners to interact with learning resources. Ever since the westward settlement, educators and learners have incorporated advances in telecommunications to improve upon the basic distance-learning model.

Distance learning entered the telecommunications age by way of educational television from public broadcasting facilities. The airwaves are used to beam instructional programs into schools and homes thereby augmenting the curriculum with visual assets. Later, satellites beamed live programming to properly equipped school buildings and, for the first time, offered feedback about the visual asset through the use of a telephone line back to the site of instruction (2).

The primary objective of advances in distance learning has been an increase in the quality and level of interactivity between learner and resource. To most educators and learners, this means fully interactive, multimedia capability that encompasses high-quality audio, high-speed data links, and full-motion video. Table 2 compares the attributes of some distance-learning delivery technologies.

The sophistication of distance learning delivery methods tracks the sophistication and availability of telecommunications infrastructure technologies. In addition, the subject of distance-learning courses tracks the shortage of teachers in specialized curriculum areas (see Fig. 4). Distance learning is one way to offer a full range of courses in small or remote schools.

Audiographics Conferencing. Audiographics conferencing is an advanced computer application in which computer interaction is augmented by two-way, real-time audio communication. Audio, data, and graphics are shared over regular telephone lines, allowing users in different locations to work on the same applications simultaneously.

Audiographics conferencing is a step up from computer conferencing inasmuch as an image component augments the audio and data resources available for teaching and learning. Audiographics conferencing is dependent on the availability of appropriate and compatible computer software, and the presence of a premises or network-based conferencing bridge into which participants can dial and in which the voice, data, and image transfers can be mediated (see Fig. 5).

Microwave Radio Systems in Distance Learning. Point-to-point microwave radio systems operate in various frequency bands. They can transmit audio, data, or video in either a simplex or duplex format. Microwave requires a clear line of sight between sender and receiver, making the signal sensitive to terrain and buildings.

In education, microwave systems are used predominantly in short-haul applications between school sites in a given district. A decline in the cost of requisite electronics has contributed to the use of microwaves for distance learning. However, crowded frequencies and difficulty in siting towers hampers the use of this distance-learning delivery medium.

Distance Learning by Satellite. Satellites are normally used to transmit audio, data, and video programming in a point-to-multipoint configuration. In education, low-altitude satellites are used for such store-and-forward applications as computer conferencing or voice mail in which immediate interaction is not required. Geosynchronous satellites, because of their fixed position, are primarily used for the delivery of real-time video programming.

Educational services are increasingly migrating to Ku-band satellites that operate at 14/12 GHz, mostly because the receive dishes are smaller and less

TABLE 2 Comparison of Distance-Learning Delivery Technologies

	Audio Graphics (Telephone)	Microwave Radio	Satellite	Cable TV	Analog Optical Fiber*	Digital Optical Fiber*
Start-up Cost	Low	Low	High	Medium	Medium	High
Expansion Potential: Channels	High	Low	Low	Low	High	High
Maintenance costs	Low	Medium	Low	High	Low	Low
Special concerns/difficulties	Limited visual interaction	FCC licenses difficult to obtain because of frequency congestion in urban areas	Microwave interference (C-Band) Rain Fade (Ku-Band)	Not designed for interaction	Right of way	Right of way
Expansion potential: Sites	High	Low-Medium	High	Low-Medium	High	High
Voice, video, data potential	Medium	Medium	High	Medium	High	High

*Voice, data, and/or video signals can be transmitted over fiber-optic cable using digital or analog techniques. The decision to use analog or digital fiber optics needs to be made with the following considerations: distance, future technologies, and integration with existing telephone networks.

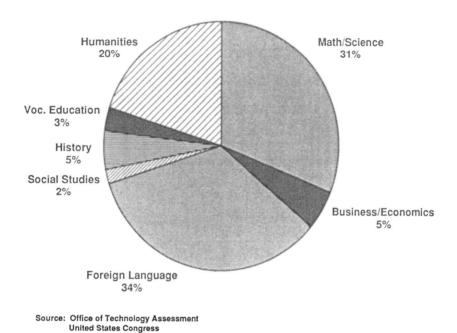

Humanities
20%

Math/Science
31%

Voc. Education
3%

History
5%

Social Studies
2%

Business/Economics
5%

Foreign Language
34%

Source: Office of Technology Assessment
United States Congress

FIG. 4 Distance-learning courses.

expensive and because interference with terrestrial microwaves operating in the C band is reduced. Data from the 1988–1989 school year indicate that approximately 7% of school districts had satellite dishes.

In distance-learning applications, a program-origination facility uses a satellite uplink to transmit programming. The signal is then transmitted down to Ku-band, C band, or Very Small Aperture Terminals (VSAT) dishes at the school. Students in the receiving classroom often use a telephone link to communicate with the instructor. This voice link is sometimes augmented by an audio bridge that enables all participants to hear the audio interaction. However, students in different locations cannot see each other.

Terrestrially Based Distance Learning: Cable Television and Fiber-Optic Cable. Terrestrially based telecommunications systems provide a range of alternatives for distance learning. Cable television systems are widely available but are generally limited by their inability to switch a signal from point to point as required. Cable television systems are predominantly one-way transmission vehicles with limited interactive potential. They are, nonetheless, important additional sources of educational programming along the educational-television model.

Increasingly, educators and learners are exploring the public telephone network as the optimal vehicle for distance learning. The network's ubiquity and standards form the basis for a fully interactive, multimedia infrastructure that can connect any learner with any resource in any format at any time.

The public network is presently limited by bandwidth restrictions in the

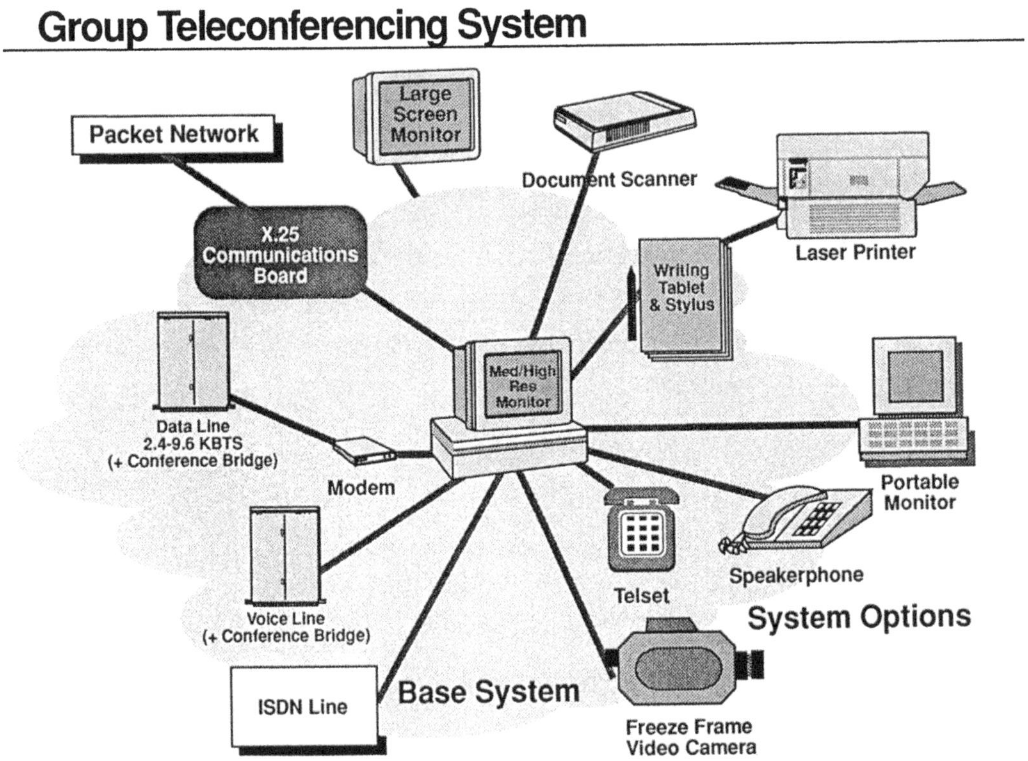

Group Teleconferencing System

FIG. 5 Base audiographics conferencing system and options. The basic and optional elements of an audiographics conferencing system are shown. Note that the system can be connected to the network by way of a basic voice line, a higher speed data line, an X.25 protocol line to a packet switch, and/or an ISDN basic-rate access line. The available transmission speed determines the rate at which data and images can be interchanged by the conferees.

local loop. Improvements in digital technology have enabled a migration from narrowband to wideband capabilities of the public network. Yet, even in the wideband environment, transmission rates available for distance learning are presently constrained by the 1.5 Mbps (T-1) limitation of copper distribution facilities.

Improvements in signal processing and encoding/decoding algorithms have resulted in the use of the public network for distance learning using T-1 or fractional T-1 circuits to transmit and receive video. Compressed interactive video, augmented by voice and data links, is an emerging method of distance learning and a prelude to broadband distance learning.

Fiber optics underpin advanced broadband networks capable of carrying high-speed data and full-motion video at transmission rates of 45 Mbps or greater.

In recent years, several applications of fiber optics in distance learning have been piloted. Some involve analog fiber (i.e., transmission of video without converting the analog signals to digital). More recent pilots involve fully digital transmission of integrated voice, data, and video over a single fiber-optic circuit. The more advanced pilots incorporate broadband switching of 45 Mbps (DS-3) signals. Figure 6 illustrates a distance-learning network that uses fiber optics, and Fig. 7 shows a distance-learning classroom.

Telecommunications Networks in Education

Specialized telecommunications networks for educational purposes have existed since the 1970s. The oldest and most developed are those sponsored by the federal government to link research facilities at major universities throughout the United States.

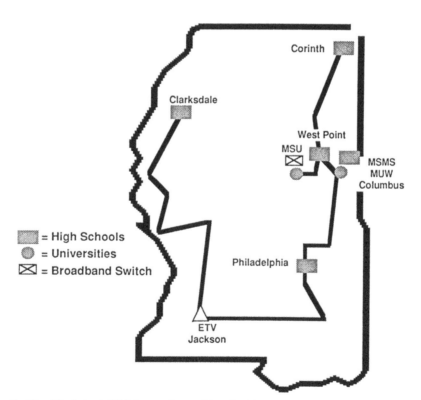

FIG. 6 The Mississippi 2000 interactive multimedia distance-learning network. This network encompasses some 700 miles of fiber-optic cable, 7 transmit/receive nodes, and a video switch capable of providing any-to-any connectivity between any combination of the 7 network nodes. This network is a prototype of the emerging broadband public telephone network. MSMS = Mississippi School for Math and Science; MUW = Mississippi University for Women; ETC = Mississippi Authority for Educational Television; MSU = Mississippi State University.

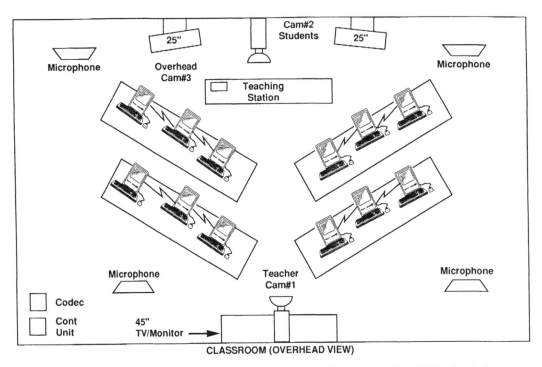

CLASSROOM (OVERHEAD VIEW)

FIG. 7 Mississippi 2000 distance-learning classroom (overhead view). Each high-school classroom on the Mississippi 2000 network is equipped with two stereo audio channels, video source equipment, video capture and display devices, and a local-area network of personal computers connected to a wide-area data network operating at 1.5 Mbps.

Research Networks

ARPANET (ARPA Computer Network) (3), the first academic research network, was started in 1969 by DARPA (Defense Advanced Research Projects Agency—called ARPA at the time). Merit, established in 1972, was the first statewide higher-education network serving colleges and universities in Michigan. CSNET (Computer Science Network), established in 1981, was intended for research in computer science and engineering, and BITNET, also established in 1981, continues to be one of the most successful and widely used academic networks.

Specialized library networks, including the On-line Computer Library Center (OCLC), established in 1971 and now international in scope, Washington Library Network (WLN), and Research Libraries Information Network (RLIN), provide bibliographic and other data to a large group of users (4).

After two decades of experience, general notions about what contributes to a successful computer network and what detracts have emerged. Users want convenience and control—networks must be accessible, easy to use, fast, and error and delay free. Academic users want to communicate with their peers and tap computing resources and databases.

The general shape of a new national data network was outlined at a meeting of National Science Foundation (NSF) networking specialists and higher-education representatives that took place soon after six supercomputing centers were established in 1985. What emerged was NSFNET—a communications backbone operating on 56-kbps lines leased from telecommunications companies and connecting the six supercomputing centers (see Fig. 8) (4). NSFNET has evolved from transmission speeds of 56 kbps to T-1. A planned upgrade to T-3 rates (44.736 Mbps) is currently underway.

Statewide Networks

Over the past two years, many states have initiated ambitious plans to construct educational networks—predominantly for distance learning, but also as a means to consolidate public-sector telecommunications traffic on a common infrastructure. Many of these plans call for the expansion of existing infrastructures to accommodate additional nodes. The most common evolution moves from networking institutions of higher learning to incorporating all elementary and secondary schools in a phased evolution.

Both Minnesota and Iowa lead the nation in the design and development of multiuser, multiagency statewide networks that will utilize the facilities of local-exchange and interexchange carriers augmented by privately owned fiber-optic and microwave facilities. Figure 9 illustrates TNT, a statewide network serving education in New York State (5).

The National Research and Education Network

A recent and important development in educational networks resulted from the 1989 introduction of the National High Performance Computer Technology Act (S1067, Gore [D-TN], 05/18/89) in both houses of Congress. The act calls for the establishment, by 1996, of a "national science and technology information infrastructure of data bases and knowledge bank accessible through the National Research and Education Network (NREN)." This network is to operate at speeds of up to 3 Gbps (gigabits per second) and is designed to accommodate the high bandwidth needs of supercomputer-to-supercomputer connectivity.

The planned NREN will link government, industry, and the higher-education community and is to be developed in close cooperation with the computer and telecommunications industries. Unlike NSFNET and earlier research networks, the NREN will encompass a broader range of applications and collaborators.

The NREN will capitalize on the latest advances in digital and fiber-optic technologies, and has been conceived as an interim capability until such time as "commercial networks can meet the networking needs of American researchers."

Issues in Educational Uses of Telecommunications

Educators and learners have developed and implemented a wide range of creative applications of telecommunications. As is the case with any evolutionary

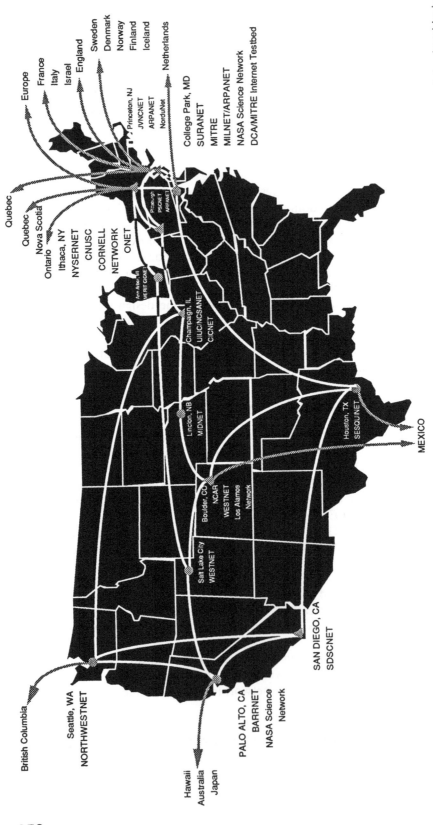

FIG. 8 NFSNET. NFSNET is a partnership of the National Science Foundation, the MERIT Computer Network (a consortium of state-supported universities in Michigan led by the University of Michigan), IBM, MCI, the state of Michigan, supercomputer centers, and midlevel regional networks.

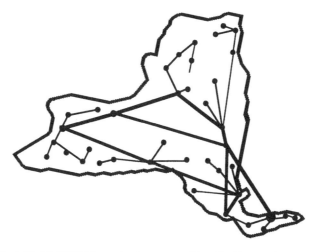

FIG. 9 New York State's TNT Network. At present, New York's TNT backbone network links the New York City Board of Education, the State Education Department, and 12 Regional Centers. The ultimate objective is to make TNT a comprehensive, integrated, and cost-effective communications network that provides all educators with the capability for electronic communications and document sharing, plus complete, easy access to resources and educational services.

technology, minor adjustments, major restructuring, and, at times, fundamental paradigm shifts result from the availability of new capabilities. This is certainly true as telecommunications becomes a foundation of the learning enterprise.

The explosion in telecommunications applications and their increasingly critical importance to the mission of educators and learners has raised several significant issues that affect the future course of developments both in telecommunications and education. Chief among these are technical limitations, equity concerns, and jurisdictional issues.

Technical Limitations

Certain technical limitations inhibit the wider use of some telecommunications applications in education. The most obvious limitation is the lack of state-of-the-art telecommunications systems and infrastructures to support teaching and learning. These limitations are a function of equity considerations, discussed below.

The most serious practical inhibitor to the fuller use of telecommunications for educators and learners is the limitation of the available radio spectrum. The shortage of radio spectrum has a particular impact on the ability of education to conduct distance learning using microwave technologies.

Closely related to the spectrum-limitation issue are limitations on the number of satellites and available transponder time available for distance-learning applications.

More globally, a primary technical limitation is the lack of adequate training available to teachers, administrators, and learners in the use of telecommunications technology to support their learning enterprises. Few schools of education have incorporated courses in instructional technology into their curricula, and even fewer focus on the potential of telecommunications applications.

Equity Concerns

Telecommunications applications are a proven means of enhancing the quality and availability of educational opportunities. Access to these applications, however, is a function of having the requisite systems and/or infrastructure in place where the application is needed. The ability to foot the capital cost of telecommunications systems and/or the expense of telecommunications services is a function of wealth and investment policies that do not necessarily take account of the educational need that might exist.

In planning the evolution of their networks, local and interexchange carriers assess the potential profit and/or expense savings to be gained from a given capital investment. These assessments generally lead to the placement of the most advanced network capabilities in areas of high commercial concentration inasmuch as business and industry is usually the early adopter of new telecommunications applications. Large cities benefit, yet smaller, rural towns and villages often wait years before these new capabilities are available locally.

Similarly, wealthier educational institutions are better able to fund the acquisition of telecommunications products or services that make these specialized applications possible. Poorer school districts or poorly endowed higher-education institutions are financially limited in their ability to embrace and exploit the possibilities.

As telecommunications applications permeate the educational environment, and as education policy makers and stakeholders increase their awareness of the potential role of telecommunications in the learning enterprise, it is reasonable to expect that an increased portion of funds will be allocated to telecommunications products and services.

It is also reasonable to expect that common carriers, with the support of enlightened regulators, will incorporate a "social benefit" criterion into their capital-planning analysis.

This criterion will assign a value to the role of telecommunications as a foundation for the learning enterprise, and should result in an accelerated and ubiquitous deployment of essential infrastructure technologies that directly address the needs of educators and learners. Clearly, the price of services made possible by this upgraded infrastructure must be within the means of all educators and learners.

Jurisdictional Issues

Formal education is predominantly a local affair. School boards and state departments of education certify teachers, curriculum objectives, textbooks, and other support materials.

Telecommunications, however, knows no boundaries. A learner in one state can tap the power of a network to take a course in another state. Using distance-learning techniques, a geometry teacher in Illinois might have students in seven other states. As telecommunications traverses geographic boundaries between learners and learning resources, the need for significant policy changes in jurisdiction over certification becomes apparent.

Conclusion

Telecommunications has permeated every aspect of our daily existence. This is certainly the case with education — perhaps the most communications intensive of human activities.

Telecommunications can be used effectively to automate the business of educational institutions. It can make electronically stored data available on demand. It can transport images and moving pictures into the learning environment. And, in its fullest effect, telecommunications can reorient education from a focus on formalized structures and teaching to a focus on self-defined, self-directed, lifelong learning enterprises.

Acknowledgment: CLASSNOTES is a registered service mark of BellSouth Corporation.

Bibliography

There are numerous papers and articles in industry and educational journals that describe creative uses of telecommunications applications. Several important works provide a primer for the uninitiated, and are highly recommended.

Bauch, J. P., The TransParent School: A Partnership for Parent Involvement, *Educational Horizons*, 187–189 (Summer 1990).

Carnevale, A. P., and Gainer, L. J., *The Learning Enterprise*, The American Society for Training and Development (February 1989).

Coates, J. F., and Jarrat, F., *Forces and Factors Shaping Education*, J. F. Coates, Washington, DC.

Colorado Educational Technology Consortium (CETC) Subcommittee on Telecommunications Systems, *Recommendations on Telecommunications Systems for Colorado K-12 and Teacher Education Programs*, Colorado Educational Technology Consortium, Denver, CO, 1988.

Planning for Telecommunications: A School Leader's Primer, U.S. West Communications, 1989.

U.S. Congress, Office of Technology Assessment, *Linking for Learning: A New Course for Education*, OTA-SET-430, U.S. Government Printing Office, Washington, DC, November 1989.

U.S. Congress, Office of Technology Assessment, *Power On! New Tools for Teaching and Learning*, OTA-SET-379, U.S. Government Printing Office, Washington, DC, September 1988.

References

1. Roesler, P., The Way We Were, *Telephony*, 13 (January 4, 1941).
2. U.S. Congress, Office of Technology Assessment, *Linking for Learning: A New Course for Education*, OTA-SET-430, U.S. Government Printing Office, Washington, DC, November 1989.
3. McKenzie, A. A., and Walden, D. C., ARPANET, the Defense Data Network, and Internet. In: *The Froehlich/Kent Encyclopedia of Telecommunications*, Vol. 1 (F. E. Froehlich and A. Kent, eds.), Marcel Dekker, New York, 1990, pp. 341–376.
4. IBM, *NSFNET—The National Science Foundation Computer Network for Research and Education*, IBM, Milford, CT, 1990.
5. State University of New York at Albany, *Long Range Plan for Technology in Elementary and Secondary Education in New York State: Using Technology to Pursue Excellence in Schools*, State University of New York at Albany, 1989.

DANIEL J. HUNT

Communication over
Fading Radio Channels*

Introduction

Considerable literature (1,2) has been devoted to phenomenological and statistical characterization of the rapid "wave interference" fading associated with time-varying propagation multipath. These changes in the channel transfer function occur within a time scale that ranges typically from fractions of a second to several seconds. The effects of this rapid fading influence the selection of appropriate methods of modulation, diversity, and coding, all conditioned by statistical models of the fading. These models and the predictions derived from them agree sufficiently well with experiments that they are widely and confidently accepted for system performance analysis, and laboratory "channel fading simulators" based on these models are generally accepted for flexible testing of new concepts or designs. This article provides an introductory review of these channel models and their parameters, and of how they are used in system design and analysis.

The effects of rapid fading on communications performance are usually described by the error-rate performance of specific modem and receiving-system designs as a function of mean signal-to-noise ratio (SNR) and of the linear distortions implied by the channel models. However, the channels also commonly undergo longer term variations in which both the actual structure of the channel (the kind of distortions that it causes) and the mean path attenuation change significantly. These changes are often relatable to solar or meteorological influences, with time scales of minutes, tens of minutes, or hours; in many cases, the long-term variations include components that are diurnal, seasonal, annual, or even involve the 11-year sunspot cycle. The wave-interference fading and the effects of channel structural changes are often called *short-term* and *long-term fading*, respectively, an inelegant and imprecise but intuitively convenient dichotomization of time scales of the changes being described. Both kinds of variation are of course continually in process. However, the distinction in time scales is extremely useful for engineering because, for most fading channels, only the short-term fading variations affect the details of received waveform structure and the interrelationships of errors within a message, while the longer term variations determine in effect the "availability" of the channel. Mathematically, the statistical model of the short-term fading usually can be regarded to be conditional upon the "instantaneous" values of parameters described by the longer term statistics.

The post–World War II flowering of statistical communication theory fit extremely well with the intense 1950s' interest in exploiting multipath fading

*Based on Fading Channel Issues in System Engineering by Seymour Stein, which appeared in *IEEE Journal on Selected Areas in Communications*, SAC-5(2): 68–89 (February 1987). ©1987 IEEE.

channels, particularly troposcatter and ionospheric high-frequency (HF) sky-wave, for long-distance radio communications. The result has been almost four decades of sophisticated modeling and analysis of fading multipath channels, and of widespread applications. Theory and experiment have progressed into a fairly mature and well-documented discipline that is routinely used by communications engineers to support new applications, enhanced performance, expanded requirements, and sometimes new designs to fit "newly discovered" fading channels.

The next section of this article outlines the best-developed part of this theory, the short-term, statistical, fading-channel models. The third section describes the ensuing understanding of reception effects for individual symbols in a message sequence and for messages as a whole. The third section also indicates the most significant recent advances for operation over fading channels, the use of adaptive equalization to eliminate intersymbol interference in high-speed serial data transmission, and the use of sophisticated coding schemes for error control and equivalent diversity effectiveness. In both cases, system design significantly involves questions of message durations and of channel parameters other than SNR, most importantly the rapidity of fading. Finally, the last section of this article discusses the issue of overall system performance in relation to the longer term variations, an aspect of meaningful system design that has primarily been honored by intuitive judgments, largely because it is so difficult to quantify. Among the difficulties are the lack of adequate databases (and the major costs of acquiring such data) to support extensive analysis or simulation. For HF skywave communications, for example, one would need a database on spatial (almost global) as well as temporal dependences in order to achieve realistic understanding of communication possibilities as related to opportunities for propagation connectivity.

Short-Term Characterization of Fading Multipath Channels

Channel Characterization as a Stochastic Time-Varying Filter

The usual line-of-sight propagation channel is characterized simply as a nondistorting two-port filter described by a gain A (attenuation) and by a propagation time delay τ_o. This is summarized as a delta-function impulse response of the form

$$h(\tau) = A\delta(\tau - \tau_o)$$

and a transfer function

$$H(f) = A \exp(-j2\pi f \tau_o).$$

This represents an all-pass filter over an infinite domain of frequencies, obviously an overidealization of any physical channel, but one that can be a perfectly adequate and correct description over the more limited bandwidth occupied by a communication signal. A multipath channel is simply one where energy arrives via several such paths, usually as a result of reflections, or of inhomogeneities in the physical medium that produce ray-splitting or ray-scattering effects. Continuous physical changes in the channel (motions or internal turbulence within the channel, or motion of either terminal of a link) cause small changes in the individual path lengths, but these may nevertheless equate to large electrical phase changes for the radio frequency. The variations between constructive and destructive interference resulting from the random phase changes comprise the effect called multipath fading, defined strictly speaking for unmodulated transmission (pure single-frequency sine wave). In phasor terms (see Fig. 1), the

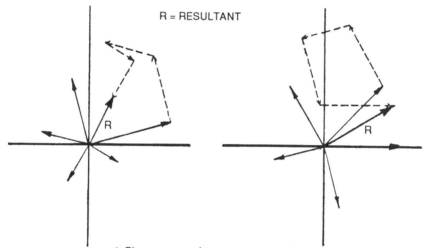

a) Phasors at one frequency, at two different times

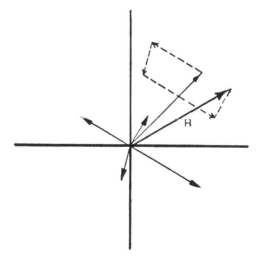

b) Phasor at second frequency

FIG. 1 Phasors due to multipath.

observed received phasor is a vector sum of several phasors, with the phase of each varying individually and randomly over a full $(0, 2\pi)$ range. Even with only a few paths of roughly equal intensity contributing, central-limit theorem arguments lead to the conclusion that the received waveform has all the characteristics of a very narrow band of stationary Gaussian noise, in other words that it consists of Gaussian quadrature components characterized by a power spectral density of nonzero width, and with a corresponding Rayleigh distribution of the received envelope. Fading that fits to this model is generically called *Rayleigh fading*.

In another view, the fact that a single tone transmitted gives rise to a received signal with a spectrum of nonzero width implies that the channel behaves like an agglomeration of frequency shifters. This should not be surprising, since any dynamic change in the channel transfer function (in particular, any phase modulation) can be interpreted as due to path-length changes, hence as having Doppler shift. In fact, one can quantitatively identify the power spectrum with an intensity distribution of apparent Doppler shifts. The width of the power spectrum of the Gaussian fading process is closely related to the rate at which fading is perceived on the received tone.

The physical channel most likely consists of paths that have considerable persistence and changing delay. For HF skywave, the ray path delays may even vary with frequency, hence are described as individually frequency selective when wide bandwidths are of interest. The mathematical model that is used more often is one of fixed delays with varying gain and phase at each delay. The ultimate justification for this model is that it is valid within the resolvability of delays corresponding to bandwidths of practical interest (1). It can also be justified by the sampling-theorem-based tapped delay line model discussed below. The discrete-path model is also often used to support channel simulations that do not include a full complement of uniformly spaced taps.

As mentioned above, the dynamic effects in many channels (HF skywave, troposcatter) are a characteristic of nonhomogeneous and time-changing natural media, but another significant cause of fading can be motion of the terminal in a static multipath environment. Line-of-sight propagation in the very high frequency (VHF) and ultra-high frequency (UHF) bands is an excellent example of the latter, with multipath due to multiple terrain reflections and to building reflections in urban or suburban areas. While some slow natural fading may occur due to foliage motion at the points of reflection, the bulk of the fading of interest in these channels occurs because of motion of one terminal during the course of a message, which again alters the constructive or destructive interference relationships among the multiple-path contributions.

The statistical model for the short-term fading assumes stationary statistics. That is, it assumes that the long-term variability is sufficiently slow that the channel statistics can be regarded to be sensibly fixed over some interval of engineering interest (typically minutes). For the mathematical purist, this defines a process that is "locally stationary" in time, realistically described by stationary process statistics. Accounting for the longer term variations is a less expert art. In some cases, analysts have convolved the two sets of statistics, or have treated one or more of the statistical parameters of the rapid fading as itself subject to slow change during time intervals of interest. Generally, however, the two sets of statistics relate quite differently to user requirements and are handled independently (see the final section of this article). Usually, only the short-term statistics are of detailed interest to the modem or coder designer, and fortunately the locally stationary model fits the actual behavior of most channels of interest quite well.

The channel model for the short-term fading is that of a linear filter whose parameters are time varying. The concept of a linear time-varying filter is simple so long as the variations occur slowly compared to the duration of any waveform being considered. We use complex envelope notation throughout (e.g., see Ref. 1). For the linear time-varying filter model of the channel, the channel impulse response $g(\tau; t)$ is represented relative to some nominal band center frequency f_c by its equivalent low-pass (complex) impulse response $h(\tau; t)$

$$g(\tau; t) = 2 \, \text{Re} \left[h(\tau; t) \exp \left(j2\pi f_c \tau \right) \right].$$

In these expressions, τ represents the usual filter response (propagation delay) variable, while the t-dependence indicates that the very structure of the impulse response changes with time (see Fig. 2). Correspondingly, the channel has a time-varying equivalent low-pass transfer function, which is the transform of $h(\tau; t)$ relative to its τ-dependence

$$H(f; t) = \int h(\tau; t) \exp \left(-j2\pi f\tau \right) d\tau$$

where frequency f represents offset from the nominal band center frequency f_c. At each frequency denoted by f, $H(f; t)$ is specifically the time-varying envelope and phase that is received with a single tone transmitted at that frequency, hence is the quantity characterized earlier as a zero-mean complex Gaussian process as an interpretation of central-limit behavior.

If a transmitted waveform has a complex envelope $u(t)$ and a corresponding

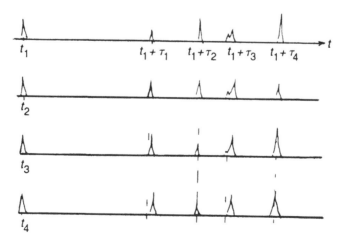

FIG. 2 Example of successive responses for discrete multipath.

spectrum $U(f)$, the waveform received after propagation over the time-varying channel has the complex envelope

$$z(t) = \int h(t - \tau; t) u(\tau) d\tau$$
$$= \int \exp(j2\pi ft) H(f; t) U(f) df.$$

That is, the filter is linear and superposition applies for all signal components. However, note that the time dependence of the channel structure enters into the time dependence of $z(t)$. The transform of $z(t)$, defined over an infinite time span, would include the effects of the channel variations. On the other hand, with the channel varying slowly, we can identify meaningfully a "short-term" spectrum $Z(f; t)$ given by

$$Z(f; t) = H(f; t) U(f).$$

"Meaningful" means that this expression can be used in analysis of waveforms that extend only over intervals within each of which there is in fact negligible structural change in the transfer function because of fading. Otherwise, only the integral interpretation has realistic meaning.

Since we can describe the channel's structural variations only as a stochastic process, the time dependence in characterizing the channel as a time-varying filter really describes a particular realization of the process that will enter into mathematical analyses only in the form of statistical correlations. Specifically, as already noted, $H(f; t)$ is the multipath channel response to a pure sine wave at relative frequency f, for the case where $U(f)$ is a delta function. The same argument that led to characterizing $H(f; t)$ as a complex Gaussian process is easily generalized to the statement that if multiple values of f are considered, the transfer function defined over multiple values of f is a higher dimensional, joint complex Gaussian process.

It is important to note that the central-limit characterization is strongest around the value zero for each of the quadrature components, namely, around the region where the value of the envelope is zero. The first-order probability density function (pdf) of a stationary complex zero-mean Gaussian process (complex envelope)

$$z = x + jy$$

is described by the bivariate Gaussian distribution (see Fig. 3-a)

$$p(x, y) = \frac{1}{2\pi S} \exp\left(-\frac{x^2 + y^2}{2S}\right)$$

where the expected value

$$S = E\left[\tfrac{1}{2} \mid z \mid^2\right]$$

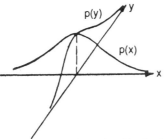

a) Bivariate Gaussian Distribution

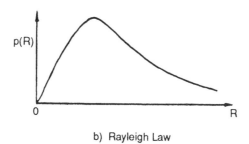

b) Rayleigh Law

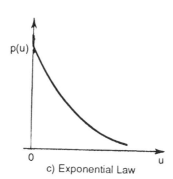

c) Exponential Law

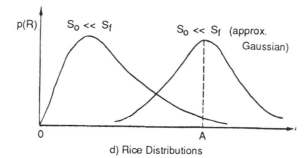

d) Rice Distributions

FIG. 3 Fading distributions.

is the mean power in the random waveform. The envelope and phase of the waveform are

$$R = \sqrt{x^2 + y^2} \qquad \theta = \tan^{-1} \frac{y}{x}$$

whose joint first-order pdf is immediately observed to decompose into independent pdfs

$$p(\theta) = \frac{1}{2\pi} \qquad 0 \le \theta < 2\pi$$

(uniform distribution)

$$p(R) = \frac{R}{S} \exp\left(-\frac{R^2}{2S}\right) \qquad 0 \le R < \infty$$

(Rayleigh distribution, see Fig. 3-*b*).

Associated with the distribution of the envelope is the pdf for its mean-squared value or *instantaneous power* as defined by $u = R^2/2$,

$$p(u) = \frac{1}{S} \exp\left(-u/S\right) \qquad 0 \le u < \infty$$

(exponential distribution, see Fig. 3-*c*).

The statement above about strong approach to the central limit corresponds to a statement that the Rayleigh law for the envelope is likely to be most correct for the low values of R, near $R = 0$. As it happens, this also characterizes the domain of most concern for signal and receiver design, describing the signal envelope when it fades deeply below its mean or median. For this region, a valid approximation to the Rayleigh pdf is

$$p(R) \approx R/S \qquad 0 \le R \ll \sqrt{S}$$

or

$$p(u) \approx 1/S \qquad 0 \le u \ll S.$$

This approximation in fact is sufficient to describe the asymptotic (low error rate) performance of most systems, as given below. The mathematical model of the channel as a complex Gaussian process is less credible at large values of instantaneous envelope (where it predicts an infinite range, whereas the physical process is usually bounded by inverse square law attenuation), but since large values of signal (large values of instantaneous SNR) are not stressful to performance, the discrepancy is not significant to the issues that drive communication system design for these classes of channels.

When there is a single, dominant, nonfading component in the received signal along with a fading process, the envelope statistics follow the Rice pdf (see Fig. 3-*d*)

$$p(R) = \frac{R}{S_f} \exp\left(-\frac{S_o}{S_f}\right) \exp\left(-\frac{R^2}{2S_f}\right) I_o\left(\frac{R\sqrt{2S_o}}{S_f}\right)$$

$$0 \leq R < \infty$$

where

S_o = power in nonfading component

S_f = mean power in fading component.

Correspondingly, the phase is no longer uniformly distributed, but rather is more concentrated around that of the nonfading component. However, if the latter is also changing (e.g., if there is a Doppler shift from the reference frequency relative to the phase being defined), a propagation experiment with statistics accumulated over a long enough interval may show an apparent uniform phase distribution with a Rice envelope distribution.

Some researchers have suggested that envelope statistics on some fading channels fit better to a Nakagami-*m* distribution with $m \neq 1$ than to a Rayleigh or Rice distribution (Nakagami-*m* is a generalization of the Rayleigh, with Rayleigh as the special case $m = 1$). There is no intuitive explanation for the claimed distributions; moreover, all the original claims for the Nakagami distribution were based on fitting envelope statistics around the mean or median, rather than the near-zero region that we assert to be fundamental to system performance over fading channels. Therefore, the practical utility of results based on a Nakagami-*m* pdf appear questionable for modem design purposes, and they are not further discussed in this article.

Covariance Functions

Given that $H(f; t)$ is a complex Gaussian process, a *spaced-tone complex covariance* can be defined to characterize the relationships of values of the transfer function at different frequencies and at different times:

$$R_F(f_1, f_2; \Delta t) = E\left[\tfrac{1}{2}H(f_1; t)H^*(f_2; t + \Delta t)\right].$$

In writing this covariance as a function of only the time lag Δt, we are explicitly assuming that the short-term fading is a wide-sense stationary process. At the same time, one expects physically that, over any radio band, the spaced-tone relationship will depend only on the frequency difference $\Delta f = f_2 - f_1$, and not on the actual value of frequency, that is, that the channel is also wide-sense stationary across frequency. (This is a "narrowband" assumption that eventually breaks down as one considers use of very wide bandwidths on any channel. The impulse response characterization that follows below provides perhaps a more intuitive indication of what might cause a breakdown in the assumption.) Assuming that the spaced-tone covariance depends only on Δf, we can write

$$R_F(\Delta f; \Delta t) = E\left[\tfrac{1}{2}H(f; t)H^*(f + \Delta f; t + \Delta t)\right].$$

Its normalized value is

$$\rho_F(\Delta f; \ \Delta t) = \frac{R_F(\Delta f; \ \Delta t)}{R_F(0; 0)}.$$

The selectivity of fading (i.e., the extent to which a realization of the channel transfer function is frequency dependent) is well characterized by $\rho_F(\Delta f; 0)$. Over a bandwidth B for which $\rho_F(B; 0)$ is near unity, all signal components fade "together," and $H(f; t)$, in effect, is an all-pass filter (for that bandwidth) with a varying amplitude and gain that applies equally to all the signal spectral lines

$$\rho_F(B; 0) = 1 \Rightarrow$$

$$H(f_1; t) = H(f_2; t) = \alpha(t) \text{ for all } f_1, f_2 \text{ such that } |\ f_1 - f_2\ | \le B$$

where $\alpha(t)$ is a complex Gaussian process (in its time variations). This has sometimes been termed the *flat-fading* or *nonselective fading* channel since it implies no frequency selective distortion by the channel. For a complex envelope $u(t)$, transmitted over such a channel, the received complex envelope is

$$z(t) = \alpha(t)\ u(t).$$

This kind of channel has also been called a *multiplicative fading* channel. Note in contrast that for totally independent fading at spaced frequencies, as is desirable for frequency diversity (see the third section of this article), one would want to use a wider spacing Δf such that $\rho_F(\Delta f; 0)$ is near zero.

The time-varying impulse response $h(\tau; t)$, as the transform of a wide-sense stationary Gaussian process, is also wide-sense stationary. It can be shown (1) that the wide-sense frequency stationarity of $H(f; t)$ is *precisely* equivalent to a statement that the impulse response changes independently at each different time delay

$$R_T(\tau_1, \tau_2; \Delta_t) = E\left[\tfrac{1}{2}h(\tau_1; t)h * (\tau_2; t + \Delta t)\right] = R_T(\tau_1; \Delta t)\ \delta(\tau_2 - \tau_1)$$

where

$$R_T(\tau; \Delta t) = \int \exp\left(-j2\pi\tau\Delta f\right)R_F(\Delta f; \Delta t)d(\Delta f).$$

The value $R_T(\tau; 0)$ is the mean strength (intensity) of the channel versus delay time, a "multipath profile." The notion that fading is independent at each delay (actually, at each resolvable delay) is physically reasonable for most channels. The transform relation between R_F and R_T, involving the variables τ and Δf, shows that the selectivity of the channel as measured by the width of R_F versus

Δf is inversely reciprocal to the width of the channel response, the value described by the multipath spread T_M. This is a significant rule of thumb for system design. If we consider simple symbol waveforms, for which bandwidth is roughly the reciprocal of the duration T, an equivalent statement for observing only nonselective fading over each symbol is

$$T \gg T_M.$$

Finally, the Δt dependence of any of these correlation functions describes the time dynamics of the processes, which can also be interpreted in terms of a power spectral density by a further transform on the Δt variable. A particularly significant physical characterization arises from the transform of R_T,

$$S(\tau; \lambda) = \int R_T(\tau; \Delta t) \exp(-j2\pi\lambda\Delta t) d(\Delta t).$$

$S(\tau; \lambda)$, called the *scattering function*, is the power spectrum of all the contributions at delay τ [or in the delay interval $(\tau, \tau + d\tau)$], hence is the relative strength at delay τ of all scatterers that cause a relative frequency shift in the range $(\lambda, \lambda + d\lambda)$. This is a direct physical characterization of scattering that can be used to predict the nature of the fading when the physical motions (delay and velocity distributions) of reflectors or scatterers, and their scattering or reflection cross sections, can be estimated from physical principles. For example, the scattering function for reflection is the primary measurement objective in radar astronomy using active probing signals, where it is described in range/ Doppler shift terms.

For an unmodulated sine wave transmitted at relative frequency f, for which the received complex envelope is $H(f; t)$, the complex autocorrelation of the waveform received at frequency f is $R_F(0; \Delta t)$, and its power spectrum is

$$V_F(\lambda) = \int R_F(0; \Delta t) \exp(-j2\pi\lambda\Delta t) d(\Delta t)$$

$$= \int R_T(\tau; \Delta t) \exp(-j2\pi\lambda\Delta t) \, d\tau \, d(\Delta t)$$

$$= \int S(\tau; \lambda) \, d\tau.$$

That is, the observed power spectrum is a sum of the spectra contributed by all the delays. This spectrum, for a sine wave transmitted, is the so-called *Doppler spectrum* of the channel, and its nominal width is termed the *Doppler spread* B_D. The rate of Rayleigh envelope fading across the median envelope level is actually proportional to a specific definition of *rms bandwidth f_N* of $V_F(\lambda)$:

$$f_N = \left[\frac{\int \lambda^2 V_F(\lambda) \, d\lambda}{\int V_F(\lambda) \, d\lambda} \right]^{1/2},$$

fade rate $= 1.475 f_N$ (average rate of downgoing crossings of median level).

The fade rate is frequently measured in propagation studies, since it can be easily instrumented or visually extracted from chart recordings. Whether the power spectrum of interest is $V_F(\lambda)$ as integrated over all delays, or whether the signal's bandwidth is used to resolve the delay spread, obviously depends in detail on signal characteristics.

For the single sine wave transmitted, coherent integration in reception is possible only over time spans Δt for which $R_F(0; \Delta t)$ remains near unity. The width in Δt of $R_F(0; \Delta t)$ is therefore loosely termed the *coherence time* of the channel. Clearly it is reciprocally related to the Doppler spread B_D, the nominal width of the corresponding power spectrum. For a simple symbol waveform of duration T to be processed coherently requires

$$T \ll 1/B_D.$$

In order for it to be reasonable to process a symbol waveform with a filter matched to the waveform as transmitted, two separate criteria must be satisfied: one, that there be essentially no loss of coherence over the symbol as received; the other, that there be essentially no frequency selective distortion of the symbol. This leads to a double inequality

$$T_M \ll T \ll 1/B_D.$$

In turn, both inequalities can be satisfied simultaneously only if the *spread factor L* defined by

$$L = T_M B_D$$

is very small

$$L \ll 1.$$

Most natural channels in use are underspread, satisfying this inequality, which allows relatively simple receive waveform processing provided the symbol keying rate used, $R_s = 1/T$, satisfies both inequalities. A roughly optimum choice of keying rate from this viewpoint is

$$R_s = \frac{1}{T} = \sqrt{\frac{B_D}{T_M}}.$$

This does not negate system design for data rates that violate one or the other of the inequalities. In those cases, useful signal design choices are more restricted, and receive system processing generally cannot realize matched filter performance. Indeed, some channels are *overspread*, $L \geq 1$, or are sufficiently close to unity spread factor that there is no choice but to accept such limitations.

Another system implication of the spread factor is that unless the channel is underspread, the channel changes state (in a time of order $1/B_D$) during the T_M seconds that it takes for an energy packet to traverse the channel. This has implications for system designs that attempt to use estimates of the channel state to optimize receive processing and performance, again not precluding the possibility of making or using such estimates, but complicating the process and reducing the performance advantages from those available for an underspread channel.

The characterization given above describes the degree of distortion of an isolated symbol. Continuous communication involves sequences of symbols and side-by-side channel-frequency allocations so that we must also be concerned with the time smear that may cause intersymbol interference or the frequency smear that may cause adjacent channel interference, respectively.

The same criterion that defines nonselective fading $T \gg T_M$ also serves to make intersymbol interference negligible, in the sense that smear of symbol energy extends at most over a small leading portion of the next symbol interval. For each received symbol, there is an interval of duration $T - T_M$ that contains most of that symbol's energy and that is free of intersymbol interference. The simple expedient sometimes adopted in a modem design is to use filters, prior to the decision processes, that have an impulse response selected to be shorter than $T - T_M$; with proper synchronization, this assures decision voltages free of intersymbol interference. This approach is sometimes described as using a *guard time* of duration exceeding T_M in the demodulation processing. On the other hand, designs that can cope with the intersymbol interference that occurs when T is smaller than T_M have also become prominent in fading-channel modem design, motivated by considerable other benefits that are attributed to using single high-speed serial streams to carry the desired traffic rather than multiple parallel slower speed streams. This has led, first in troposcatter and more recently in HF modems, to the development of adaptive equalizer systems that can mitigate the effects of considerable intersymbol interference in operation over continuously fading channels.

The dual problem of adjacent-channel interference caused by the frequency spread implied by the Doppler spread B_D has been of considerably less concern in system design primarily because the values of B_D tend to be quite low (in the fractional hertz to the tens-of-hertz range) compared to keying rates (signal bandwidths) of practical interest. As a result, the guard spacing between channels allocated for independent traffic is controlled largely by the spectral shaping of the transmitted signals rather than by Doppler spread. Note that this is a statement on Doppler spread, not on the allowances that might be needed to compensate for overall mean Doppler shift in systems involving high-speed moving platforms.

Parameters of Typical Channels

Ionospheric Skywave HF

Multipath arises from paths with different numbers of multiple reflections between earth and ionosphere (multiple-hop paths), from paths at multiple elevation angles connecting the same endpoints that occur because of the electron-density variation versus altitude (high or low rays), from magnetic-ionic effects causing polarization-dependent paths (ordinary and extraordinary rays), and from the natural inhomogeneity of the ionosphere layers. The natural inhomogeneities cause typically 20–40 μs multipath spread of each path or mode, and the high/low and ordinary/extraordinary rays result in paths spread by 100–200 μs, although all such paths in a given mode tend to coalesce when operation is at frequencies very close to the actual maximum usable frequency (MUF) for that mode. For single-hop links over medium ranges (800–2000 km), a maximum multipath spread of the order of 100 μs is not unusual. Because all paths in that case are via the same reflection area, there tends to be no major Doppler spread differences and the Doppler spread often is under 0.01 Hz (very slow fading, with time stabilities exceeding 100 s). The more notorious multipath spreads for HF, usually in the 1–2 ms range, occur either for short ranges (near vertical incidence) under about 800 km due to delayed energy arrivals via repeated earth-ionosphere reflections, or over long paths, 2000–10,000 km, that require two or more hops. In the latter range, a variety of modes can occur, involving different numbers of hops and including reflections between layers; for very long paths, even longer multipath spreads can be encountered, out to 5 ms or more. Moreover, because different Doppler shifts are often encountered at the different reflection points, the overall Doppler spectrum is often multi-lobed (channel scattering function, see Fig. 4). The overall Doppler spread, controlled by the Doppler shift differences, can range up to 1–2 Hz. These long skywave paths are the ones for which most HF modems are designed, usually for multipath spreads out to about 3 ms and Doppler spreads out to 2 Hz. However, paths with reflection points in or near the auroral belt can have greater multipath spread (5–10 ms) and significantly greater Doppler spread (10–20 Hz). Predictions on propagation via skywave after a high-altitude nuclear detonation give similar (or larger) values. Note that these latter paths have a large spread factor, although still below unity; they are likely to require significantly different designs for worst-case operation than the more traditional HF

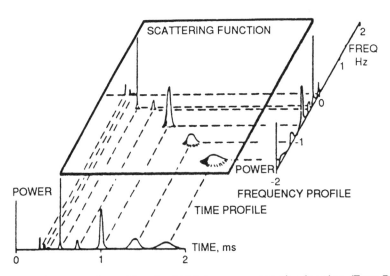

FIG. 4 Time-frequency plane of long-haul high-frequency scattering function. (From Ref. 3.)

modem designs intended for use on channels with a spread factor of order 0.01 or less.

High-speed aircraft terminals can contribute additional Doppler effects on all paths (Doppler shift up to 10 Hz at 10 MHz for Mach 1 speed) and will cause some broadening of Doppler spectra because of the different takeoff angles for different ray paths. However, neither geometric calculations nor reported experience suggest a significant change in Doppler spread, except perhaps for paths that otherwise would exhibit extremely slow fading (values of B_D under 0.1 Hz).

As indicated above, high Doppler spreads in ionospheric skywave propagation tend to be associated with large multipath spreads, and thus with longer distance propagation. Low Doppler spreads can also occur with high multipath spreads, but there is no apparent mechanism for the converse (high Doppler spread but low multipath spread). There is also little evidence of any significant association of changes in path loss (signal attenuation) on a particular link with changes in multipath parameters on that link. Intuitively, attenuation conditions on any single propagation path are not likely to be coupled to the existence or nonexistence of other paths reflecting additional energy. Thus, one might expect at best a weak monotonic coupling between SNR and multipath spread. Moreover, any such effects might well be masked operationally by the common use of frequency-changing procedures as diurnal changes occur.

If we consider a "typical" conservative HF modem design, for $T_M \leq 4$ ms and $B_D \leq 5$ Hz, the optimum keying rate suggested above for distortionless reception is

$$R_s \approx \sqrt{\frac{5}{0.004}} = 35 \text{ baud.}$$

Early HF operation (Morse, low-speed telegraphy) is indeed consistent with such a value. However, demands for digitized speech and higher speed data channels lead to requirements for values like 1200 or 2400 bits/second (bps), and desires that extend to as high as 9600 bps. Since the early 1960s, the 1200/2400 bps requirement has been filled (with various views of success in meeting requirements for high availability) by using multiple parallel tone modems within a 3-KHz allocation. The best known of these uses 16 parallel tones, spaced 110 Hz apart, and keyed synchronously at 75 baud with four-phase keying to realize 2400 bps transmitted (often, however, operated at a lower throughput rate, with the redundancy utilized for coding or diversity as discussed in the next main section of this article). A major limitation in use of such a modem, aside from its complexity vis-à-vis a comparable speed phone-line serial modem, is the requirement for a linear-transmitting amplifier to avoid intermodulation effects among the multiple streams. The peak-to-rms ratio for 16 tones is about 12 dB, and it is common to operate with an rms 7–10 dB below peak rating (thus compromising by allowing a tiny amount of intermodulation); with respect to amplifier designs, this derating is often regarded as a real loss in available system margin. In the early 1980s, successful operation was shown for 2-, 4-, 8-, or possibly even 16-phase-keyed, constant envelope serial data at 2400 baud, with adaptive equalizer demodulation (this topic is explored further in the next main section of this article). Considerable success in designing parallel-tone modems has also been achieved by including a significant amount of error-control coding in the transmitted stream (with concomitant increase in the transmitted data rate). Coding that results in usably low throughput error rates even at channel error rates of the order of 0.01 has been shown capable of handling the errors caused by intermodulation distortion that results from clipping the composite waveform, thereby circumventing the need for much transmit amplifier derating.

Tropospheric Scatter

Troposcatter involves a more continuous, but nonsymmetric, distribution of delays. The most commonly used links range from just over the horizon (or short links with horizons locally elevated because of major terrain obstacles) to links about 200 km long. For such links, the multipath spread usually does not exceed 0.1–0.2 μs, while at microwave frequencies the Doppler spread is in the 0.1–10 Hz range. The actual spreads are controlled by the beamwidths used (always narrow) and by the actual geometry of the "common volume" intersection of the transmit and receive beams. Aircraft echoes occasionally cause higher fade rates ("flutter fading") of 100 Hz or more due to the additional Doppler shift. A typical scattering function is shown in Fig. 5.

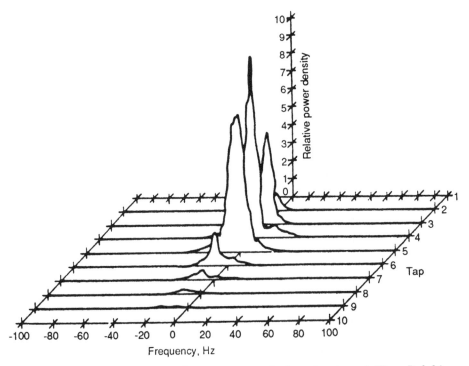

FIG. 5 Scattering function of a medium-range tropospheric scatter channel. (From Ref. 2.)

Considerably longer links have been implemented experimentally and for special applications. The multipath spreads can then extend to a few microseconds, although Doppler spreads tend to remain the same.

The attenuation mechanism associated with forward scatter is highly angle dependent. As a result, the intensity versus delay tends to be highly peaked to the earliest delays, and the effective multipath spread is usually the width of that peaked function. There is some indication that a significant weakening of the early delay return is associated with occurrence of a lowered mean SNR on the fading channel, coupled with a noticeable increase in the apparent multipath spread.

For $T_M \approx 0.1$ μs and $B_D = 10$ Hz, an optimum keying speed would be about

$$R_s \approx \sqrt{\frac{10}{0.1(10)^{-6}}} = 10 \text{ kbaud}$$

This is well below the keying rates desired in use of troposcatter, which has always tended to be regarded as handling the traffic at least equivalent to a low-capacity microwave line-of-sight relay. Fortunately, because of the very low spread factor, it is possible to key without distortion at much higher rates—hundreds of kilobaud—when T_M is realistically no more than about 0.1 μs. Up to the 1980s, almost all troposcatter used FDM/FM (frequency division multiplexed basebands, with frequency modulation of the carrier), with analog multichannel telephony basebands. More recently, fully digital troposcatter radios have appeared, including some that use adaptive equalizer designs to allow keying speeds up to 6.5 Mbaud, 4-phase (4–6), or, in another instance, using data-rate adaptation (downward in factors of two) from maximum speeds of about 2.5 Mbaud (7).

Scintillation on Earth/Satellite Channels

Unusual conditions in the natural ionosphere can result in significant scintillations on earth/satellite links. These effects are highly frequency dependent $(1/f^3)$, hence, most significant at VHF and low UHF frequencies. Primarily, they involve paths with ionospheric penetration in the local nighttime equatorial zone (latitude below 15°), although occasional effects also occur in the geomagnetic polar zone. In mild scintillation, the channel probably follows a Rice model (although many propagation researchers believe that the envelope distributions are better fit by Nakagami-m distributions) but has been observed to be purely Rayleigh fading at times of strong scintillation. The Rayleigh fading component has a narrow multipath spread, perhaps as small as 0.01 μs (hence a large coherence bandwidth), while fading rates are in the fractional hertz to few hertz range.

The fade rates appear to be governed primarily by movement of the path's ionospheric penetration point across the ionosphere, resulting in a fade cycle that corresponds to movement over a scale distance roughly about 1 km. Hence, fade rates during scintillation events depend on whether the satellite is in low orbit versus nearly geostationary, and whether the terrestrial link is fixed or on an airplane.

Considerable military interest in the scintillation phenomena results from predictions that a high altitude nuclear detonation would result in considerably more intense phenomena of the same type in a large region surrounding the detonation in the hours following the blast. At sufficiently high frequencies, as ionospheric absorption effects decrease enough to allow renewed link operation, considerable scintillation is projected, with the fade rates themselves gradually slowing down. In many cases, fully developed Rayleigh fading is projected over time spans of interest, with fade rates varying downward from the KHz range.

Terrain Multipath in VHF/UHF Channels

The VHF and lower UHF bands are used extensively for both military and civil nondirectional communications, most recently in the introduction of mobile telephone networks in the 800–900 MHz band (in the United States). Terrain multipath, including built-up structures in urban areas, can be considerable. The result is a kind of spatial, stochastic standing-wave pattern with a correlation width ranging from one-half of a radio-frequency (RF) wavelength to a few or at most a few tens of wavelengths, depending on details of geometry. Motion of a receiving platform through this pattern results in a fading received signal, typically showing the full characteristics of Rayleigh fading. In some high-resolution propagation tests (0.1 μs resolution) at 1 GHz (8) over a variety of terrains, the multipath spread was found to be up to 10 μs with 1–3 μs most common, and with upward of 10 significant resolvable paths often found. Similar results have been found in tests to support mobile telephony (9,10), with an indication that suburban or rural areas tend to have lower multipath spreads, sometimes under 1 μs.

The fade rate in these applications is directly related to the speed of the moving terminals, with the Doppler spread a fraction of the Doppler shift as determined by the angular spread of the multipath. For a velocity of $V = 20$ meters per second (mps) (about 45 mph) and a frequency of 800 MHz ($\lambda = 0.4$ m), the peak shift (V/λ) is about 50 Hz. One set of results (9) suggests a minimum coherence time (correlation coefficient of 0.5) of 1.3 ms, which is roughly consistent with this estimate.

Underwater Acoustic Communications

Underwater acoustic communications links have been suggested for tactical naval applications over ranges of 10–100 miles. At the lower frequencies (under 5 KHz) that allow such ranges, multipath spreads (typically direct path plus bottom bounce or multiple bounce) are several seconds, while the Doppler spread often exceeds 1 Hz. This is generally an overspread channel, requiring signal designs that involve noncoherent waveform processing (e.g., frequency shift keying with energy detection).

Microwave Line of Sight

High-speed microwave line-of-sight digital links with 20–40 MHz bandwidths (symbol keying rates of 15, 22.5, and 35 Mbaud are common with high-order alphabets) were widely introduced in the 1980s. In temperate latitudes, meteorological conditions can occasionally introduce refractive multipath (i.e., splitting of the direct ray into two or three rays) within a few nanoseconds overall spread, with performance degraded below toll quality unless the effects are compensated. Generally, the SNR margins on these systems are sufficiently high that the primary correction needed is use of adaptive equalization to eliminate intersymbol interference, rather than to modify the modulation to be more fade resistant. While one also encounters diversity techniques on such links, the diversity is largely used to select the least-distorted branch rather than for any optimization of SNR.

Laboratory Simulation

A challenging practical issue in the engineering of equipment to cope with fading channel distortions is testing to validate the ability to meet performance requirements. The use of on-the-air testing is unsatisfactory for at least two reasons:

1. It is at best difficult to determine the actual characteristics on a channel simultaneous with on-the-air modem testing over that channel.
2. The worst-case distortions against which a modem is specified may occur only a small fraction of the time, and are not available on demand for scheduled testing.

The flexibilities needed are achieved instead by the use of laboratory simulators of fading channel conditions. In addition to testing engineered equipment, such simulators allow experiments to test new concepts under controlled statistical conditions.

The usual laboratory channel simulator expresses realizations of a short-term process with fixed (stationary-process) statistics — in other words, a run is usually specified by a multipath intensity versus delay, Doppler spectrum versus delay, and by a signal level or SNR. Probably most of the simulators built to date have been aimed at the narrowband HF channel (up to 4 KHz bandwidth). Such simulators can readily be implemented for real-time operation using current microprocessor technology. To test a modem, the modulator output is presented at some convenient intermediate frequency (IF), conditioned, and digitized at some rate usually under 25 kilosamples/s. All the channel simulation is then performed digitally, and the output is converted back to an analog waveform for presentation at IF to the demodulator circuitry. An example block diagram of such a simulator is shown in Fig. 6-*a*. Controlled levels of additive Gaussian noise or interference are frequently included within the digital circuitry, but sometimes are also added externally as analog waveforms. Simulation systems may also include the ability to generate selected waveforms digitally. The inclusion of separate magnetoionic components at each path delay

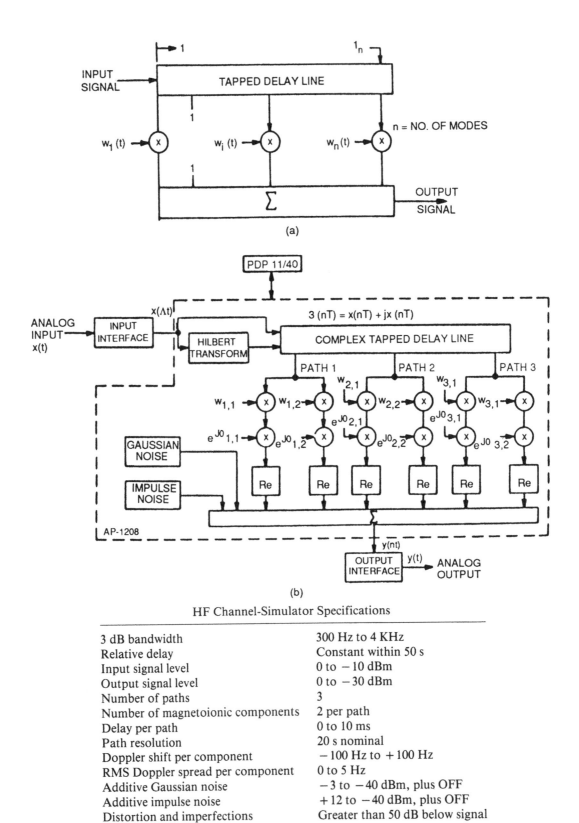

HF Channel-Simulator Specifications

3 dB bandwidth	300 Hz to 4 KHz
Relative delay	Constant within 50 s
Input signal level	0 to −10 dBm
Output signal level	0 to −30 dBm
Number of paths	3
Number of magnetoionic components	2 per path
Delay per path	0 to 10 ms
Path resolution	20 s nominal
Doppler shift per component	−100 Hz to +100 Hz
RMS Doppler spread per component	0 to 5 Hz
Additive Gaussian noise	−3 to −40 dBm, plus OFF
Additive impulse noise	+12 to −40 dBm, plus OFF
Distortion and imperfections	Greater than 50 dB below signal

FIG. 6 Fading channel model and simulator: a, block diagram of validated channel model; b, high-frequency channel simulator implementation. (From Ref. 11.)

shown in Fig. 6-*a* is a nicety for HF often not included for narrowband HF channel simulation.

Most of the simulators provide only single-channel operation. To test diversity modes of system operation (i.e., modes that require two or more independently fading signals, as discussed in the next main section) requires multiple simulators, adroit use of recordings, and the like.

For any significantly wider bandwidths, as might be of interest for troposcatter or for spread-spectrum operating modes, real-time operation often is no longer consistent with use of digital processing, although the limitation obviously changes with time. Analog simulators may be used (with all the additional problems of precision in control), or, as an expedient, the channels simulated may be very restricted in nature (while still hoping to mimic all the important effects), or recordings may be used on input and output to allow the simulator itself to be run much slower than real time. With digitally generated signals (and perhaps even a digitally simulated receiver), time scaling is relatively easily introduced.

Two approaches are used to model the multipath channel itself and, in fact, many practical simulators use a kind of compromise of the two for economic implementation. One approach is to model a specific set of relatively delayed paths with independent gain and phase controls for each path. This obviously fits well with the physical intuition about the HF skywave channel. For example, many of the early simulators used a fixed gain at each delay, and only a phase modulation to mimic the dynamically varying path length. For limited signal bandwidths that do not resolve the individual paths, the result is Rayleigh fading by the same central-limit arguments used earlier. However, the more modern simulators of this type introduce a multiplicative, independent, Rayleigh fading process on each path delay that more suitably simulates the fading caused by the multipath in a narrow delay spread around the nominal path delay for each mode.

The second approach, intuitively more consistent with representing a multipath continuum, as in troposcatter, uses a sampling theorem representation of the channel in the form of a tapped delay line with uniform tap spacings, a finite impulse-response (FIR) filter in signal-processing terms (see Fig. 6-*b*). Basically, if the channel response of interest is confined to some bandwidth W, the representation of the time-varying channel transfer function $H(f; t)$ can be confined to that bandwidth; in turn, the corresponding time-varying impulse response can be represented by complex samples taken $1/W$ apart

$$\hat{h}(\tau; t) = \sum \hat{h}\left(\frac{k}{W}; t\right) \delta\left(\tau - \frac{k}{W}\right),$$

and the response to a transmitted complex envelope $u(t)$ is given by

$$v(t) = \int u(t - \tau)\hat{h}(\tau; t)\, d\tau$$

$$= \sum u\left(t - \frac{k}{W}\right) \hat{h}\left(\frac{k}{W}; t\right).$$

The last equation is clearly reflected in the tapped delay-line model shown in Fig. 6-*b*. The time-varying complex tap weights

$$\alpha_k(t) = \hat{h}\left(\frac{k}{W}; t\right)$$

are each complex Gaussian in the Rayleigh-fading model. Although these tap weights are not strictly uncorrelated unless the multipath profile is slowly varying, the correlations are small and usually ignored.

In experiments with such simulators to test modem designs, it has generally been reported that performance measures (almost always bit error rate) are found to depend on SNR, overall multipath spread, and overall Doppler spread, but that they are generally not sensitive to the number of taps used to achieve those spreads so long as at least three taps are involved. For economy of implementation, use of three taps has tended to be the common practice (11,12).

When a channel is simulated by a tapped delay-line filter with taps spaced by T_s, represented by a set of coefficients $\{\alpha_k\}$, the impulse response and transfer function respectively are represented by

$$\hat{h}(\tau) = \sum \alpha_k \delta(\tau - kT_s)$$

$$H(f) = \int \exp(-j2\pi f\tau)\, h(\tau)\, d\tau$$

$$= \sum \alpha_k \exp(-j2\pi f kT_s).$$

The channel represented is clearly periodic with period $1/T_s$, and this periodicity persists even when the $\{\alpha_k\}$ are time varying (although the details of the transfer function obviously change). An interesting interpretation of such a filter is that the coefficients $\{\alpha_k\}$ represent the Fourier coefficients of the expansion of $H(f)$ over a fundamental bandwidth such as $(-W/2, W/2)$. Truncation of the filter to K coefficients thus implies that the fastest variation across frequency is at a rate K/T_s. When only three taps are implemented, the effective fundamental of the periodicity is the reciprocal of the greatest common divisor of the two successive spacings (using integer factors). Thus, if three taps are equally spaced by $T_M/2$ over a multipath spread T_M, the transfer function will be periodic over only $2/T_M$ Hz. While this could be adequate for testing single-tone modems with bandwidth considerably smaller than $1/T_M$, it could be inadequate for realistic testing of a multiple-tone modem. Hence, more likely, one would try to use relative tap spacings that have a low common divisor.

As a result of increasing interest in wideband HF designs, wideband simulators are likely to be built in the near future, and probably will be of the first type identified above (i.e., with selected tap spacings). In wideband HF, it is additionally pertinent to try to model the fact that skywave path length (delay) is itself frequency dependent (frequency dispersive) and this fits most easily with the selected tap approach. On the other hand, concerns about inadvertently using a channel model with periodicity within the bandwidth being tested are real, and will also have to be considered.

In modem procurements, acceptance testing of new modems for fading channel operation is usually based on operating a simulator at specific preestablished parameters.

Signal, Modem, and Receiving-System Design Issues

Error Rates in Single-Channel Fading

The complex Gaussian process model of the fading channel is a theoretician's delight because the received waveforms are complex Gaussian (although usually nonstationary because of the signal modulation) and theoretical analysis of receiving-system operation involves the relatively simple mathematics of a Gaussian signal in Gaussian noise. Indeed, direct derivations of probability of error are often much simpler for the fading channel than for the nonfading channel, a fact that has been exploited in many detailed analyses. Nevertheless, because results existed first for the nonfading channel, the more common presentation of theoretical performance results starts with probability of error for a nonfading signal in noise and averages it over the additional effects of fading.

A premise underlying the statement just made, but more importantly underlying much of the approach to design for fading channel operation, is that it is desirable, if at all possible, to select signal parameters (symbol bandwidth, keying rate) that allow essentially distortionless reception of individual symbol waveforms. As pointed out above, this is substantially possible for underspread channels by selecting a symbol waveform with roughly unity time-bandwidth product and for which the symbol duration T satisfies

$$T_M \ll T \ll 1/B_D.$$

For more complex situations, the approach is generally one of introducing adaptive circuitry that attempts to undo the effects of the distortions (for underspread channels), or that uses modem schemes that are substantially tolerant of the distortions, by accepting some "penalty" in the performance achievable versus SNR.

Keyed streams that do satisfy the inequality above have performance that can be well understood in terms of a slow, flat-fading channel where an individual symbol waveform is sufficiently narrowband (sufficiently long in duration) so that frequency selectivity is negligible in the fading of its spectral components, yet the waveform is also sufficiently short so that fading variations cause negligible loss of coherence within each symbol waveform as received. In consequence, the receiving system can be designed on the basis of optimal processing of the transmitted waveform (matched filters or some suitable approximation), just as in the nonfading case.

Such a receiving system makes its decisions based on the output of a linear filter processing an undistorted symbol waveform. The probability that a symbol error is caused by stationary additive Gaussian noise depends only on the instantaneous SNR associated with each symbol (i.e., the ratio of squared signal voltage at the filter output to the mean-squared value of the stochastic noise output). For the simplest cases of binary keying, these probabilities of error are expressed for the nonfading case in terms of the SNR γ at the filter output by one of two forms

$$P_b^{(1)}(\gamma) = \tfrac{1}{2} \exp(-\alpha\gamma)$$
$$P_b^{(2)}(\gamma) = \tfrac{1}{2} \operatorname{erfc} \sqrt{\beta\gamma}$$

where
   $\alpha = \tfrac{1}{2}$ for noncoherent orthogonal symbols such as noncoherent frequency shift keying (FSK)
   $\beta = \tfrac{1}{2}$ for coherent orthogonal symbols (e.g., coherent FSK)
   $\beta = 1$ for ideal coherent antipodal symbols (e.g., ideal phase-shift keying [PSK], coherently detected)
   $\alpha = 1$ for differentially detected antipodal symbols (e.g., differentiated PSK [DPSK])

Both forms result in an asymptotic exponential dependence $\exp(-\alpha\gamma)$ or $\exp(-\beta\gamma)$ at SNRs high enough to result in low error rates of practical interest. Similar dependences are found for higher order alphabets, but we use the simple binary cases for our examples. For the nonfading, additive Gaussian noise channel, the result is that a relatively narrow range of SNRs (from about 7 to 14 dB) covers the error rates of practical interest (10^{-3} to 10^{-6}) for all signal types, with small further reductions in SNR per bit (to perhaps as low as 3 dB) achieved by use of elaborate high-order alphabets or error-correction coding.

For the slow, flat-fading channel, the signal voltage for each symbol includes a Rayleigh-fading factor, constant over the symbol, hence appearing as a multiplication factor in the output voltage. Because of this fading factor, the per-symbol SNR γ is now itself a stochastic variable over the total ensemble of symbols received, and the previous nonfading formulas can simply be regarded as conditional probabilities of error, conditioned on the instantaneous SNR γ. In turn, the Rayleigh fading implies an exponential pdf for γ

$$p(\gamma) = \frac{1}{\Gamma} \exp(-\gamma/\Gamma) \qquad 0 \leq \gamma < \infty$$

where Γ is the ratio of mean signal power at the filter output (averaged over fading) to the mean noise power; this is just a restatement of the exponential distribution for the squared envelope, stated above. The result is that the overall averaged probabilities of error are given for the binary cases cited earlier by

$$P_b^{(1)} = \int_0^\infty P_b^{(1)}(\gamma) \frac{1}{\Gamma} \exp\left(-\frac{\gamma}{\Gamma}\right) d\gamma = \tfrac{1}{2}(1 + \alpha\Gamma)^{-1}$$

$$P_b^{(2)} = \int_0^\infty P_b^{(2)}(\gamma) \frac{1}{\Gamma} \exp\left(-\frac{\gamma}{\Gamma}\right) d\gamma = \tfrac{1}{2}\left[1 - \sqrt{1 + \frac{1}{\beta\Gamma}}\right]^{-1}$$

Asymptotically, for low error rates of practical interest, both formulas exhibit an inverse algebraic dependence between error rate and mean SNR, of the form $1/2\alpha\Gamma$ or $1/4\beta\Gamma$, respectively, instead of the exponential dependence on SNR found for the nonfading channel (see Fig. 7). If one were to try to operate a system in this manner, 10^{-3}–10^{-6} error rates correspondingly require roughly 30–60 dB mean SNR, a considerable impact on system design. However, it is readily shown that the reason for these results is the nonzero probability in Rayleigh fading of occurrence of signal values well below the rms (*deep fades*), for which the instantaneous SNR is small and for which the instantaneous probability of error approaches 0.5. In fact, the asymptotic results on error rate can be obtained by approximating $p(\gamma)$ just by its form for $\gamma \ll \Gamma$, $p(\gamma) \approx 1/\Gamma$. The occupancy of these lowest levels is inverse with the mean SNR, which is the inherent reason for the inverse dependence of error rate on mean SNR. As is

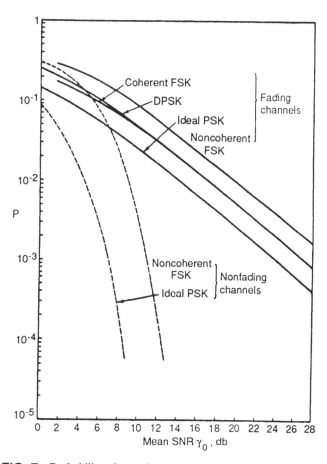

FIG. 7 Probability of error for several systems in Rayleigh fading.

shown in the section that follows, reducing the effective occupancy at low levels is at the heart of the techniques of diversity.

A series of papers (13–15) examined the over-idealizations inherent in the slow-flat fading model by analyzing the Gaussian process more rigorously. For example, it can be shown that instants of deep fading are accompanied by quite rapid and highly frequency-selective changes in channel amplitude and phase, such that the undistorted symbol model is not applicable during such instants even if it applies to the bulk of all symbols. Nevertheless, when interest lies in error rates like 10^{-3}-10^{-6}, distortion effects that occur on only 10^{-3}-10^{-6} of all symbols can all by themselves provide an undesired irreducible error rate. The analyses show that the probability of error involves additional distortion-related factors, such that the probability of error approaches a constant value for very high SNR (i.e., becomes independent of SNR), so-called error-rate bottoming. Typically, the additional factors involve the Doppler spectrum width or the multipath spread, normalized to the transmitted symbol duration. It should be noted that fading is not the only source of error-rate bottoming effects. Another familiar source is distortion caused by equipment non-idealities, which again leads to occasional errors independent of additive noise. Fortunately, the use of diversity also tends to ameliorate these bottoming effects (the irreducible error rate decreases significantly).

Diversity

In simplest form, diversity rests on the observation that if the same information is received redundantly, over two or more independently fading channels (*diversity branches*), there is a very much lowered probability that any individual symbol will be struck by a deep fade coincidentally in every one of the branches. With intelligent processing, each symbol decision can then be based primarily on the versions received at high SNR, thereby greatly reducing the overall error rate. For Mth-order diversity with equal mean SNR in each branch, the probability of error is essentially the Mth power of the probability of error for a single no-diversity fading channel. Hence, with Mth-order diversity, the probability of error varies with the mean SNR Γ on each individual branch as $1/\Gamma^M$. There are constants involved (functions of M) that make the improvement not quite so great. Nevertheless, the result is that for error rates of 10^{-3}-10^{-6}, where the single channel requires $\Gamma = 30$-60 dB, dual diversity ($M = 2$) brings the requirement down to about 15–30 dB per diversity branch, and quadruple diversity ($M = 4$) brings it down to the 8–15 dB range.

A variety of detailed methods has been used for diversity combining. While the essence of diversity lies in selecting the single best branch for each symbol decision, the decision SNR can be additionally improved by appropriately combining signal contributions from the various branches. Generally, with independent additive noise on each branch, the combiner weighting factors should include terms that are equivalent to using gain in every branch that brings the mean noise power in each branch to a common value. In bands where receiver noise is dominant and there is a similar receiver in every branch, this is often accomplished by using a single common automatic gain control (AGC) voltage for all receivers (*AGC lock*). When the channel fading is slow enough to obtain an estimate of the channel complex gain on each branch, the optimum combining in stationary, spectrally flat additive noise (after equalizing the noise levels) is to form a weighted sum at a common IF, in which channel phase is compensated (all branches are phase locked) and, additionally, the weighting amplitude is proportional to the channel voltage gain. This has been termed *maximal-ratio combining*, since it results in a decision SNR that is the sum of the SNRs on the individual branches (it is a direct extension of the matched-filter principle). It has been shown that just phase locking without the additional amplitude weighting is almost as effective (so-called *equal-gain combining*) when the contributions from the individual channels are of roughly equal mean strength. Phase locking necessarily is a form of predetection (IF) combining. At the other extreme is selection of the instantaneously best channel, as judged by some criterion such as AGC voltage being derived in each receiver, or by monitoring errors in known patterns embedded in the datastream (e.g., framing). Such selection often is implemented as *switching diversity,* in which rather than continually switching to the best, a branch is used until it falls below some quality criterion, and then another (better quality) branch is selected (if available). Selection or switching combining is often implemented after detection (*postdetection combining*); but if there is benefit to the designer in using a single modem, selection can be implemented at IF with the proviso that if the modem uses any phase-sensitive processing, the branches must be phase locked before combining (suggesting that a gradual switching action between the two branches may then be implemented). Neither the phase-locking predetection techniques nor the selection or switching techniques depend on signal structure, hence they can be used for analog signals as well, and indeed they have been the basis for implementing diversity with FDM-FM transmissions, notably in troposcatter.

For diversity combining of digital streams, matched-filter processing is often first carried out independently in each diversity branch using locked synchronization (adding delays in each branch if necessary), and the combining is then accomplished on the matched-filter voltage outputs, retaining amplitude and phase, or complex (*I and Q*) representation. For phase-coherent detection, as in ideal PSK using phase-locked loop carrier recovery, the branches must again be phase locked by compensating for the channel-phase variations, and again amplitude weighting may also be used for slight additional improvements in performance. For orthogonal waveform detection, like noncoherent FSK, phase-coherent combining of branch outputs is still optimal, if available. A separate matched filter is used for each waveform in the symbol alphabet, with separate coherent combining for each set of voltages that correspond to one of the symbols, and decisions are then based on envelope (magnitude) comparison of the combined values. However, orthogonal waveform keying is often used in fading channels because of the fear that fading may sometimes be rapid enough to preclude any phase-locked loop type of carrier recovery (estimation of channel phase). In such an application (or simply when the additional circuit complexity of phase-coherent combining is deemed not to be needed or warranted), phase-coherent combining across diversity branches is no longer applicable. Instead, it is assumed that there is no knowledge of individual channel phase. It has been shown for most cases that (again, with equal noise levels in all branches) the optimum combining then involves squared-envelope detection of each matched-filter output, and addition of the set of squared envelopes for each symbol to form decision voltages with the decision based on the largest resultant. This is so-called *square-law combining*. Its performance is intermediate between maximal-ratio (when available) and selection diversity. Figure 8 illustrates these relationships for binary FSK for $M = 1, 2, 4$ with equal strength diversity branches, including the asymptotic error probabilities (coherent FSK with maximal-ratio combining has exactly 3 dB poorer performance than ideal coherent PSK). For square-law combining with significantly unequal mean signal contributions in the several branches, the weighting factors involve these relative mean SNRs. This can be especially important in applications where many of the diversity branches often contain no useful signal, so that their inclusion without weighting can significantly degrade performance (cf. Ref. 10). Indeed, if the branch mean levels are varying dynamically in these cases, and cannot be tracked but only described statistically, the optimum diversity combining may involve nonlinear functionals of matched-filter output voltages in each branch. We also note the recent significant addition of the use of coding plus interleaving to achieve diversity performance with digital streams, which is discussed below.

Modern modem designs for fading channels, where phase-locked loop-carrier recovery is not expected to function well, often will use differential PSK (DPSK) rather than independent orthogonal-symbol keying. The ideal demodulator for DPSK can be considered as using a set of filters operating over successive pairs of received symbols, matched to the set of possible phase differences between the two symbols, with comparison of envelope detected output. Alter-

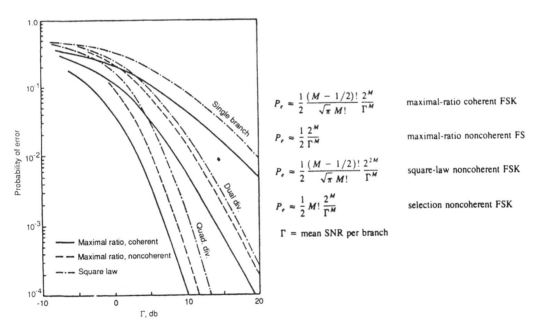

$$P_e \approx \frac{1}{2} \frac{(M - 1/2)!}{\sqrt{\pi} \, M!} \frac{2^M}{\Gamma^M} \qquad \text{maximal-ratio coherent FSK}$$

$$P_e \approx \frac{1}{2} \frac{2^M}{\Gamma^M} \qquad \text{maximal-ratio noncoherent FS}$$

$$P_e \approx \frac{1}{2} \frac{(M - 1/2)!}{\sqrt{\pi} \, M!} \frac{2^{2M}}{\Gamma^M} \qquad \text{square-law noncoherent FSK}$$

$$P_e \approx \frac{1}{2} M! \frac{2^M}{\Gamma^M} \qquad \text{selection noncoherent FSK}$$

Γ = mean SNR per branch

FIG. 8 Error rates with diversity combining; binary frequency shift keying. (From Ref. 1.)

natively, if z_1 and z_2 are the successive outputs of a symbol-matched filter, the phase of the product $z_1z_2^*$ can be shown to be an equivalent decision variable. The net result is that a combining rule for DPSK that corresponds precisely to square-law combining for orthogonal waveforms is to sum the scalar products $z_1z_2^*$ over all the diversity branches, and to make a decision according to the phase of the resultant.

Our introductory remarks on diversity emphasize that diversity primarily takes advantage of the nonconcurrence of deep fades on independently fading diversity channels. Since the deep fades are relatively rare phenomena in any case, it will be appreciated that relatively high correlations can hold for the broad course of fading on multiple channels, without losing diversity performance, because the deep fades are still not simultaneous. Thus, while theoretical analyses of diversity usually assume statistically independent fading on all branches, the practical applications do not require it. This results in a certain amount of vagueness in the statements, such as those below, about the physical criteria for establishing successful diversity.

The means for realizing diversity branches have included spaced antennas, crossed polarization, angle diversity, frequency diversity, multipath diversity, and time diversity.

Spaced Antennas

Historically (1929), the diversity principle was discovered using spaced receiving antennas for HF skywave reception. Typically, for HF and troposcatter, spacings of the order of 10–100 λ, where λ is the RF wavelength, have been found sufficient. With a single transmitting antenna, spaced receiving antennas actually increase the total mean signal energy captured, but this is not by any reckoning as significant for performance gain as the independence of occurrence of deep fades. Spaced transmitting antennas also provide diversity, but note that they cannot use the same frequency and polarization since the two signals would not be separable then into independent receivers at the receiving end. A familiar diversity configuration in troposcatter is quadruple diversity using two spaced transmitting antennas that have orthogonal polarizations, each received by two spaced dual-polarization receiving antennas. Commonly, to facilitate RF isolation at the receiver (and to minimize cross-polarization degradations), two different transmitting frequencies are used, and often it is not at all clear whether the mechanism providing diversity is the antenna spacing or, in large part, is the frequency spacing (see below). Indeed, for the two transmit–receive paths that cross at midrange, the use of different frequencies likely contributes significantly to independence of fading between them.

Vertically spaced receiving antennas are frequently used on microwave line-of-sight links to provide diversity protection against destructive interference due to ground reflections (under subrefraction conditions) or, in the newer high-capacity digital radios, to protect against the refractive-ray-splitting multipath that occasionally occurs. This latter application is sometimes termed *height diversity*.

Crossed Polarization

Most media are dielectric in nature, with little depolarization of a single transmitted signal. Correspondingly, crossed-polarization transmissions on the same path and same frequency exhibit highly correlated fading unsuited for diversity. However, on HF skywave *F*-layer paths, the magnetoionic polarization-rotation effects on individual modes tend to result in independently fading orthogonal polarizations for a single polarization transmitted. Cross-polarization reception therefore can provide dual diversity for such paths. This has not been used to any great extent at fixed sites, partly due to the difficulties of achieving high-gain beam patterns in crossed polarizations. However, polarization diversity is feasible when log-periodic or whip antennas are applicable. For a single, magnetoionic path, a transmitted linear polarization should result in receiving a slowly rotating linear polarization (*Faraday rotation*) with the appearance of periodic fading on a linearly polarized receiving antenna. A single such mode will result in anticorrelation of fading for signals received on cross-linear polarized antennas; however, this is not a form of multipath fading, and does not appear to be the polarization fading commonly observed. In microwave line-of-sight relay, the small pattern differences for horn antennas in the vertical and horizontal directions have been observed to result in sufficient independence of deep fades on the two polarizations so as to result in effective diversity against refractive multipath when both polarizations are transmitted.

Angle Diversity

In troposcatter, the angular sector for effective forward scattering (as viewed from either terminal) may be greater than the beamwidths available with large antennas, allowing use of multiple beams that traverse disjointed physical regions, hence resulting in independent scattering. The beams need be narrow only at the receiving end of the link. It is not believed that this method has been used in troposcatter other than experimentally. In the late 1980s, angle diversity was also introduced against refractive fading on line-of-sight microwave radio, with considerable success reported.

Frequency Diversity

Independently fading branches can be achieved by transmission and reception of the same information on multiple-carrier frequencies spaced far enough apart that fading is uncorrelated. For skywave HF, this often can be achieved on subcarriers spaced well within a 3 KHz allocation, where it is often termed *in-band frequency diversity*. It has been implemented either with parallel data modulation on spaced subcarriers, or in some cases by sequential (time-shared) transmission of the same information on spaced subcarriers. The use of wideband frequency hopping over fading channels, with the same information carried in different hops, also tends to result in independent states of fading at each hop frequency. The Soviet journal literature describes an elegant technique of composite or compound modulation for generating frequency diversity signals for troposcatter and other fading-channel applications (6). A periodic angle modulation is applied to the carrier, retaining its constant envelope character for transmission. However, the periodic modulation creates multiple subcarriers uniformly spaced around the carrier, with spacing equal to the fundamental of the periodicity. Superimposing information as additional angle modulation on the periodically modulated carrier implicitly imposes the angle modulation on each of the subcarriers (the technique has also sometimes been termed *stacked carrier*), while the overall transmitted waveform is still constant envelope. As described, spacings of a few megahertz would be used in troposcatter. The literature also describes an autocorrelation technique (*Akkord*) for bulk processing all the received carriers rather than separating them through individual receivers, with performance roughly akin to that of square-law combining.

Frequency diversity also is employed often in microwave line-of-sight relay in $1:N$ protection switching, where one frequency is available to provide diversity switching for any one of the N other carriers (frequencies) being used on the same link, each carrying independent traffic. When diversity is needed, the appropriate traffic is simply switched on to the backup frequency as well (usually with anticipation, to allow time for phase locking before the switching action occurs).

Multipath Diversity

When randomly keyed signals are used with a bandwidth W greatly exceeding $1/T_M$, the correlation lobe in a cross-correlation receive processor has a width of order $1/W$, much narrower than the multipath spread T_M. Multiple cross correlators, operating at uniform spacings of $1/W$, will provide multiple outputs, each of which represents only the energy arriving via its resolved $1/W$-width segment of the total multipath profile, hence fading independently from any other correlator output. The concept is perhaps most easily understood by considering a tapped delay-line model of the channel, with taps also $1/W$ apart. The most common implementation of this type uses direct-sequence spread-spectrum (pseudonoise) modulation, with a high-rate pseudorandom sequence providing the multipath resolution and a lower-rate information modulation superimposed (10). In modern implementations, instead of a correlator bank working off a tapped delay line (which gave rise to the name "Rake" for this technique), a more likely implementation is a single programmable matched filter, successively reprogrammed to match to the detailed waveform structure for each information symbol. For each symbol (symbol duration much greater than T_M), the multiple delayed contributions appear in time sequence on the matched-filter output, and are combined coherently or noncoherently by further low-pass filtering (weighted summation). In one modern troposcatter system (the AN/TRC-170), the same effect is achieved using pulselike waveforms, there called a "distortion adaptive receiver" (7).

Time Diversity

If the same data are repetitively transmitted at time spacings that exceed the coherence time of the channel, the multiple repetitions will be received with independent fading conditions, hence appropriate for diversity combining. This can be achieved simply by repeating messages or portions of messages at appropriate intervals, with storage or accumulation of diversity-combined values until the last repetition. Time diversity has probably never been implemented in this pristine form. However, it is certainly implicit at a message level in many communications protocols, where messages contain check data (error-detection codes) with provisions for repeating until receipt of an error-free version. Combining or selection at the individual bit level is not known to be used in any implemented systems, although often proposed, particularly in the form of simple majority decision for each data symbol.

The modern implementations of both time diversity and of in-band frequency diversity involve the use of error-control coding, as discussed in the next subsection. Such methods as time diversity or in-band frequency diversity involve an obvious sharing of available transmit power or energy over the multiple replications of a signal. Hence, unlike spaced receiving antenna diversity, one may regard that there is a transmit power division penalty for these forms of diversity. With the diminishing rate of performance improvement versus increasing order of diversity, one might speculate that there is an optimum order of diversity in such power-sharing techniques, beyond which net performance degrades. When ideal coherent maximal-ratio combining is available (recall its equivalence to a matched filter using all signal energy), it can be shown that there is no such intermediate optimum, only a diminishing rate of return. But with square-law combining, one obtains a set of results summarized for binary orthogonal keying by the curves shown in Fig. 9. The solid curves apply for constant transmitted energy per bit, showing that an optimum order of diversity in fact occurs when the received SNR per branch is about 3 (5 dB).

Before leaving the subject of diversity, it is useful to note other applications of the use of diversity that are different in tone from the methods and techniques described above.

1. In skywave HF, many concepts have been implemented or are being investigated for adaptive selection of the operating RF frequency and, in some cases, for adaptive selection of the path (connectivity) in an HF relay network. These concepts are sometimes called frequency or path diversity, respectively.

2. In many applications, the availability of diversity branches can be used to eliminate an interfering signal by coherent weighted processing, on the basis that the relative amplitudes and phases associated with the interference on the several branches are different from those associated with the signal of interest. The weighting used for that purpose is different from that used for optimizing the signal level (because the "noise" is now treated as coherent from branch to branch).

Coding and Interleaving

Time diversity or in-band frequency diversity, as portrayed above, can be regarded to be a brute force use of redundancy, in which each bit or symbol is repeated n times. From a coding point of view, the technique involves use of a code of rate $1/n$, with a particularly unsophisticated code. One might expect

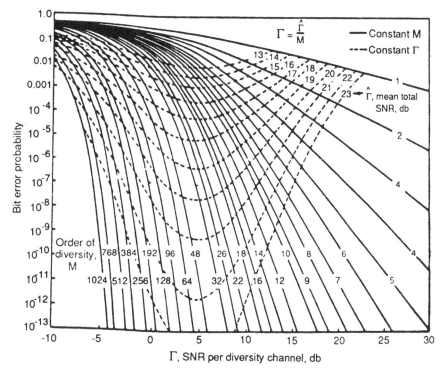

FIG. 9 Error rates for energy-sharing diversity with square-law combining, noncoherent frequency shift keying. (From Ref. 1.)

that a more adroit selection of a code will lead to a more efficient system while still maintaining the benefits of the diversity concept. This has indeed proven to be true, and coding is now well accepted as a significant component of fading-channel signal and modem design.

Some of the early thoughts were to use burst correction codes, that is, codes with sufficiently long constraint lengths to "surround" an error burst (due to a signal fade) by enough correctly decided bits that the error burst could be corrected. Efficient codes were found with this characteristic, but the concept foundered when detailed investigations showed that error bursts on typical fading channels might be excessively long and, in particular, that one could not guarantee sufficiently long error-free intervals between error bursts (7). Instead, attention turned to efficient use of codes designed for correcting random errors, by spacing the bits of a codeword sufficiently far apart so that they encountered independent fading. The technique by which this is standardly done is called *interleaving*. The simplest form is block interleaving using a memory containing *M* rows and *K* columns, with a coded stream written into memory row by row, and read out column by column. Successive bits of the coded stream then appear as every *K*th bit of the transmission stream. In reception, the coded stream is restored by using a corresponding *M* × *K* memory as a deinterleaver. Other interleaving approaches have been suggested that require smaller amounts of memory, but with the continually decreasing costs of memory devices, the interleaver design no longer appears to be a significant issue. Moreover, while the block interleaver is most naturally described by reference to a block code, the concept is equally applicable for convolutional codes.

One early coding-type theoretical result for simple diversity with binary data communications using interleaving is that with M repetitions of a bit (M odd), one can simply make a hard decision on each branch, and then take a majority vote over the branches. An error occurs when half or more of the individual hard decisions are in error, hence with an asymptotic error rate that varies as $1/\Gamma^{M/2}$. In fact, the constants are essentially the same as for square-law combining diversity of order $M/2$.

This last result extends almost immediately to hard-decision error-correction decoding of any block code using interleaving. Errors can occur when the number of errors exceeds half the minimum distance of the code set, hence the effective order of diversity is half the minimum distance of the code (bounded, then, as half the minimum weight of the codewords for linear codes). A similar statement holds true for convolutional codes, with the effective order of diversity equal to half the minimum-free distance of the code. On the other hand, soft-decision phase-coherent decoding is a form of matched-filter detection of codeword alternatives, and even soft-decision decoding based on square-law detection of individual bits (i.e., use of postdetection matched filters to form decision variables for the various codewords) comes close to the same result. For these, the effective order of diversity corresponds to the minimum distance of the block code (the minimum number of bits that are different between codewords) or to the minimum-free distance of a convolutional code. These results are illustrated in Fig. 10, which shows results for two dual-diversity FSK signals and for a Golay (24, 12) or (23, 12) rate $\frac{1}{2}$ block code using FSK signals. The $M = 2$ diversity and the rate $\frac{1}{2}$ code both involve the same degree of redundancy in the transmission, but the Golay codes have a minimum distance of eight or seven, respectively, translating into performance equivalent to eighth-order diversity for soft-decision decoding and to roughly fourth-order diversity for hard-decision decoding (as judged by the asymptotic slopes).

Figure 11 shows the results for soft-decision decoding of several binary convolutional codes. The top three curves show the effect of increasing constraint length while keeping the code rate constant at $\frac{1}{2}$, while the lower four curves show the effects of additionally using n repetitions of the coded bits, for a fixed constraint length of 5.

The effectiveness of coding within a given redundancy can be increased by increasing the length of a block code (while keeping the code rate fixed) or by increasing the constraint length of convolutional codes. There are practical limitations to how far one will go in this regard, with tradeoffs in the diminishing rate of improvement of performance with increased code length and in the increasing complexity of the decoder.* A third, subtle tradeoff at the system

---

*K. Brayer has pointed out that since successful decoding depends on a mix of few errored bits with mostly nonerrored bits within each codeword, there is a point in codeword elongation beyond which performance starts to degrade because individual codewords tend to bracket more than one error cluster. That is, the bursty nature of errors still presents a nonstationary channel to the decoder. A troposcatter channel example appears in Ref. 17, p. 434, Figs. 6–8; as the interleaver spacing increases from 6 to 576 to 1200 bits (delays of 0.015, 1.44, and 3.0 s), the decoded error rates on test runs 45, 50, 48, and 52 pass through a minimum and then increase.

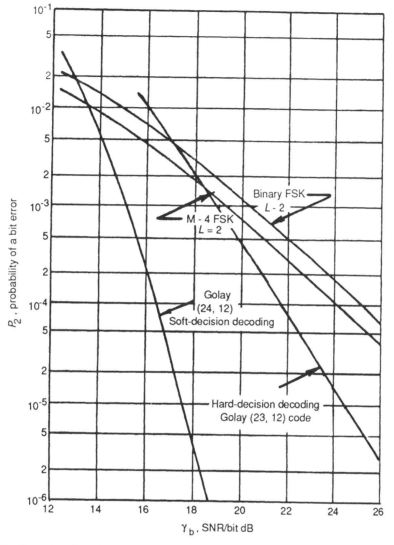

FIG. 10 Example of performance obtained with conventional diversity versus coding (dual in-band diversity versus rate $\frac{1}{2}$ code). (Adapted from Ref. 2.)

level is that the time span bracketed by an interleaved codeword increases proportionally to the increased constraint length and, hence, there can be a significant throughput delay upon receipt of a message before the deinterleaver is ready to output data. For example, if 0.5 s delay between successive codeword bits is desirable in order to operate in a fade rate of the order of 1 Hz, a Golay (24, 12) code will involve 12 s deinterleaver delay. Such delays can interfere with, or at least force modification of, network acknowledgment protocols that might otherwise be in use. As well, overall message lengths may be quite short in some applications, and stretching the transmission time to allow successful high-order time diversity, whether coding is employed or not, may imply an inefficient use of the total time-bandwidth product available for transmission

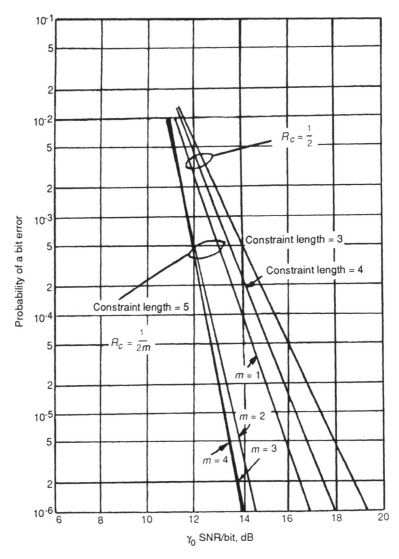

FIG. 11 Performance of binary convolutional code with soft-decision decoding (Rayleigh-fading channel). (Adapted from Ref. 2.)

(i.e., may result in not having enough message bits or coded message bits to populate an interleaver memory efficiently, implying gaps in the interleaver output that are not used for actual data transmission). One result of the complexity issues is that some fading-channel applications that involve use of a high degree of redundancy in transmission divide the redundancy between coding and repetitions. For example, consider a modem designed for 2400-bps transmission rate that is instead used to support 75-bps throughput—the 16-fold redundancy can be divided between use of a rate $\frac{1}{4}$ code and use of fourfold repetition of every transmitted bit (all with interleaving across time or across frequency in a multitone modem).

Achieving the performance of a very high order of diversity is probably of greatest interest for data communications applications that require very low throughput error rate, below about 10^{-5}. For digital voice, where error rates in the 10^{-2} range may be tolerated, there is little advantage to very high orders of diversity, although a low order is still quite beneficial.

It should be added that quite another effective use of coding on fading channels is the use of coded waveforms for error detection rather than correction. Error-detection coding can be very efficient with a relatively small amount of added redundancy. Normally, an acknowledgment type of protocol is used, with requests for repeats of unaccepted message segments (such as packets). On HF skywave, such systems are familiar as the automatic repeat request (ARQ) techniques that have been in use for four decades on telegraphy systems. When error detection is involved, there is no motivation for interleaving. Quite the opposite is true: within the same overall average rate of errors, it is beneficial if the errors encountered are clustered, since this will require fewer repeat transmissions (assuming that the code can maintain a very low rate of accepting segments with undetected errors). An interesting variation on the use of error detection with repeat transmissions recently has been described for an HF skywave packet-transmission application (18). An error-correction code is applied in forming each transmission packet, while an error-detection code is embedded additionally as part of the information content. The error-detection code determines whether to request repeats; any repeat is treated as part of an additional soft-decision decoding of the previously accumulated metrics (from processing the prior receipts of the same packets). As opposed to the independent decoding of each repetition, this procedure adheres more closely to the principle of combining all the available repetitions of the data bits, and results in considerable reduction in average time for throughput of accepted messages.

Note that in many applications, such as HF, coding also plays a major role in circumventing the effects of impulsive noise disturbances.

Use of Adaptive Equalizers

The successful development of adaptive equalizers for operation over fading channels was mentioned above. These have been adaptations of the kinds of equalizers earlier developed for voiceband data transmission over the telephone plant (19-21). The distinctive challenges of designing an equalizer for fading channels are the necessity for continual and rapid updating of the equalizer coefficients to track the changes in the channel, and a need to compensate for multiple paths with nearly equal strengths that cause deep spectral nulls within the bandwidth of interest. In application to troposcatter (4-6), use of a high symbol rate in the Mbaud range allows reception of a very large number of symbols per fade cycle (fade rate in the 1-100 Hz range), hence, a large number of updates is available for the popular data-driven, iterative adaptation scheme. The primary problem is one of processing speed. The original equalizer developed for this purpose partly used analog processing, but with currently available analog-to-digital (A/D) converters and logic speeds, a completely digital implementation can be used. Because the multipath spread extends only over a couple of symbols, the equalizer can be relatively short. Fully digital, very-large-scale-integrated circuit (VLSI) adaptive equalizers have been developed to cover the occasional refractive fading problems encountered in microwave digital lines operating at rates up to 35 Mbaud. At the other extreme, the adaptive equalizers developed for HF operation have all operated in conjunction with 2400-baud transmission, and this can sometimes provide too few symbols for data-derived updates. Instead, the most successful of these equalizers (22) interleaves learning patterns (known data patterns) with data packets, and uses the learning patterns as the basis for equalization updating (also using additional data from the data sequence as decisions are made).

The problem of multiple paths nearly equal in strength has been handled by using decision feedback in the equalizers. The equalizers are implemented as transversal filters, with a feedforward section that supplies corrections for inter-symbol interference caused by precursors or early arrivals of symbols yet to be decided, and a feedback section that corrects for intersymbol interference from symbols already decided. The feedforward section operates as a linearly weight-ed sum on the received signal plus noise. However, the presence of nearly equal paths (one of which, in effect, must be cancelled out) can result in a requirement for many taps in the feedback section. In turn, if the feedback section were to operate as a linear weighted sum, excessive noise could then appear in the equalizer output. Instead, in decision feedback, the weighting is applied to ideal noiseless versions of the decided symbol values; so long as most decisions are correct, the equalizer converges rapidly in an iterative scheme (with a learning pattern, of course, the correct data values are known). In addition, it has been found that use of equalizers with taps spaced by a fraction of a symbol duration (e.g., $T_s/2$ or $2T_s/3$) provides better ability to compensate for intricate channel transfer functions than the more traditional symbol-duration spacing. One ex-planation is that the transfer function realized by any transversal filter with uniform spacing T is periodic in $1/T$. By using $T = T_s/2$ or $2T_s/3$, the band-width that can be controlled is larger than $1/T_s$, which better conforms to the total signal spectral occupancy and allows better compensation for such artifacts as deep, in-band spectral nulls.

A side benefit from using adaptive equalizers to eliminate intersymbol inter-ference is that they provide an implicit multipath diversity benefit by also adding together (with weighting) the smeared components of each symbol.

Generally, it may be said that the effective use of adaptive equalizers in troposcatter appears to have been well demonstrated, while their effective use in HF skywave has been demonstrated on a few links but still awaits extensive testing over the wide variety of conditions found on HF links.

Understanding Long-Term System Performance

As the material above indicates, signal and receiving system design for a station-ary fading channel is a well-developed art that is continuing to evolve. New concepts continue to appear, along with improved implementations largely based on continuing speed increases and cost reductions in digital signal-processing technology. Technologically, this is all very satisfying for radio modem designers. Yet the stationary fading channel is only a time-local model of short-term fading and its statistical parameters themselves vary considerably over any longer periods of time. The designer of a telecommunications network or system has to infer, from an understanding of these long-term variabilities, how to select an appropriate set of these parameters to guide selection of those design aspects of modem, coder, diversity, and the like that are in turn conditioned upon the stationary-channel models. This selection is usually expressed as a worst-case design, where the modem design goal is specified by a mean SNR for the Rayleigh (or Rice) fading, a gross-distortion description including multipath spread and Doppler spread, and whatever finer details may be needed to substantiate performance under these conditions. The system designer's broader efforts may include tradeoffs involving other parameters and costs of a system design, such as transmitter power, antenna gains, lengths of links in a relay network, the performance required of the modem, and so on.

On the surface, one can hardly quarrel with the apparent reasonableness of such an approach. The intent appears straightforward and consistent with rational intuition about how to go about designing a system to meet user service requirements. Still, many aspects of the design procedure constitute an art, depending on customary (accepted) judgments about how the user requirements translate into technical specifications. In this section, we discuss how the use of fading channels introduces yet another dimension into this art.

One conundrum lies in identifying a "worst case" for design. Performance difficulties on fading channels are usually attributed both to fading and to multipath distortions. Therefore, the usual candidate for the worst-case design is a stationary fading channel that has the largest multipath spread and the largest Doppler spread that are reasonably expected to occur on the planned links. Against these, the modem designer is asked to achieve some specified level of performance (e.g., error rate) at a minimum mean SNR, within some cost and complexity tradeoffs. As shown in the section on signal, modem, and receiving-system design issues, a contemporary fading-channel modem design very likely would include some form of coding and interleaving to achieve the effective high order of diversity afforded by that technique and corresponding reduction in required mean SNR. But can it be claimed now that, so long as the mean SNR is above the design minimum, its performance is assured to be above that required? Unfortunately, the answer is not comforting. We have outlined a design where the signal and receiving system design is an elaborate amalgam of sophisticated use of both time diversity and in-band frequency diversity, and where performance at the specified mean SNR depends on achieving the order of diversity that is implicitly assumed to be available within the worst-case channel characteristics. That is, the designed system performs best for a given mean SNR when the multipath-distortion effects are at or nearly at their worst! But what if, perversely, some of the links have lengthy periods of time during which fading is much slower than was assumed to be the "worst" case, or the selectivity of fading is much less severe than was assumed? Much of the time diversity or frequency diversity upon which the modem's performance was predicted would no longer be available and, in those situations, the required mean

SNR on the link might have to be much higher, perhaps approaching the no-diversity values!

The problem outlined above is, in fact, a very real problem for any thoughtful system designer trying to formulate new systems for fading-channel operation. For example, in HF skywave systems, new flexibilities allow essentially automated on-line selection of good operating frequencies, instead of the older techniques of operating within limited and manually controlled frequency assignments. At the same time, the newer modem/coder techniques push back the boundaries of multipath-distortion effects and required SNR within which a desired quality of performance can be realized. Instead of the 60%–70% availability found in 1960s' testing of HF modems on individual links, projected availabilities are well into the 90th percentile. Given these improvements, can the system designer begin to promise HF networks that approach the continual availability of other media? And if so, how closely, and where are the limiting factors? The effect outlined above, that occasionally the channel may be nearly distortionless, might represent one clear limitation, where instead of the diversity benefits from clever modem and coder design that take advantage of the distortions, the system will be stranded, as it were.

Obviously, system designers can work toward incorporating additional capabilities to redress such limiting effects, whenever they can be identified. However, for example, assessing the overall impact of the conundrum above requires an extensive database on global mixes of fading channels, describing joint statistics of locally mean SNR and of the associated channel distortions. Such data are simply not available, and accumulating them implies so large an undertaking that no such engineering database is likely within any foreseeable time. The pragmatic alternative is to trust to eventual accumulation of reports and insights from individual research and engineering tests, routine usage reports, operational observations, and the like. Altogether, this is not a very satisfactory situation for the system engineer who recognizes a community of intuitions around which more could be promised, but without any database to support the claims.

Another common system-design problem posed by long-term variabilities is exemplified by a kind of performance specification common in past years for troposcatter designs to support multichannel telephone networks, which read "channel test tone-to-noise ratio (or bit error rate) shall remain above X dB (or below 10^{-y}) for all but eight hours of the worst month." The eight hours was to be interpreted as a cumulative eight hours, in effect a requirement that the specified performance would be realized with at least 99% probability at any randomly selected instant within that worst month. In more recent times, the concept of the errored-second has crept into use (certainly in the microwave line-of-sight relay practice), giving a more refined notion to the testing of bit error rate. A specification might then read "no more than 28800-error seconds during the worst month," where error counts are taken over successive intervals of one second each, and an error-second is simply defined as one in which the count exceeds some threshold value (e.g., at 8.448 Mbps, one might state a count of 85 errors in a second instead of stating an error probability of 10^{-5}). Whatever the detailed criterion, such system performance specifications have sometimes been promulgated as the governing specifications to equipment developers or manufacturers. Differing interpretations by bidders on the long-term effects could lead to different detailed equipment performance specifications, hence to different cost bases that could not fairly be compared.

The long-term fading statistics for any of the fading channels are, at best, an accumulation of multiple-year statistics. Most often, they are empirical accumulations from many test links and different time frames scaled, as best as someone can judge, to equivalent values. There is no way that the statistics during any single year, or for that matter during any few-year period, can be expected to fit tightly to those long-term statistics, especially at the tails of the distributions. Therefore, there is no meaningful way to test performance of equipment for long-term availability, within any realistic interval for deciding acceptability. Instead, for fading channels, any long-term performance specification must be regarded as largely a set of objectives, whereby the system planner draws upon the empirical database (the long-term statistics) to gauge the equipment requirements of an overall system design. That these are objectives and not specifications is intrinsically understood by all who are involved in the detailed design process, but often not explicitly so stated (often left stated as if they were quantitatively meaningful system specifications). The judgments properly made by the system planner will include some appreciation of whether user requirements are best met by erring on the side of pessimism or optimism, and usually will find quantitative expression by the amount of additional margin allocated in the link budgets.

In any case, it is clear that detailed technical specification of equipment for use on fading channels should be precisely that, and that any performance objectives relating to fading-channel operation should be stated in terms of specific tests that can be run with different sets of conditions on a simulated stationary channel, in other words, on a channel-fading simulator. The burden must fall on the system planner to specify these critical sets of conditions that the equipment must fulfill. To the extent that the system planner can foresee difficulties (with considerable cost impact) in specifying future equipment requirements because of inadequacies or uncertainties in the database of long-term statistics, per the examples above, the planner must convince his or her own organization of the importance of mounting long-term data collection and analysis efforts to acquire the missing information.

An ideal long-term database for understanding availability of a system using fading channels would probably involve at least a three-dimensional joint histogram, with hourly mean SNR, Doppler spread, and multipath spread as the variables. By contrast, the existing long-term statistics are generally only of mean SNR, supplemented by qualitative comments on variations in the other parameters and how they relate to mean SNR. System design purposes probably could be well served by providing only a detailed estimate of the nature of the tails as defined by extremal values of any one of the parameters that could stress performance. On the other hand, further evolutions in design concepts may result in requiring more detailed characterizations of the fading channels in order to project performance. For example, future fading-channel system design may require not only an understanding of how frequently outages occur (i.e., failure to achieve throughput), but of how rapidly the channels change in a long-term sense (i.e., how long the time may be before the system is again providing reliable throughput at the rates desired). This notion of speed of service, or its converse of delay, has not obviously been a criterion in design of such systems in the past, but it is a very obvious concept to put before the user in asking him or her to accept a communications system or network that cannot promise 100% instantaneous availability.

Whenever more detailed databases are available on the long-term fading, they also allow intelligent comparisons of alternate approaches to system design (signal design, demodulation, message lengths, coding) where one approach is not uniformly better than another, for example when one design copes better with fast fading but requires higher SNR.

References

1. Schwartz, M., Bennett, W. R., and Stein, S., *Communication Systems and Techniques*, Part 3, McGraw-Hill, New York, 1966. (This book contains extensive references to the pre-1966 literature.)
2. Proakis, J. G., *Digital Communications*, McGraw-Hill, New York, 1983.
3. Lomax, J. B., HF Propagation Dispersion, *AGARD Conf. No. 33, Phase and Frequency Instability in EM Wave Propagate*, Technivision Services, United Kingdom.
4. Grzenda, C. J., Kern, D. R., and Monsen, P., Megabit Digital Troposcatter Subsystem, *Proc. NTC 75* (December 1975).
5. Gruber, P., and Ellins, V., Evolution of Troposcatter Radio Transmission, *Proc. NTC 78*, (December 1978).
6. Ehrman, L., and Monsen, P., Troposcatter Test Results for a High Speed Decision Feedback Equalizer Modem, *IEEE Trans. Commun.*, COM-25:1499–1504 (December 1977).
7. Unkauf, M., and Tagliaferri, O. A., An Adaptive Matched Filter Modem for Digital Troposcatter, *Proc. ICC 75* (June 1975).
8. Neilson, D. L., Microwave Propagation Measurements for Mobile Digital Radio Applications, *Proc. EASCON 77,* paper 14-2.
9. Gupta, S., Visvanathan, R., and Muammar, R., Land Mobile Radio Systems — A Tutorial Exposition, *IEEE Commun. Magazine*, 34–45 (June 1985).
10. Turin, G. L., Introduction to Spread Spectrum Antimultipath Techniques and Their Applications to Urban Digital Radio, *Proc. IEEE*, 68:328–353 (March 1980).
11. Ehrman, L., et al., Real-Time Software Simulation of the HF Radio Channel, *IEEE Trans. Commun.*, COM-30:1809–1816 (August 1982).
12. International Radio Consultative Committee (CCIR), Use of High Frequency Ionospheric Channel Simulators, *Recommendation 520, HF Ionospheric Channel Simulators*, Rep. 549-1, 1974–1978 (1978).
13. Bello, P. A., and Nelin, B. D., The Effect of Frequency Selective Fading on the Binary Error Probabilities of Incoherent and Differentially Coherent Matched Filter Receivers, *IEEE Trans. Commun. Sys.*, CS-11:170–186 (June 1963). [Corrections in CS-12:230–231 (December 1964).]
14. Bello, P. A., and Nelin, B. D., The Influence of Fading Spectrum on the Binary Error Probabilities of Incoherent and Differentially Coherent Matched Filter Receivers, *IRE Trans. Commun. Sys.*, CS-10:160–168 (June 1962).

15. Bello, P. A., and Nelin, B. D., Predetection Diversity Combining with Selectively Fading Channels, *IRE Trans. Commun. Sys.*, CS-10:32–42 (March 1962).
16. Gusyatinskiy, I. A., and Nermirovskiy, A. S., A System for Interference Fading Prevention in Tropospheric Telecommunications Links, *Electrosvyaz* (Telecommun. Radio Eng.), part 1, 27(2) (1973).
17. Brayer, K. (ed.), *Data Communications in Fading Channels*, IEEE Press, New York, 1975.
18. Chase, D., Code Combining – A Maximum-Likelihood Decoding Approach for Combining an Arbitrary Number of Noisy Packets, *IEEE Trans. Commun.*, COM-33:385–393 (May 1985).
19. Giordano, A. A., and Hsu, F. M., *Least Square Estimation with Applications to Digital Signal Processing*, Wiley, New York, 1985.
20. Monsen, P., Adaptive Equalization of the Slow Fading Channel, *IEEE Trans. Commun.*, COM-22:1064–1075 (August 1974).
21. Monsen, P., Feedback Equalization of Fading Dispersive Channels, *IEEE Trans. Inform. Theory*, IT-17:56–64 (January 1971).
22. Perkins, F. A., and McRae, D. D., A High Performance HF Modem, paper presented at the Int. Defense Electron. Expo., Hanover, West Germany, May 1982.

SEYMOUR STEIN

Communication Printed Wiring Interconnection Technology

Introduction

Interconnection technology encompasses the physical and mechanical design considerations and the medium and means for making electrical connection between functional electronic components and planar or flexible conductive patterns supported on an insulating substrate. The resultant printed wiring or hybrid integrated-circuit assemblies, either alone or fabricated to a larger assembly, are mounted on equipment frames for external connection.

Types of Printed Circuits

The electrical connection is made between an electronic component and one of many fabricated interconnection circuits. These circuits include

- Thick and thin films
- Hybrids
- Printed wiring
  Rigid
  Flexible
  Multilayer
  Molded

These interconnection circuits have met the major objectives of telecommunications interconnection technology, namely

1. Reduced weight and size by taking advantage of improvements in plastic materials and graphic reproduction
2. Greater reliability due to uniformity of product
3. Lower cost by improvements in manufacturing process for fabrication
4. Greater speeds in signal processing

Printed circuits are features of all types of electrical and electronic equipment whose connectivity and reduction in weight and size is paramount. From consumer electronics with 1–10 years service life to high-reliability military missile electronics, thermally stressed automotive control systems, telecommunications, and computer equipment, printed wiring functions as the primary connectivity route.

Thin-Film Circuits

Thin-film circuits are predominately formed by the deposition of the condensed vapor of a liquid or solid on a solid inert substrate. Evaporative and sputtering techniques are employed. Chemical formation methods such as anodization create dielectric films (oxides of metals) by electrochemical oxidation, which is employed in fabrication of both capacitors and resistors. Electrodeposition or electroless deposition is employed to build up metal thickness after vapor deposition of the primary thin film. By appropriate choice of metals deposited and masking, conductive traces, resistors, and capacitors can be formed on the inert substrate.

Thick-Film Circuits

The fabrication of thick-film circuits has its origin in ceramic technology. A mixture of metal particles, temporary organic binder, and glass frit is applied to a high-temperature ceramic. When this mixture is carefully sintered at high temperature, a 25,000–250,000 angstrom thick film is obtained. By screen application through masks, conductor traces and circuits can be developed after sintering.

Hybrid Integrated Circuits

Hybrid integrated circuits (HICs) consist of several components, passive and/or active, integrated or appliquéd on a ceramic substrate. Hybrid integrated circuits provide small-sized, high-performance, fully functional, low-cost precision circuits with high interconnection capability. There are three basic types:

- Thin-film HICs—a combination of interconnected circuit elements supported on an inert substrate. Conductors and resistors are formed on a ceramic substrate using such thin-film processing techniques as vacuum deposition, photolithography, and anodization.
- Thick-film HICs (conductors, resistors, and capacitors)—formed on a ceramic substrate using thick-film processing techniques, first printing the pattern with the appropriate paste, drying, and then firing (sintering)
- Multilayer HICs, multilayer interconnection system and, specifically, multilayer hybrid circuits (MHCs)—permit the connection of complex electronics beyond a single-plane geometry. Multilayer interconnection structures by definition comprise two or more metalization layers separated by an insulating layer. In thick-film MHC structures, the metalization layer can be conductor tracks separated by a dielectric layer (i.e., crossovers) or a full multi-

layer structure of alternative printing and firing metalization patterns on a ceramic dielectric substrate. This forms a layered structure with connectivity established between layers with small-diameter, plated through holes (PTHs), also called vias. The complexity of the MHC structure places stringent demands upon the materials of construction and their relation to the design requirements.

The following components comprise hybrid packages:

- crossovers
- resistors (thin, thick, appliquéd)
- capacitors (thin, thick, appliquéd)
- inductors (appliquéd)
- silicon chips (beam leaded, wire bonded)
- diodes
- ceramic chip carriers

Printed Wiring

Whereas typical thin- and thick-film circuits comprise a ceramic layer upon which conductor traces are deposited and developed, printed wiring traditionally is thought of as the development of planar or flexible conductive traces on a rigid or flexible plastic substrate.

Lamination of copper metal foil to typical plastic substrates (i.e., epoxy glass-reinforced plastic or polyimide substrate) forms a single-sided or double-sided rigid or flexible copper laminate. This is followed by various subtractive and additive metalization process technologies that create an interconnection circuit. The density of the circuit (wiring) is dictated by the limitations of the manufacturing process. The poor high-temperature performance of many of the plastic-substrate materials places restrictions on the feasibility of performing some of the traditional processes for generating metalization and fabrication of passive devices as conducted on thin- and thick-film circuits on ceramic substrates. Printed wiring with plastic substrates provides complex structure shapes. These are obtainable from molded and flexible printed wiring.

As in ceramic-substrate printed wiring, plastic laminates can be fabricated to form different types of printed wiring with such specific utility as:

- Single-sided rigid circuits, the simplest circuit to fabricate with low cost. These contain a single layer with conductive tracks and are limited in applicability to low-density circuit interconnection.
- Double-sided rigid (DSR) circuits, which contain two opposite layers of conductive tracks with interconnection established through PTH vias. DSR boards provide a low-cost interconnection medium that, combined with advanced processing technology, can increase conductor track density to produce a printed circuit with connectivity competitive with some multilayer structures.

- Flexible circuits, which contain conductive tracks on both sides of a flexible dielectric base material. General applicability permits saving in weight, size, space, and cost, and these circuits can conform by bending to equipment tolerances. Protective coating is generally employed over circuitry.
- Bonded flex circuits. A flexible circuit is bonded to a rigid-support dielectric plane, providing a low-cost interconnection medium permitting support of flexible circuitry where required for attachment.
- Multilayer boards (MLBs). When circuit density makes a double-sided rigid board insufficient, an MLB is required. The multilayer structure comprises a multilevel sandwich with DSR rigid boards separated by a dielectric (i.e., reinforced) thermosetting polymer layer (prepreg) or adhesive. The sandwich is assembled with inner layers of the double-sided rigid board having printed conductive tracks. Once assembled, the interconnections, between inner layers alone or with external top and bottom surfaces of the multilayer structure, are established with buried and PTH vias.
- Molded circuits. Molded printed wiring is composed of a thermoplastic resin substrate that has been injection molded, extruded, or thermoformed into a three-dimensional shape. Molding offers many opportunities for design configuration not offered in traditional planar printed wiring. Conductive interconnects or tracks are formed by printed conductive inks or employment of subtractive and additive technology on copper film plated on the molded substrate.

Printed Wiring Interconnection Selection

This section only addresses printed wiring circuit interconnection. It is beyond the scope of this article to detail further the materials, design, and fabrication of ceramic- or metal-substrate printed circuits.

Materials

The choice of materials for a printed wiring design is dictated by the end-point and reliability requirements. All printed wiring materials have a dielectric substrate to which conductive tracks are attached. As mentioned above, these tracks can be either subtractive (developed from copper foil originally laminated to the dielectric substrate) or formed on the substrate additively. Dielectric substrates can be either flexible or rigid materials comprising a thermoplastic or thermosetting resin (alone or as a reinforced material containing an inorganic or organic web, matt, or random fibers). Typical flexible printed wiring substrates are tetrafluoroethylene (Teflon®), polyamides (nylon), polysulfones, polyimides (Kapton®), and polyesters (mylar). Rigid printed wiring dielectric materials are phenolic-resin or epoxy-impregnated paper, and epoxy-impregnated fiberglass cloth or fiberglass cloth impregnated with triazine-based, high-temperature application specialty resins. Other reinforcements for flexible or rigid printed wiring are Kevlar® fibers.

Recently, printed wiring products have been categorized by CLASS. In effect, the applicable CLASS in which a product belongs dictates the type of laminate-dielectric to be fabricated into a printed wiring circuit. Such factors as dimensional, thermal, and mechanical and physical integrity during fabrication and service life are considered in the choice of the printed wiring circuit material. The printed wiring classes are:

- *CLASS I—Consumer Product.* Used in noncritical applications where extended life is not the major criterion but cost/reliability is of importance.
- *CLASS II—General Industrial.* This class includes high-performance commercial and industrial products in which extended life is desired/required. Examples of instances where high performance and reliability are desired are automotive computer electronics, modems, terminals, and computers. High-performance computers and telecommunications equipment require a type of industrial printed wiring where reliability is of prime importance.
- *CLASS III—High Reliability.* In this category are printed wiring products that are used for life-support systems, military electronics, undersea cable repeater electronics, satellites, and space vehicles.

Fabrication Process

Single-Sided Printed Wiring Circuits

The simplest printed wiring circuits are single-sided circuits that are fabricated from a single-sided, copper-clad laminate. The simplest and most common fabrication sequences are depicted in Fig. 1. In these low-cost methods, circuit tracks may be developed by creating a pattern from the original copper foil by etching away areas not protected with a resist coating (see Fig. 1-*a*). Alternatively (Fig. 1-*b*), the original copper foil may have areas segregated with a developed organic-resist coating by reverse printing. The exposed copper areas are electroplated with an etch-resist coating such as tin or tin/lead alloy (solder). The protective organic-resist coating is then removed and the exposed laminate copper is removed by chemical etchants to produce the circuit. The placement of devices on this circuit is facilitated by preferentially punching holes or drilling holes for insertion of component leads. The placement of holes can be conducted either at the beginning or end of the circuit fabrication.

Double-Sided Rigid Printed Wiring Circuits

A typical fabrication sequence of a DSR printed wiring circuit is shown in Fig. 2. As diagrammed in the illustration, the drilled hole and copper surfaces are coated additively with electroless copper, followed by application of an organic-resist coating that is developed to define circuit tracks and pads for component attachment. The copper plating is typically followed by electroplating of a

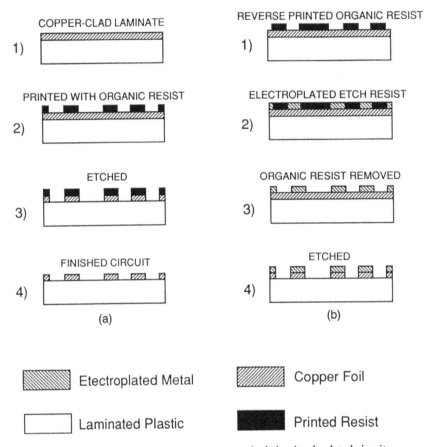

FIG. 1   Basic printed wiring types: *a*, etched circuits; *b*, plated circuits.

tin or tin/lead alloy as an etch resist or solderable coating on the exposed copper tracks and pads. This is followed by the removal of the organic protective resist (Steps 5 and 6). The etch resist can be retained or removed in the remaining steps of fabrication.

Single- and Double-Sided Flexible Printed Wiring Circuits

The fabrication sequence for flexible printed wiring circuits is a typical subtractive process (shown in Fig. 1) similar to that for double-sided rigid circuits. However, for single- and double-sided flex, the fabrication process for low-cost, large-volume circuit production is a continuous roll-to-roll tandem process that begins with a roll of copper-clad flexible material that is then drilled or punched. The circuits are then passed through electroless and electroplating baths and, where applicable, this is followed by plating of a tin or tin/lead etch-resist coating. Traditionally, the tin or tin/lead etch-resist coating is condensation

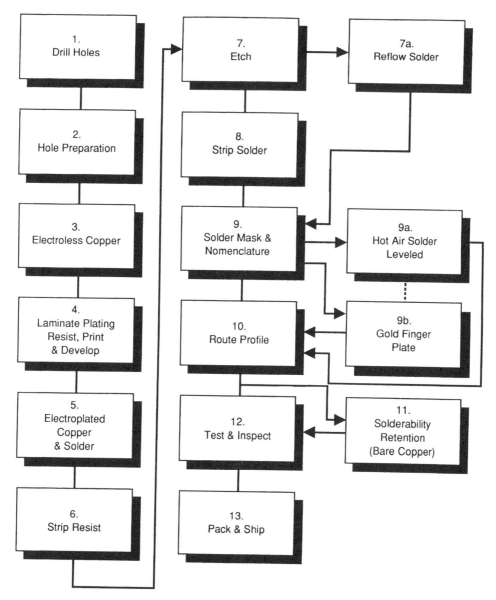

FIG. 2 Typical double-sided rigid printed wiring circuit manufacturing sequence.

reflowed in a separate operation prior to application of a solder-mask protective coating.

The tin or tin/lead etch-resist coating plated on conductor tracks attached to a flexible circuit typically cannot be hot-air solder leveled. The thermal dimensional instability of common flexible-base laminates prohibits the extended exposure to the hot-air solder leveling. Condensation reflow is the preferred method of reflowing the etch-resist tin or tin/lead coating.

Multilayer Board (MLB) Printed Wiring Circuits

The MLB printed wiring fabrication sequence is presented in Fig. 3. A significant component of the MLB is the inner-core layers, which are typically fabricated from thin (0.005 in thick) flexible epoxy-glass, copper-clad, or polyimide-laminate circuits. Individual circuits are then combined with bonding piles of

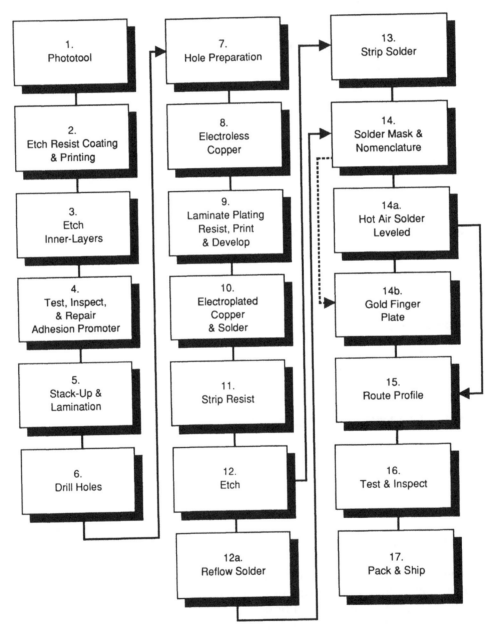

FIG. 3 Typical multilayer board printed wiring circuit manufacturing sequence.

uncured epoxy-impregnated glass cloth or adhesive-coated polyamide film to form an assembly. This assembly is placed in a press and bonding is affected between the uncured epoxy-impregnated glass cloth (prepreg B stage) and the thin copper-clad laminate circuits (see Fig. 4). The remaining manufacturing process sequences are similar to those for rigid printed wiring circuits but with significant attention paid to the integrity of the drilled hole and plating-process sequence, shown in Steps 6–11 of Fig. 3. Alternative materials (i.e., polyamide film, copper-clad laminate) are employed in the fabrication of the inner and outer layers of the MLBs.

Reliability

Of major concern in printed wiring reliability is maintaining the desired continuity and isolation of the circuit. Some seminal work on the reliability of printed circuit boards done at AT&T Bell Laboratories was reported at the IEEE Reliability Physics Symposium in 1976 (1–7).

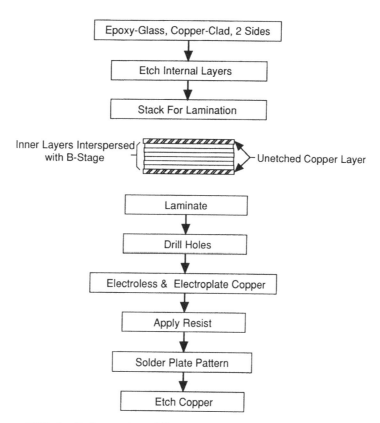

FIG. 4 Basic steps in multilayer board printed wiring fabrication.

Plated Through Hole Reliability

Plated through hole (PTH) integrity is a significant factor affecting double-sided and multilayer circuit reliability. As the vehicle for establishing interconnections between circuit planes, increased impedance will result if the PTH is compromised by fracture of the barrel, inner-layer foil cracks, and edge cracks of the barrel pads.

These PTH failure modes occur due to differential thermal expansion characteristics of the printed wiring structural materials, preparation of the drilled hole for electroless/electroplating generating, reduction in the adherence and ductility of the plated-copper layers, as well as the copper foil employed in circuit board preparation. The failure of the PTH arises during fabrication operations where thermal shock occurs, such as during solder plate reflow or solder coating (hot-air solder leveling) and circuit board assembly operations, including solder wave or condensation soldering.

Electrical testing of the PTH resistance by four-point measurements can be employed to obtain evidence of a predisposition to cracking. Test coupons having PTHs are placed on fabricated circuits, the circuits are thermally stressed, and measurements of the electrical resistance of the PTH are made. In addition, coupons are microsectioned for subjective visual analysis of hole cracking. Employment of test coupons on fabricated circuits is a form of manufacturing process control.

Insulation-Resistance Reliability

A compromise in the intrinsic high impedance of the dielectric substrate affects the design performance of the printed wiring circuit. This will occur when the insulation resistance between conductive tracks is lowered sufficiently to create a conductive path. The insulation resistance of a dielectric material is inherent with its structure. Moisture sorption, degrees of cure of a thermoset material employed as a laminate or solder-mask protective coating, and entrapped residual ionic contaminant can influence the value observed.

The bulk and surface leakage current between conductive traces or PTHs will, under high humidity, increase in the presence of ionic contaminant. Under electrical stress testing with temperature/humidity aging, ionic contamination can initiate electromigration of metalization and formation of metal dendrites between biased surfaces. Ultimately, with sufficient contamination present, the metal migration will propagate to bridge surfaces and create an electrical short-circuit failure.

Reliability testing for printed wiring circuits to assess insulation-resistance failure takes the form of exposing test circuits to electrical stress, humidity, and temperature. Insulation resistance is measured as a function of time elapsed to failure. Typical Arrhenius behavior has been observed and medial lifetimes determined for voltages below 100 V for telecommunications interconnection printed wiring.

Various structural-material failure modes have been characterized to account for insulation-resistance failures. These structural effects arise from residual ionic contamination in raw materials, such as solder masks, conformal coatings, and base plastic laminate, providing leakage paths. The effect of induced ionic contamination arising from plating residues in fabrication of printed wiring or residual high-activity flux residues after printed wiring assembly has been highlighted by dramatic changes of insulation resistance. Insulation resistance decay over time under temperature, humidity, and voltage testing has been observed, and is followed by hard electrical shorts.

Printed Wiring Assembly

Printed wiring assemblies can be classified by their end-product use. As with printed wiring, these assemblies are represented by three classes:

- *CLASS I—Consumer Products.* Includes noncritical applications that shall be reliable and cost effective but whose service life is limited (consumer electronics such as videocassette recorders, televisions, radios, and residential telephones)
- *CLASS II—General Industrial.* High-performance commercial and industrial products in which extended life is required. High-performance computers and switching telecommunications equipment are representative examples.
- *CLASS III—High Reliability.* High reliability exemplifies this class. Examples include life-support electronics, satellites, missile systems, and undersea-repeater electronic assemblies.

The assembly of printed wiring encompasses the mounting of electronic components, securing them by hand, wave or reflow soldering, removal of residual corrosive flux residues arising from the soldering process by solvent or aqueous cleaning, testing of the electrical integrity of the circuits and for the presence of residual ionic contamination, and repairs, if necessary.

The mounting of passive and active components is one key element in the success of the printed wiring assembly process. Two broad categories of components are (1) those that have axial or radial leads (leaded components) that allow positioning of the components through plated and nonplated holes in the printed wiring and (2) surface-mount devices, components that are leaded or leadless and mounted on circuit pads on the planar surface of the printed wiring (see Fig. 5). The subject of this section is the manner of mounting and the securing and repair in the assembly of through-hole and surface-mount components.

Through-Hole Mounted Component Mounting

Two major classes of through-hole components are axial leaded and radial leaded. The axial-leaded components are more readily assembled and mounted

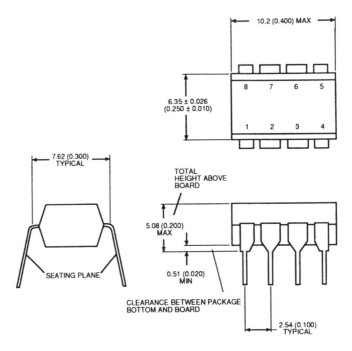

a. 8-Pin DIP

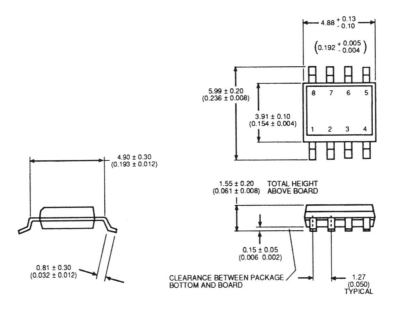

b. 8-Lead SOIC

FIG. 5 Types of surface-mounted components — comparison of dual-in-line package (DIP) and small-outline integrated circuit (SOIC) package sizes. (All dimensions are in millimeters and parenthetically in inches.) (Illustration courtesy of Texas Instruments, Inc.)

by automatic (robotic) insertions as these components can be supplied in a taped reel. Radial-leaded components, because of the variety in body shape and lead shape and size, are conventionally supplied in delivery tubes for automatic insertion.

Soldering

Once the component is inserted, it is typically secured in place by hand or wave soldering. Before soldering, it is necessary to precondition the printed wiring and component leaded metallic surfaces with a "flux." The purpose of the flux is to remove metallic oxides and other extraneous contamination on the exposed circuit surfaces. This presents a clean metallic surface for soldering by the molten tin/lead eutectic that secures the leads to the printed wiring circuit. Wave soldering is the preferred method of securing through-hole leaded devices to the printed wiring. Wave soldering is a process by which a solder fountain flows against the undersurface of the preassembled printed wiring as it is conveyed past the solder. Those copper surfaces that do not contain secured leads are afforded added environmental protection by the tin/lead metalization.

Assembly Cleaning

Cleaning of the soldered printed wiring assembly is conducted to remove flux residues either for cosmetic reasons or to assure electrical reliability of the circuit by removing induced contamination between biased circuitry. The cleaning regimen is dictated by the corrosiveness of the flux residues, the hermeticity of the components, and compatibility of the component packaging with the cleaning medium.

Repair

Subjective visual, x-ray, or cross-section microsection examination is made at selected test-site areas of soldered component lead joints to the printed wiring through holes to assure the integrity of the assembly. This inspection is generally done on a sample of assemblies. The level of inspection is dictated by process control established for the assembly process. The inspection for solder joint integrity is accompanied by an electrical testing of the circuit. Repair of the circuit is conducted with reworking of the solder joints and replacement of components damaged during the assembly process. Care must be taken not to compromise the integrity of the resoldered surface by excessive heat exposure.

Surface-Mount Component Mounting

The advent of high-density interconnection technology with the necessity of conserving printed circuit wiring space by reducing both the distance between conductor tracks and conductor width has been accompanied by the reduction in the size of active and passive elements. The comparative size of through-hole and surface-mount devices with equivalent electrical characteristics is shown in Fig. 5. A significant electrical advantage in employment of surface-mount devices in electronic assembly is that the short lead length offered by surface-mount devices improves the electrical characteristics (inductance) of the assembly.

Surface-mount components have been assembled in the past on ceramic substrates in the fabrication of thick-film hybrid circuits. This assembly technology has now been adapted to populate printed wiring circuits with surface mount components. With many surface-mount components, leads are eliminated and such passive devices as inductors, resistors, and capacitors have metalization at ends of the planar package that function as the interconnect to the interior element. The metalized ends of these "chip" devices are directly attached to pads on the printed wiring boards during assembly. Some typical leadless surface-mount components are chip capacitors, chip inductors, and chip carrier packages. Another family of surface-mount devices are small-outline components, such as the small-outline integrated circuits (SOICs) shown in Fig. 5, which are surface mountable with shortened leads. Chip carriers that contain packaged hybrid circuits are low-profile, square packages with connections on all four sides. These surface-mount packages come in leaded and leadless configurations.

Although surface-mount devices have made tremendous inroads in replacing conventional through-hole-mounted devices, the availability of one-to-one replacements in electrical characteristics has not been achieved. Consequently, many printed wiring assemblies are mixed assemblies.

Simple Mixed Assembly

In a simple mixed assembly, the small size of the surface-mount components permits high-speed automatic placement of devices on printed wiring pads. To secure the devices in place prior to soldering, a thermosetting adhesive is positioned in the area between the printed circuit pads. The device is then placed on the adhesive. The adhesive is cured in a separate operation, securing the device to the printed wiring substrate; the device is then soldered to the printed wiring pads.

Resistors, inductors, capacitors, and small-outline packages can be positioned on the underside of printed wiring in a wave-solder assembly process. The underside also contains the projected leads of through-hole-mounted devices earlier positioned on the upper side of the printed wiring. Wave soldering then is performed and both the surface-mount devices and through-hole leads are secured to the printed circuit wiring surface. A single cleaning step follows, with care taken to assure removal of flux contamination from beneath surface-mounted devices.

Complex Mixed Assembly

With the high-density printed wiring circuits associated with complex mixed assembly, surface-mount devices are secured on both surfaces of the printed wiring. Through-hole devices are also positioned on the top surface of the printed wiring.

The solder assembly can be many faceted. In a typical mixed-assembly surface-mount device, "leads" contact the pads of the printed wiring on the top surface. These pads contain a solder-paste coating that has been preferentially placed on the printed wiring pads in a separate dispensing operation. *Solder paste* is a finite dispersion of solder spheres in a binder fluid predominantly containing flux. The function of the solder paste is to add bulk solder to affix the device to the pad. Solder-paste dispensing and ensuing technology were developed in the fabrication of ceramic surface-mount device assemblies. The printed wiring subassembly containing surface-mount devices resting on the solder-paste-coated pads is then reflowed by employing high-temperature inert solvent vapors (vapor-phase condensation reflow soldering) or by convection and infrared heating. A cleaning step to remove residues from flux binder of the solder paste can be imposed if the binder is corrosive. The assembly at this stage contains the soldered-in-place surface-mount devices on the top surface; through-hole-mounted devices are then positioned by hand or automatically with extended leads crimped to bottom side pads. The printed wiring circuit is then populated with adhesively bonded surface-mount devices on the board's underside, followed by wave-solder assembly.

Component Packaging for Printed Wiring Assembly

The printed wiring circuit, in conjunction with assembled components, makes up the electronic assembly product. Inherent in the ultimate reliability of the product is the compatibility of the printed wiring and components to the assembly process. In simple assembly of surface-mount and through-hole devices, such hermetically packaged devices as chip resistors and capacitors mounted on the underside of printed wiring and plastic, metal or glass packaged devices are designed to survive the thermal rigors of the wave-solder assembly process. In the case of HICs, the surface-mount devices had to be hermetic to survive solvent cleaning. The postcleaning to remove flux residues of the underside of the printed wiring having through-hole-mounted devices with chlorofluorocarbon or chlorinated solvents was the preferred method to achieve assembly cleanliness and reduce induced contamination of circuit tracks. Plastic packaging for these devices was not specifically designed to be immersed in solvent during flux removal.

With the advent of mixed assembly, devices on both the top and bottom side of the printed wiring are solder assembled. The component packaging has to be compatible with total-immersion cleaning to assure complete removal of residual corrosive-flux residues. Recent worldwide interest in removal of chlorofluorocarbons and halocarbons from use because of their contamination of the atmosphere has introduced aqueous and semiaqueous total-immersion cleaning during circuit-pack assembly. Component packaging is evolving to accommodate the requirements of total-immersion exposure to solvents and to aqueous environments.

Recent developments in flux technology have resulted in the introduction of "non-clean" fluxes and soldering in inert atmospheres. These fluxes have sufficient activity to permit solder assembly without a post-assembly cleaning. The flux residues remain on the printed wiring without compromising circuit reliability.

Reliability of Assemblies

The guidelines that follow can assist in assuring the reliability of a printed wiring product.

1. Compliance with acceptable physical and mechanical design concepts
2. Stringent new assembly and manufacturing processes. Samples of the products should be qualified initially to assure compliance with all design requirements for the class of product. These tests shall include environmental stress tests, encompassing temperature, humidity, and bias, to establish failure modes and acceptable failure levels.
3. Establishment of a quality assurance program. Establish inspection procedures to qualify components and printed wiring for the assembly process. Manufacturing control should be performed by a statistical process.
4. Product qualification and field performance. In-process testing should be done to determine presence of ionic contamination by solvent-extract conductivity if corrosive fluxes are employed in assembly. Visual examination for coating/laminate delamination and solder-joint integrity should also be performed. Documentation of field performance failure is necessary to assure corrective action.

Acknowledgment: Teflon, Kapton, and Kevlar are registered trademarks of E. I. du Pont de Nemours.

Bibliography

Combs, C. F., Jr., (ed.), *Printed Circuit Handbook*, 3rd ed., McGraw-Hill, New York, 1987.
The Institute for Interconnecting and Packaging Electronic Circuits, 7380 North Lincoln Avenue, Lincolnwood, IL 60646. A major source of updated information on interconnection technology. In addition, the institute provides guidelines for industrywide standards and specifications for printed wiring.
Kear, F. W., *Printed Circuit Assembly Manufacturing*, Marcel Dekker, New York, 1987.

References

1. Boddy, P. J., Delancy, R. H., Lahti, J. N., Landry, E. F., and Restrick, R., Accelerated Life Testing of Flexible Printed Circuits, *14th Annual Proceedings, IEEE Reliability Physics Symposium*, 108–117 (1976).

2. Ammann, H. H., and Jocher, R. W., Measurement of Thermo-Mechanical Strains in Plated-Through-Holes, *14th Annual Proceedings, IEEE Reliability Physics Symposium*, 118–120 (1976).
3. Oien, M. A., A Simple Model for the Thermo-Mechanical Deformation of Plated-Through Holes in Multilayer Printed Wiring Boards, *14th Annual Proceedings, IEEE Reliability Physics Symposium*, 121–128 (1976).
4. Oien, M. A., Methods for Evaluating Plated-Through-Hole Reliability, *14th Annual Proceedings, IEEE Reliability Physics Symposium*, 129–131 (1976).
5. Hines, J. N., Measurement of Land to Plated-Through-Hole Interface Resistance in Multilayer Boards, *14th Annual Proceedings, IEEE Reliability Physics Symposium*, 132–134 (1976).
6. Rudy, D. A., The Detection of Barrel Cracks in Plated-Through Holes Using Four Point Resistance Measurements, *14th Annual Proceedings, IEEE Reliability Physics Symposium*, 135–140 (1976).
7. Ammann, H. H., and Farkass, I., Simulation of Thermal Stress Stimuli in the Testing of Printed Wiring Products, *14th Annual Proceedings, IEEE Reliability Physics Symposium*, 141–146 (1976).

W. BERNARD WARGOTZ

Communication Protocols for Computer Networks: Fundamentals

Introduction

The ultimate objective of a communications system is to allow entities to exchange information. The entities may reside within the same system, or may be physically located in different environments. We use the word *system* to denote a collection of data-handling devices and peripherals that are capable of information processing and transfer. As the geographical distance between the communicating entities increases, the complexity of designing such a communications system increases remarkably. Several functions need to be performed in order to allow the orderly exchange of information between the communicating entities. These functions include the management of logical channels, data transfer at different levels, flow and congestion control, routing, deadlock avoidance, and concurrency control. These functions fall into two categories, namely, communication functions and network resource-allocation functions. Due to the physical separation, the communication entities must conform to a mutually acceptable set of rules for generating and interpreting the messages they exchange. The set of conventions governing this exchange of data between the communicating entities is referred to as a *protocol*. The key elements of a protocol include syntax, semantics, and timing. The *syntax* of a protocol is concerned with such issues as data format, encoding techniques, and signal levels. The *semantics* of the protocol define the control information involved in the coordination of the data exchange and the management of errors. Finally, the *timing functions* ensure the synchronization and speed matching between the communicating entities. Stated differently, the key elements of the protocol define what is communicated, how it is communicated, and when it is communicated.

Protocols have been part of human practices since the early stages of history, as people have devised various techniques for communicating their thoughts, needs, and desires to others. Written messages and fire signals were then the basic vehicles for these people to achieve communication adequately. The discovery of the telegraph in 1838 ushered in a new epoch in communications and paved the way to the developments and implementations of increasingly more sophisticated communications systems. Over the last few decades, the revolutionary changes in chip size to performance ratio brought about by the very-large-scale-integrated (VLSI) design techniques and the resulting improvement in processing storage and transmission link capacities saw the merger of the fields of data communications and computer science. The successful integration of data, voice, and video is driving the technology and the technical-standards organizations to a unified treatment of the information, making virtually all data and information sources easily accessible. It is the purpose of this article to provide a unified view of the fundamentals involved in computer communications.

Network Architectures and Protocols

Based on the techniques used to transfer data, communications networks can be classified as switched networks or broadcast networks. The three main types of switched networks are circuit-switched networks, message-switched networks, and packet-switched networks. In circuit-switched networks, the physical path between the source and destination must be established before data can be transmitted. Upon establishing the connection, the circuit remains exclusively and continuously dedicated to the ongoing communication until completion. Message switching does not require a dedicated physical path between the sender and the receiver. The message first travels from its source to the next node in the path. When the entire message is received at the intermediate node, the message is temporarily stored until the link to the next node becomes available. This store-and-forward procedure continues until the packet reaches its destination. Packet switching is basically similar to message switching. The only difference between the two strategies is that packet switching decomposes messages into smaller packets to overcome the long transmission delays inherent in message switching. Message decomposition allows many packets to be transmitted simultaneously, thereby creating a pipeline effect.

Broadcast systems have no intermediate switching nodes. All stations share a single transmission channel. Packets transmitted by one station are received by all other stations. An address field within the packet specifies the destination of the packet. Packets that are intended for other stations are ignored. Two similar types of broadcast networks are packet radio and satellite networks. Another common instance of broadcasting is the local-area network.

The task of designing a communications network is too complex to be handled as a monolithic unit. An alternative to the monolithic design of a communications protocol is a structured approach that aims at dividing the communication task into manageable parts. The approach describes the communication functions in terms of an architecture. The architecture defines the relationship and interactions between network services and functions through common interfaces and protocols. This viewpoint has been adopted by the International Organization for Standardization (ISO) in their recommendation for a standard network architecture. The proposed model is depicted in Fig. 1. The model referred to as the Open System Interconnection (OSI) defines a framework for the specification of protocol standards for connecting heterogeneous computers. The model defines the rules and conventions for various functions within each layer, specifies the general relations among these functions, and determines the constraints on the types of functions and their relations. Note that since the model serves as a framework for standards, it does not specify either the details of the implementation or the definition of the interlayer interfaces. The OSI model provides the basis for two open systems that conform to the reference model and the associated standards to exchange information. Notice that the communicating systems need not implement the same interlayer interfaces, since these are not visible from the outside.

The ISO reference model defines seven layers: the physical layer, the data-link layer, the network layer, the transport layer, the presentation layer, the

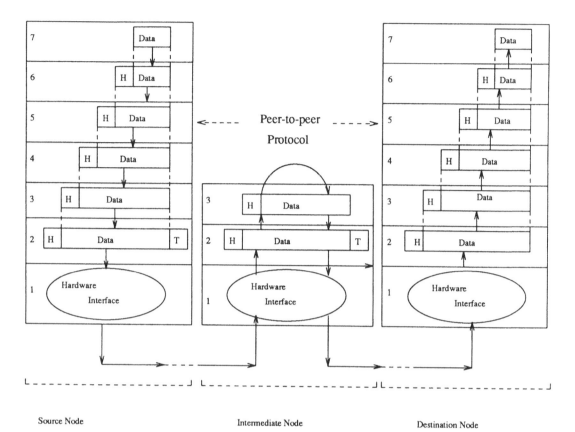

Intermediate Node Destination Node

FIG. 1 International Organization for Standardization/Open System Iinterconnection communication model: *1*, physical layer; *2*, data-link layer; *3*, network layer; *4*, transport layer; *5*, presentation layer; *6*, session layer; *7*, application layer; H = header; T = trailer.

session layer, and the application layer. Each layer performs a related subset of functions required for communication, and adds value to the services provided by lower layers so that a user at the highest layer is offered the full service needed to interact with other users in the network. Direct communication between peer layers in two different machines is only achieved at the physical layer. Because of its widespread use and attractive architecture, the OSI model is adopted as an organizational tool for the discussion of network communications protocols in this article. Notice, however, that an alternate approach, which resulted from the extensive research and practical experience of the Advanced Research Projects Agency Computer Network (ARPANET, currently ARPA Internet), is increasingly gaining importance in the field of communications. The fundamental principle of the ARPANET architecture is based on the concept that communication between local and remote processes can be accomplished by first identifying the host where the remote process resides, and then identifying the remote process within the remote host. Both actions can be handled independently. Consequently, the task of the network in the ARPANET model is reduced mainly to routing data between hosts. The network need not be concerned with how the data is directed to the remote process within the host. Based on the concept above, the ARPANET architecture was conceived as a hierarchical ordering of protocols that can be organized into four layers: the application layer, the transport layer, the Internet layer, and the network layer. Since the model is hierarchically organized, entities at higher layers may bypass an adjacent layer and use the services of a lower layer directly. Notice that both approaches recognize that the communication task is too hard and too diverse to be carried out by a single unit. Nevertheless, philosophical differences between the ARPANET hierarchical approach and the ISO layered approach exist. Advocates of the modular and hierarchical structure claim that the ISO layered approach tends to be prescriptive rather than descriptive, thereby reducing the freedom of the designer to develop efficient cost-effective and rich protocols. While the practical differences between the two approaches is still being debated, several protocols of the ARPA Internet suite are gaining wide acceptance. These protocols are discussed within the organizational framework of the ISO model.

Physical-Layer Protocols

In its simplest form, data communication takes place between two devices that are connected by a transmission medium. The devices, which may include simple terminals as well as large computer systems, are generically referred to as data termination equipment (DTE). A DTE accesses the transmission medium through the mediation of data circuit terminating equipment (DCE). DCEs at each side of the transmission medium are responsible for providing the functions required to establish, maintain, and terminate a connection. In addition, the DCE may be required to perform signal conversion and coding that is needed to interface the DTE to the communications system. Every DTE interacts with the DCE on its side by exchanging both data and control information over a set of interchange circuits. The interactions between the DTE and DCE are governed by a complementary interface. The exact nature of the interface is specified by a set of standards that define the mechanical, electrical, functional, and procedural characteristics of the interface.

RS-232 Standard

Several standards have been developed to describe the exact interface between the DTE–DCE pair. The first specification is known as RS-232 (see also Ref. 1). It is the C-level version of the interface, RS-232-C, issued in October 1965 and reaffirmed in June 1981, that gained a widespread use in the community. Notice, however, that a D-level version of the interface was issued in January 1987, but its use is growing slowly as vendors are making their equipment compatible. The C-level standard calls for a 25-pin connector with a specific arrangement of leads. The electrical characteristics specify a voltage more negative than −3 V with respect to the common ground to represent a binary 1, and a voltage more positive than +3 V to represent a binary 0. Typically, however, the voltage of a binary 1 bit is −5 to −10, and a binary 0 bit is +5 to +15. The functional specifications of the RS-232 circuits are summarized in Table 1.

As can be seen from Table 1, the transmit data and receive data circuits of the RS-232 interface are used by the DTE and the DCE to transmit data. The other lines collectively perform the control and timing functions required for setting up and clearing the connection. The DTE expresses its wishes to send data by asserting the Request To Send circuit. The modem responds by asserting Clear To Send, thereby allowing data to be transmitted over Circuit BA. Notice that a Request To Send can only be attempted if the Data Set Ready circuit is asserted. The Carrier Detect circuit allows the DCE to inform the local DTE that the remote DTE is transmitting. Timing circuits are only used during synchronous transmission. They provide a clock signal for bit-synchronization purposes. The clock signal is supplied by the DCE to the DTE on the transmit and receive signal element timing control circuits. These circuits are assigned Pins 15 and 17, respectively.

In some situations, the two DTEs are close enough to signal each other without the need for a DCE. However, when no DCE is provided, the interconnection wiring must be modified to emulate the role of the DCE. The resulting wiring configuration, often referred to as *null modem*, is depicted in Fig. 2. As can be seen in the illustration, in addition to reversing the transmit and receive circuits, some of the control circuits are also reversed.

Although the distance over which the signal may travel is not specified by the RS-232-C interface, the maximum amount of capacitance is limited to 2500 picofarads. This limitation determines the maximum distance for signal propagation supported by the interface. Over the years, the RS-232-C interface has been very adequate for the speeds and distances usually encountered in the communications environment.

RS-449 Standard

The need to achieve higher data rates over longer distances and the necessity to provide for a better control of the DTE over the DCE called for a new set of standards to perform the new functions. The new specification, identified as RS-449, was issued in 1975 and was supposed to supersede RS-232. RS-449 is both a mechanical and functional specification that also contains two subspecifications, RS-422-A and RS-423-A, which identify the electrical characteristics of the interface standards. The subspecification RS-422-A defines a balanced electrical interface and uses a differential amplifier to reduce the sensitivity to noise. On the other hand, RS-423-A defines an unbalanced electrical interface. In essence, RS-449 retains all the interchange circuits of RS-232-C with the exception of the protective ground, and introduces 10 new circuits to provide greater DTE control over the DCE. The additional circuits are Send Common (SC), Receive Common (RC), Terminal in Service (IS), New Signal (NS), Select Frequency (SF), Local Loopback (LL), Remote Loopback (RL), Test Mode (TM), Select Standby (SS), and Standby Indicator (SI). If an RS-422-A balanced

TABLE 1   RS-232-C Interface

Pin	RS-232-C Name	Direction	Circuit Name	Function
1	AA	Both	Protective ground	Conductor attached to the equipment frame and may be connected to external grounds as required by local codes
7	AB	Both	Signal ground	Conductor to establish a common ground reference potential for all interchange circuits except AA
2	BA	To DCE	Transmitted data	Data signal generated by DTE
3	BB	To DTE	Received data	Data signal received by DTE
4	CA	To DCE	Request To Send	Signal the wish of the DTE to transmit. Control the direction of data transmission on half duplex.
5	CB	To DTE	Clear To Send	Signal that DCE is ready to transmit data. Response to clear to send.
6	CC	To DTE	Modem Ready	Signal indicates the status of the local DCE
20	CD	To DCE	Terminal Ready	Signal indicates the status of the local DTE
22	CE	To DTE	Ring Indicator	Signal indicates that DCE is receiving a ringing signal of the communication channel
8	CF	To DTE	Carrier Detect	Signal indicates DCE is receiving carrier signal
21	CG	To DTE	Signal Quality Detector	Signal indicates high probability of an error on the received data
23	CH	To DCE	Data Signal Rate Selector	Asserted to select the higher of two possible data rates
	CI	To DTE	Data Signal Rate Selector	Asserted to select the higher of two possible data rates. If secondary channel is not used then CI is on pin 12.
24	DA	To DCE	Transmit clock (DTE source)	Providing timing information relative to transmitted signals
15	DB	To DTE	Transmit clock (DCE source)	Provide DTE with timing information relative to Circuit BA
17	DD	To DTE	Receive clock (DCE source)	Provide DTE with receive timing information relative to BB
14	SBA	To DCE	Secondary Transmit Data	Performs the same function as Circuit BA for the secondary channel in reverse channel modem
16	SBB	To DTE	Secondary Receive Data	Performs the same function as Circuit BB for the secondary channel
19	SCA	To DCE	Secondary Request To Send	Performs the same function as Circuit CA for the secondary channel
13	SCB	To DTE	Secondary Clear To Send	Performs the same function as Circuit CB for the secondary channel
12	SCF	To DTE	Secondary Carrier Detect	Performs the same function as Circuit CF for the secondary channel

Data Terminal Equipment   Data Terminal Equipment

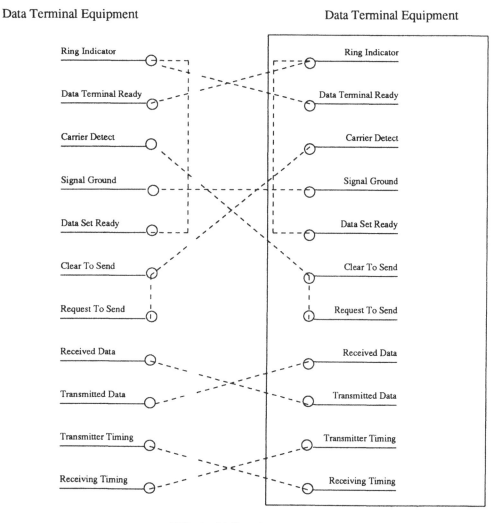

FIG. 2 Null modem.

circuit is used, a data rate of up to 100,000 bits per second (bps) can be supported over a distance of 4000 feet. As the distance decreases, the data rate of the interface increases and reaches 10 million bps over a distance of 40 feet.

X.21 Interface

It is clear from the discussion above that new functionalities of the interface were introduced at the expense of extra circuits and connections. An alternative way to provide the functionalities sought by RS-449 is to upgrade the logical functions of the DTE–DCE pairs and reduce the number of circuits required. This approach was used to design the X.21 interface. This interface, which represents a major improvement over RS-232 and RS-449, specifies a 15-pin connector and provides for balanced and unbalanced modes. Of the specified pins, only eight are currently in use. The X.21 interface is depicted schematically in Fig. 3.

As can be seen from the illustration, the Transmit (T) circuit carries the data from the DTE to the DCE during the data-transfer phase. However, the T circuit may also carry control information from the DTE to the DCE during the call set up or the call clear. Similarly, the Receive (R) circuit carries the data and control information from the DCE to the DTE. During the data-transfer phase, the Control (C) circuit remains on. The C circuit is controlled by the DTE, and is used to indicate to the DCE the meaning of the data being transmitted. The DCE uses the Indication (I) circuit to indicate to the DTE the type of data being transmitted on the R circuit. Notice that during the control phases, both C and I signals may be either On or Off, depending on the current state of the protocol.

In the X-21 interface, two states are allowed for the DCE and three for the DTE. When the DCE is in the Ready State, the R signal is 1 and I is Off. In this state, the DCE is ready to operate. When the DCE is in the Not Ready State, the T signal is 0, I is Off, and the service is not available. The DTE may be in the Ready State, Uncontrolled Not Ready State, or Controlled Not Ready State. In the first state, the readiness of the DTE to operate is indicated by a steady 1 on the T circuit and a control Off on C. The Uncontrolled Not Ready State indicates that the DTE is unable to enter operational phases owing to an abnormal condition such as a fault condition. In this state, the T control circuit

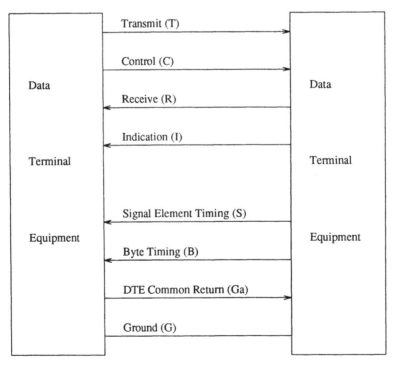

FIG. 3 X.21 interface.

is 0 and C is Off. Finally, in the Controlled Not Ready State, the DTE is operational but is not ready to accept calls at the current time. This is indicated by a pattern of alternating binary 0 and binary 1 on the T circuit, while maintaining the C circuit in the Off state.

The Signal Element Timing provides the DTE with timing information. This information is used to receive the incoming data over the R circuit properly. Byte Timing provides the DTE with 8-bit byte element timing. The signal is normally On, but changes to Off at the same time the S circuit signifies the last bit of an 8-bit byte. The service is optional, and may be agreed upon during the data-transfer phase by the DTE and the DCE.

Data Encoding

The capability of moving information between a sender and a receiver relies on the manipulation of an electrical signal. Two types of signal, analog and digital, are commonly used. Analog signals are represented by continuously varying electromagnetic waves that may be propagated over a variety of media, whereas digital signals are categorized as a sequence of voltage pulses. Note that both analog data and digital data could be encoded into either form of signal.

Digital data can be transmitted by encoding each bit of the data into an element of the digital signal. Non-Return-to-Zero (NRZ), return-to-zero, biphase, delay modulation, and multilevel binary techniques are the most commonly used encoding techniques. The various techniques are usually evaluated based on their synchronization and error-detection capabilities. Although non-return-to-zero techniques are still widely used in data communications systems, biphase techniques are rapidly gaining widespread recognition. The differential Manchester, a biphase encoding technique, has been adopted in local-area networks.

Digital data may also be transmitted using an analog signal. The process, known as *modulation*, involves the manipulation of one or more characteristics of the carrier signal to encode the elements of the digital data. The characteristics of the carrier signal include amplitude, frequency, and phase, resulting in three modulation techniques, namely, amplitude-shift keying, frequency-shift keying and phase-shift keying. More efficient use of bandwidth could be obtained if an element of a signal carried more than one bit of data. Based on this observation, a common encoding technique known as quadrature phase-shift keying uses multiple phase shifts to encode more than one bit into a signal element. This scheme achieves a higher data rate, expressed in bps, for the same signal modulation rate, expressed in bauds.

A process known as *digitization* can be used to encode analog data into digital signals. The most commonly used techniques to achieve the digitization of analog data include Pulse Code Modulation (PCM) and Delta Modulation (DM). The PCM scheme is based on sampling the signal at regular time intervals and at a fixed rate. The samples are then represented as narrow pulses whose amplitude is proportional to the value of the original signal. This phase is usually referred to as Pulse Amplitude Modulation (PAM). The PAM samples are quantized, approximated by an integer, and encoded as a digital signal. The basic approach of the DM scheme is to approximate the analog signal by a binary staircase function. At each sampling time, the approximating stair function moves either up or down, resulting in the generation of either 1 or 0, respectively. The accuracy of the scheme depends on the sampling rate. The principal advantage of the DM scheme over the PCM scheme is the simplicity of its implementation. In general, however, the PCM scheme exhibits superior performance in handling noise for the same data rate.

The transmission media commonly used in communications networks can be categorized as conducted or radiated. The conducted media include telephone and telegraph wires, private wires, coaxial cables, and fiber optics. Radiated media include radio broadcast, microwave radio transmission, infrared transmission, and satellite transmission. Each of these communications media has unique performance characteristics that make their suitability for a specific situation depend on the environment, the desired speed and security, the distance covered, the susceptibility to error, and the overall cost.

Regardless of the design or the type of the communications medium, however, the transmitted signal is always subject to degradation due to the imperfect response of the medium and to the presence of noise. Noise is usually caused by thermal agitation of electrons in a conductor, intermodulation, cross talk, or impulses. The performance of the transmission medium may be improved by the use of an adequate encoding scheme. Nevertheless, errors are still likely to occur and a mechanism to improve the reliability of the medium is required. This is the objective of the data-link layer.

Protocols of the Data-Link Layer

The basic purpose of the data-link layer is to establish communication between two ends of the data link, maintain and provide for orderly transfer of data frames during the data-transfer phase, and finally terminate the connection upon data-transfer completion. Unlike the physical layer, which is concerned basically with the transfer of signals over a communication link, the data-link layer aims at achieving reliable data transmission over a potentially unreliable transmission link. The basic unit of communication at the level of the data-link layer is referred to as a frame. A *frame* is defined as a serial block of bits containing data and control information. The basic structure of the data-link protocol is derived from the nature of the transmission medium being used. However, in order to accomplish the basic requirements, a data-link-layer protocol usually performs the basic functions of link management, frame segmentation, frame synchronization, error checking, and flow control; these functions are discussed below.

The *link-management function* establishes an active connection over an already existing transmission link, maintains the sustained exchange of data, and initiates the termination process to relinquish control of the link following the transfer of information. Functions at this level may also be required to cope with abnormal conditions and to allow for the detection and recovery from such situations as loss of response or link failure.

The *frame-segmentation process* consists of breaking long streams of data into more manageable, less error-prone data frames.

The process of *frame synchronization* refers to the capability of the receiver to correctly receive and interpret transmitted data frames. One way to achieve this synchronization is by inserting timing gaps between frames. However, equally spaced timing gaps between frames cannot be guaranteed by networks. Several methods have been proposed as an alternative to timing-based frame synchronization. The most commonly used techniques include character count, character-based frame delimiters, and flag-based frame delimiters. Character-count techniques reserve a specific field in the header of the frame to denote the number of characters in the frame. The receiving station uses the information above to determine the end of the current frame. The major drawback of this method is its dependency on the correctness of the character-count field. It may easily happen that noise affects the character count and causes the receiver to lose track of the frame delimiters.

Character-oriented schemes achieve synchronization by forcing each frame to start and end with a specific character sequence. The most common sequences used are DLE (Data Link Escape), STX (Start of TeXt) to start a frame, and DLE ETX (End of TeXt) to end a frame. To allow the transparent transmission of the sequences above as part of the data field, the method uses a technique called character stuffing, which precedes every DLE with a DLE. The major shortcoming of this method is its dependency on the type of code. As networks developed, the limitations of the character-oriented methods became obvious and a new method was required. The new method, usually referred to as bit oriented, uses a unique configuration of bits, called a *flag*, as a frame delimiter. The most commonly used flag is 01111110. The occurrence of the flag in the data field is prevented by using bit stuffing. The method breaks the pattern by automatically inserting a zero after five consecutive ones. This method guarantees that the boundary between two frames can be unambiguously recognized by the flag pattern. Flag-based framing methods allow faster recovery from loss of synchronization, since the only pattern the receiver has to search for is the flag.

As stated above, transmission impairments usually result in data errors. *Error control* requires the implementation of an error-detection and correction mechanism. This is usually achieved by appending to the original frame of bits an additional set of bits that constitute an error-detecting code. The code itself does not convey any information, but is used by the receiver to detect errors. Several schemes based on a certain degree of information redundancy have been proposed in the literature. The most commonly used methods include parity bit, longitudinal and vertical redundancy check, and cyclic redundancy check. The data-link layer must be capable of detecting and correcting errors to achieve a high degree of information integrity. The most commonly used procedure to control errors is for the receiver to detect the error and request retransmission of the corrupted frame.

The term *flow control* refers to the regulation of traffic on an individual connection between a source and a destination. The main goal of flow control is to prevent overflow of the buffers dedicated to an individual connection. It may be perceived as a bipartisan agreement between the source and destination that locally limits the amount of traffic contributed to the overall network traffic.

The simplest form of control is known as Stop-And-Wait. In this scheme, the sender transmits a frame and waits for the receiver to acknowledge the reception of the frame. Notice that if the two stations are simultaneously exchanging data, acknowledgments may be piggybacked in data frames. The acknowledgment may be either positive or negative. Positive acknowledgments (ACKs) are issued by the receiver upon receiving an error-free frame. These are invitations from the receiver to the sender to transmit the next frame. Negative acknowledgments (NAKs) are sent by the receiver upon detection of an error and are indications to the sender that the frame needs to be retransmitted. This retransmission technique is usually referred to as Automatic Repeat Request (ARQ). To account for possible losses of the acknowledgments, this method uses timers. At the occurrence of a time out, the sender retransmits the unacknowledged frame. The main advantage of this method is its simplicity. The major drawback of the method, however, is the time spent by the sender waiting for acknowledgments. The waiting time increases with the link propagation delay, causing poor performance of the scheme.

An alternate solution, known as the *sliding-window protocol*, allows multiple frames to be in transit at one time. Prior to data transfer, the sender and the receiver agree on an initial window size that determines the number of frames the sender can send without waiting for an acknowledgment. The sender maintains a sending window of the frames it is allowed to send, whereas the receiver maintains a receiving window of the frames it is allowed to receive. Each frame is uniquely identified by a sequence number (modulo the size of the window). The receiver acknowledges frames by sending an acknowledgment containing the sequence number of the next frame expected. Note that this scheme can be used to acknowledge multiple frames, as an acknowledgment containing a sequence number $n + 1$ may be used to acknowledge all frames including n. The efficiency of the sliding-window protocol depends on the size of the window and the link propagation delay.

Two variants of the sliding-window protocol are proposed: Go-Back-N ARQ and Selective-Repeat ARQ. Both schemes aim at improving the throughput of the link, especially if the propagation delay is not negligible compared to the frame transmission time. Go-Back-N ARQ allows the sender to transmit a series of data frames, within the sending window, without waiting for an acknowledgment. The receiver uses the multiple acknowledgment scheme to positively acknowledge error-free frames. When the receiving station detects an error in a frame, it responds with a NAK containing the sequence number of the frame in error, and discards all further incoming frames that the transmitter may have sent. Upon receiving a NAK, the sender retransmits the frame in question and all succeeding frames. As in the case of Stop-And-Wait ARQ, the Go-Back-N protocol uses timers to detect time-outs. In contrast with the Go-Back-N scheme, the selective repeat ARQ uses a more refined approach, which requires the sender to retransmit only the negatively acknowledged frame. Although more efficient, the scheme requires that the receiver contain enough buffer space to hold the out-of-sequence frames. In addition, more complex logic for sending and receiving frames out of sequence is required at the sender and the receiver side, respectively.

The general principles introduced above have been used extensively in practice to implement data-link protocols. Based on the framing technique used, these protocols can be classified into three categories: character oriented, byte-count oriented, and bit oriented.

Character-Oriented Protocols

The most popular character-oriented scheme is Binary Synchronous Communication (BSC), developed by IBM. This scheme uses specified control characters from a given character set to delimit frames and to control data interchange. The set of control characters can be divided into three groups. The first group, used to delimit messages and blocks within a message, includes SYN, SOH, STX, ETX, ETB, and ITB. The SYN character, Synchronous Idle, is used as a time fill in the absence of any data or control character to maintain synchronization. The SOH (Start of Heading) character is used at the beginning of a sequence of characters that constitutes a header. The STX (Start of Text) character denotes the beginning of a sequence of characters that is to be transmitted entirely as an entity. The sequence is referred to as text. The STX character may also be used to terminate a sequence of characters started with SOH. The character ETX (End of Text) indicates the end of a message, or terminates the last block of the message if the message contains multiple transmission blocks. In the latter case, ETB (End Transmission Block) is used to terminate preceding blocks. The ITB (End of Intermediate Transmission Block) character is used to terminate an intermediate block. Notice that the block check character for error control is sent immediately following ITB, but no turnaround occurs.

The second group of characters, used to control the half-duplex exchange of information, include ACK0 (Acknowledge Even-Number Message), ACK1 (Acknowledge Odd-Number Message), ENQ (Enquiry), WACK (Wait before Transmit Positive Acknowledgment), NACK (Negative Acknowledgment), EOT (End of Transmission), and RVI (Reverse Interrupt). ACK0 and ACK1 indicate that the preceding transmission block has been achieved successfully, and the receiver is ready to accept the next block. The alternate use of ACK0 and ACK1 provides a sequential checking control for a series of replies. WACK provides a positive acknowledgment of the preceding transmission block, but indicates also that the receiver is not ready yet for the next block. NACK indicates a negative acknowledgment of the last transmitted block. EOT signals the end of transmission and resets all stations on the line to control mode. The RVI character is used in BSC to request premature termination of the transmission in progress. RVI is usually sent instead of ACK0 or ACK1. Consequently, the character also acknowledges the last transmitted block.

The third group of characters, represented by a combination of the DLE character with other control characters, is used to achieve a transparent mode of transmission, which allows control characters to be included in the data field of the frame.

The BSC protocol suffers several shortcomings that limit its use in computer networks, and has led to the introduction of new schemes. The limitations include a heavy dependency on the character code being used, a lack of sequence numbers that allow better performance in a long-distance, high-speed environment, the incapability to handle full-duplex channels, and the unnecessary complexity required to handle transparent and nontransparent transmission modes.

Although BSC brought about a major advancement in data communications, it has significant shortcomings that helped motivate the development of other data-link control protocols. The major weaknesses of BSC are due to the fact that it can be used only in half duplex, and would require major modifications to operate in full duplex. The protocol operates in a select-hold mode, in which a pair of stations hold the data link until the completion of their communication. This mode of access may cause excessive delays for the other stations. The variety of options and special features make the implementation difficult. In addition, the handling of transparent text is clumsy and cumbersome.

Byte-Count-Oriented Protocols

The second class of protocols, byte-count-oriented protocols, were developed as an attempt to reduce the complexity of the character-oriented data-link protocol. The most commonly used protocol of this class is Digital Data Communication Message Protocol (DDCMP). The basic principle of DDCMP is to include a byte count in the frame header to keep track of the number of characters in the data field, thereby achieving transparency without the use of control characters. Due to the importance of the byte-count information, the protocol dedicates a special cyclic redundancy check to protect the frame header. DDCMP aimed at being very general. The protocol allows handling different environments, including half duplex, full duplex, synchronous, and asynchronous serial and parallel transmissions. The protocol also handles different configuration environments including multipoint or point-to-point systems. The protocol uses frame sequence numbers to allow a sender to transmit multiple frames and the receiver to acknowledge a group of frames simultaneously. The principal advantage of the protocol is its simplicity and its generality. However, DDCMP has several major drawbacks. The first is due to its heavy dependency on the reliability of the byte-count information. A wrong byte count may cause the receiver to go out of synchronization. Several frames may be discarded before resynchronization is achieved. The second drawback is due to a relatively short header compared to the maximum frame length. It may easily happen that, because of the relatively short processing time required by the header, the system may be required to handle a large amount of data on short notice.

Bit-Oriented Protocols

Character-oriented protocols have been extensively used in many applications. However, their shortcomings made them unsuited for modern interactive applications, which led to the development of bit-oriented protocols. In addition to satisfying a wide range of data-link requirements, bit-oriented protocols aimed at achieving code independence, adaptability, high reliability, and high efficiency. Several bit-oriented protocols exist and are known under a variety of names and mnemonics. Most of the proposed protocols derive from the Synchronous Data-Link Control (SDLC) protocol developed by IBM. In their effort to produce international standards, the American National Standards Institute (ANSI) and ISO proposed modified versions of SDLC. ANSI proposed Advanced Data Communication Control Procedures (ADCCP), while ISO produced High-Level Data-Link Control (HDLC). The International Telegraph and Telephone Consultative Committee (CCITT) adopted and modified HDLC to produce the Link Access Procedure (LAP) as part of the X.25 packet-switched network standard. The LAP protocol was later modified to produce a balanced version (LAP-B), which is more compatible with HDLC. All these protocols are based on the same principles and the differences between them are very minimal. Consequently, the remainder of this discussion focuses on HDLC.

In order to satisfy a wide variety of configurations, HDLC defines three types of stations: primary, secondary, and combined. Secondary stations operate under the control of primary stations, which have the responsibility of maintaining separate logical links for each secondary station. Dialogue between the primary and secondary stations is established by the exchange of commands and responses over an unbalanced-link configuration. A combined station assumes both roles of primary and secondary stations. Communication between two combined stations is achieved through a balanced link. The transmission may be half or full duplex. Three types of frames, information (I frame), supervisory (S frame), and unnumbered (U frame), are used to describe different operations of the protocol. The I frame is used to carry the exchange of data between two stations. The bit poll in this frame specifies whether it is a command or a response frame. The S frame is used for flow and error control. The scheme allows for both go-back-N and selective repeat ARQ. The U frame enforces such other control functions as mode setting and recovery.

The main functions provided by HDLC include link management and data transfer. Prior to any exchange of information, a logical connection between the two communicating stations must be established. This is accomplished by two unnumbered frames. In a multidrop link, the primary station first sends an SNRM (Supervisory Normal-Response Mode) frame containing the address of the secondary station. The purpose of the frame is to poll the secondary station for an eventual data transmission. The secondary station responds with a UA (Unnumbered Acknowledgment) frame containing its own address. The procedure used to set up a point-to-point link is similar to the one used for multidrop links, except that an Asynchronous Balanced Mode (ABM) is selected and hence an SABM (Set Asynchronous Balanced Mode) frame is sent first. In this mode, both sides may initiate the transfer of data independently. After all data have been transferred, a DISC (Disconnect) frame is sent to terminate the connection.

The two most important aspects of the data-transfer phase of HDLC are the error-control and flow-control techniques the protocol provides. Error control is provided by a continuous ARQ procedure with either a selective repeat retransmission or a go-back-N retransmission strategy. Flow control is achieved by a basic window mechanism. The basic operation of these procedures is similar to the mechanisms described above.

Local-Area Network Protocols

In many local-area networking environments, communication among stations is usually provided by means of a unique channel. It is characteristic of this channel that only a single station can transmit a message at any given time. Therefore, shared access of the channel requires the establishment of a protocol among the network stations.

The difficulty in designing an effective multi-access protocol arises from the spatial distribution of the stations. To reach a common agreement, the stations must exchange some amount of explicit or implicit information. However, the exchange of coordinating information requires the use of the channel itself. This recursive aspect of the multi-access problem increases the complexity of the protocol and the overhead of the channel. This issue is further complicated by the absence of an instantaneous state of the system. The spatial distribution of the system does not allow any station on the network to know the instantaneous status of other stations on the network. Any information explicitly or implicitly gathered by any station is at least as old as the time required for its propagation through the channel.

The factors that influence the aggregate behavior of a distributed-multiple-access protocol are the intelligence of the decision made by the protocol and the overhead involved. These two factors are unavoidably intertwined. An attempt to improve the quality of decisions does not necessarily reduce the overhead incurred. On the other hand, reducing the overhead may result in lowering the quality of the decision. Thus, a tradeoff between these two factors has to be made. A distributed-multiple-access protocol potentially can benefit from the globally available knowledge about the status of the critical resource. An update of the status of the transmitting station may be included in the transmitted information. The new status then becomes known globally to all to other stations in the network.

Determining the nature and extent of information used by a distributed-multiple-access protocol is a difficult task, but potentially a valuable one. An understanding of exactly what information is needed could lead to an understanding of its value. Most of the proposed distributed-multiple-access protocols operate somewhere along a spectrum of information ranging from no information to perfect information. Three types of information, predetermined, dynamic global, and local, can be readily identified. Predetermined information is known to all stations. Dynamic-global information is acquired by different stations during the evolution of the protocol. Local information is known to the individual station. The use of local information may result in a lack of coordination among stations and eventually failure to achieve the objectives of the protocol. Predetermined and dynamic-global information may result in perfect coordination among the stations, but usually exacts a price in terms of wasted resource capacity.

Over the last decade, several approaches to solve the multiple-access problem were proposed. These solutions attempt, by various mechanisms, to strike a balance between the amount of information provided to the stations and the overhead in providing it. Multiple-access protocols divide broadly into two classes: contention-based protocols and control-based protocols.

Contention-based protocols can be characterized as partitioning processes in which the set of contending stations is gradually reduced, according to some predetermined and globally known rules, until the set contains exactly one station. The remaining station successfully transmits its message. At the end of the transmission, a new partition is found. Several mechanisms have been proposed to determine the partitioning process. These mechanisms can be broadly divided into three groups. The first group is based on a probabilistic partitioning, in which stations in the contending set randomly reschedule their transmission. The amount of coordination among different stations varies from one scheme to another. The second group uses an address-based mechanism to achieve partitioning of the set of the contending stations. At the occurrence of a collision, the set of contending stations is reduced to those stations whose addresses fall within a specific range. Further collisions will reduce the range of addresses until a single station is uniquely determined. The third group of contention-based protocols uses the generation times of messages to partition the contending set of stations. A time window is determined, and only stations holding messages generated during the time window have the right to transmit. The window is recursively narrowed until it contains only one station.

Control-based protocols are characterized by collision-free access to the channel. Collisions are prevented by imposing an explicit or implicit ordering of the stations in the network. Explicit ordering requires a predetermined channel allocation; the channel is statically allocated to different stations on the network. Demand-adaptive protocols dynamically adjust in an attempt to satisfy the immediate requirements of the stations consistently. Demand-adaptive protocols can be divided further into two classes: the reservation class and the token-passing class. In the reservation scheme, stations reach a consensus about which station is next to transmit using a reservation procedure that typically precedes the successful transmission of a message. The token bus access scheme relies either on an explicit or implicit token that is passed among active stations. The active stations form a logical ring on which the token circulates. The token gives its holder the right to transmit.

Contention-based schemes are simple to implement. They do not require active channel interfaces and do not rely on global information circulating on the bus. Consequently, they are robust and reliable. Control-based access mechanisms reduce the degree of concurrency but guarantee a limit on the maximum time any given station must wait in order to access the channel.

Three standard protocols have been proposed to handle local-area network traffic: Carrier Sense Multiple Access with Collision Detection (CSMA/CD) (IEEE Standard 802.3); token bus (IEEE Standard 802.4); and the token ring (IEEE Standard 802.5). Note that these three standards are all defined in the data-link layer or, more precisely, in the medium-access-control (MAC) sublayer.

Carrier Sense Multiple Access with Collision Detection
(CSMA/CD) Bus Network

CSMA/CD is a random-accessing scheme in which a network station transmits only after sensing an idle channel. Should the channel be busy, the station must wait until it is clear before it can attempt transmitting again. Once it transmits its message, the station continues to listen to the channel to detect any collision with messages being sent simultaneously by other stations on the network. CSMA/CD methods ensure that collision can be detected by requiring message length to be greater than a predetermined minimum. In the event of a detected collision, all transmitting stations abort their operations and wait for a random amount of time before attempting to retransmit. This random backoff time ensures that one station will subsequently gain access to the channel. However, the total number of times a station must try before successfully accessing the channel is strongly influenced by such factors as network traffic volume, message length, and network physical length. Consequently, the CSMA/CD-based protocols perform optimally under light conditions, but the performance of such schemes is very sensitive to the traffic load. As the load increases, the scheme performance degrades significantly. The conflict that occurs due to random access can cause waste of bandwidth and, most of all, may result in an unbounded access delay. This last shortcoming makes the scheme unsuitable for real-time applications. In addition to its lack of determinism, the standard CSMA/CD does not support a priority function necessary to handle different classes of traffic characterized by different timing constraints.

Several variations of IEEE 802.3 have been implemented. The most commonly used is the Ethernet system, known as IEEE 802.3 Type 10 Base 5. The specifications allow for up to 1024 stations to achieve baseband transmission over a coaxial cable at 10 megabits per second (Mbps) over a distance of 2.5 kilometers (km). The cable is connected in 500-meter (m) segments. A second implementation, IEEE 802.3 Type 10 Base 2 (Thinnet), aims at providing a reduced version of the full capabilities of the Ethernet Standard. The Thinnet standard results in a simple, low-cost, and user-manageable network. A third implementation of the IEEE 802.3 standard was developed by AT&T. The specification, known as Type 1 Base 5 (StarLAN), uses unshielded twisted-pair cable. Stations are connected to the network by means of a hub, which acts like a switching system. The resulting network configuration is hierarchical.

Although CSMA/CD schemes were initially developed and implemented with coaxial cable, several attempts have been made to implement the scheme over fiber optics. Most of these schemes differ in the way they achieve collision detection. Two basic implementations, passive star and active star, are usually adopted. The passive-star approach uses either power sensing, pulse width, time delay, or directional coupling to allow stations to transmit. However, most of these schemes are either unreliable in detecting a collision or are hard to implement. Active-star based implementations usually use a miniature coaxial backplane bus to detect collision. The receiver converts the optical input signal into an electrical signal and places it on the receiving bus. When the control module detects a collision on the coaxial bus, it transmits a jam signal; otherwise it transmits the bit stream.

Token-Passing Bus Network

Token passing involves circulating a unique bit sequence (the token) among the network stations in a specific order. Only the station currently holding the token has access to the channel. Token bus incorporates the advantages of the token ring (see below). Stations are connected to the bus via taps. Since the taps are passive elements, as opposed to the active interfaces of a ring, bus networks are less susceptible to failure.

Token passing guarantees access to each network station within a prescribed period of time. The highly deterministic nature of this scheme is a critical quality for distributed real-time applications. In a standard token-passing protocol, three bits in the packet are used to represent up to eight different priority classes (PCs). However, only four priority classes are used in the current proposal. These four classes of service are synchronous (PC = 6), asynchronous urgent (PC = 4), asynchronous normal (PC = 2), and asynchronous time available (PC = 0). In this priority scheme, the lower priority packets are deferred when the network is heavily loaded. Network loading is computed at each station by measuring the time elapsed between two consecutive tokens. When the network is lightly loaded, a station passes the token and receives it again in a short period of time. As loading increases, the time for the token to return to the station increases. If the time exceeds a predetermined threshold value, low-priority traffic is deferred until the network load decreases. A separate threshold value exists for each of the three lower priority classes. The highest-priority class of packets can always be transmitted when the station receives the token and the token-holding timer is not expired.

The major drawback of token-based networks is the overhead involved in passing the token. This overhead has a significant impact on the performance of the network, especially in a light-load situation or when the traffic distribution among stations is asymmetric. Since the token-circulation path is static, the token may visit idle stations, resulting in a waste of bandwidth and an increase in the overall delay of the network. This effect becomes drastic when priority schemes are implemented. As described above, the object of the priority is to allocate network bandwidth to the higher priority class messages and to send lower priority messages when there is enough bandwidth. Consequently, a station receiving the token may not be able to transmit any messages due to the constraints imposed by the priority scheme. The token must be passed to the next station, resulting in a waste of bandwidth and a degradation of the overall performance. A closer look at the token bus priority scheme reveals also that the priority scheme lacks fairness and fails to meet hierarchical independence requirements. Since the token-passing sequence is not necessarily transferred in the order of the addresses for any priority class except the one with the highest priority, the order of channel access is dependent on the traffic in other classes and different values of thresholds. In the worst case, it is possible that the packets of a lower priority class at a station get blocked, while the packets with the same priority at another station get transmitted.

Note finally that the reliable operation of token-based networks relies on the integrity of explicit information, such as a unique token, or on the integrity of the active stations. In addition to degrading the overall performance of the network, improper behavior of the stations destroys the deterministic properties of the network. The resolution of exceptional events, such as the addition of a new station to the logical ring, may not be collision free, violating thereby the main objective of the protocol: collision elimination. Similarly, it is difficult to predict what impact exceptional events, such as the election of a new control unit or the recovery from a token loss, might have on access delays.

Token-Ring Network

In a ring-topology network, the token is passed along from one station to another in one direction. Data is sequentially transferred in a bit-by-bit fashion from station to station on the ring. Each station on the ring introduces at least one bit delay during which it regenerates and repeats a received bit. The token received can be either a free token or a busy token. A station can transmit a packet when it receives a free token. The station converts the free token to a busy token and transmits its packet. The packet travels along the ring and is removed by the sending station. At the end of a packet transmission, the station will convert the busy token back to the free token and pass the free token to the next station on the ring. The token-ring scheme is less sensitive to throughput degradation than the CSMA/CD scheme. The delay performance of the scheme is not strongly affected by the number of stations in the ring. Further, since the signal is regenerated at each node, greater distance can be covered.

The token-ring network provides a priority handling mechanism. In this scheme, the start delimiter (SD) field (1-byte long) of the packet format assigns three priority mode (PM) bits to designate up to 8 different priority classes for the packet. When a station receives a token packet, it compares the priority bits of the SD field with the priority bits of the packet to be transmitted. If the comparison is favorable, the station transmits the packet; otherwise, the packet is held and the station immediately forwards the token to the next station. The SD field also has three priority reservation (PR) bits. A station can reserve its priority request in the PR field of a packet if the station's priority is higher than any current reservation request. The current transmitting station examines the PR field and releases the next free token with the new priority-mode indication, but retains the interrupted priority class for later release. A requesting station uses the priority token and releases a new token at the same priority so that other stations assigned that priority also can have an opportunity to transmit. When the station that originally released the free token recognizes a free token at that priority, it then releases a new free token at the level that was interrupted by the original request. Thus, the lower priority token resumes circulation at the point of interruption.

In general, the priority scheme adopted in the token-ring network performs well. However, the scheme is unfair and favors stations located downstream and closer to the most active high-priority stations. As a result, some stations may cut off the ring if they are located farther away from stations with high-priority-class traffic.

IEEE Standard 802.5 evolved from the extensive research work undertaken by the IBM Research Laboratory in Zurich, Switzerland. The original implementation is based on copper media. However, a fiber-based implementation of the token ring has been considered. The design, known as Fiber Distributed Data Interface, allows for a maximum configuration of 500 stations connected through 100 km of optical fiber.

Network-Layer Protocols

Fundamentals

The main purpose of the network-layer protocol is to relieve the transport layer from dependency on the operational characteristics and topology of the transmission facilities. In essence, the network layer provides functionalities that complete the definition of the interaction between host and network. The layer defines the basic unit of transfer across the network, provides services to achieve routing and congestion control, and allows for internetworking to enable the connection of multiple networks and provide a unified cooperative interconnection of networks that supports a universal communications service.

The network layer may provide two types of services, connection-oriented service or connectionless service, to route packets from source to destination. Connection-oriented service, also referred to as virtual circuit by analogy with the telephone circuit, requires the establishment of a reliable logical path along which data is transferred. In connectionless service, also referred to as datagram service, packets are considered self-contained units that carry sufficient information to allow their routing from the source DTE to the destination DTE independently. The type of connection service provided by the network layer has been the object of extensive debate among network protocols developers. Advocates of connectionless service argue that providing a specific functionality can only be fully realized at the source. Duplication of higher level functionalities at the network layer can only improve the efficiency of the network. On the other hand, those who favor connection-oriented services argue that the network should provide for a reliable transfer of packets between the source and the destination. Initially, the ISO model called for connection-oriented services. However, ISO eventually modified its service definition to include both services.

Routing

The principal task of the network layer is to route packets from the source DTE to the destination DTE. In connection-oriented service–based networks, routing decisions are made only when a new virtual circuit is being set. In datagram-based networks, however, the decision is made anew at the arrival of every new packet. Routing algorithms can be classified into two categories: adaptive and static. Adaptive algorithms aim at reflecting the current traffic and topology of the network into the routing decision. Decisions in static algorithms, however, are made based on off-line, precomputed routing information that is usually downloaded into the intermediate nodes.

Based on the information used in the routing decisions, adaptive algorithms may be classified as centralized, isolated, or distributed. Centralized algorithms rely on the availability of a routing control center that collects network status information, determines the optimal routes between each pair of intermediate nodes, and distributes information to the nodes of the subnetwork. Isolated routing algorithms aim at relieving the network from the dependency on a single node. In this scheme, nodes make their routing decisions based on their local status information. Distributed algorithms attempt to strike a balance between the two other algorithm schemes. This is achieved by having intermediate nodes exchange routing information with their neighbors. Routing decisions are then made based on the collected information.

Congestion Control

The second important task of the network layer, closely related to routing, is congestion control. Congestion control is primarily concerned with controlling traffic in the network to avoid overloading the finite resources within the network. More specifically, congestion control refers to a global procedure that is exercised by such network components as bridges and routers to prevent network congestion. The procedure may be perceived as a social agreement between all nodes in a network that dictates how congestion prevention is accomplished or how it is relieved once it occurs. The controlling action may be applied to many source–destination pairs indiscriminately or simultaneously. The congestion-control scheme must provide mechanisms to detect the occurrence of congestion, criteria to select and inform the sources of traffic that are causing the occurrence of congestion, and techniques to recover from congestion by restoring the network to its normal operation.

Congestion control is critical in regulating access to network resources and preventing degradation of the network effective throughput, ensuring thereby a fair share of the resources among users. In an unconstrained network, unrestricted competition for network buffers may cause the intermediate nodes to drop packets. When a packet is discarded, the sender eventually times out and retransmits the lost packet. This situation may occur several times, preventing the sender from releasing a buffer that would have normally been freed. Furthermore, the relative position of a user in the network or the particular selection of network traffic parameters may result in one user capturing more resources than other users, thus having an unfair advantage of use of the network. An extreme case of congestion may result in a deadlock.

Achieving congestion control requires the establishment of control procedures to regulate random demands for resources. The main task of these procedures is to pass control information necessary for access regulation. Two factors make this task difficult in a distributed-system environment. First, the control information may itself be subject to random delays, which usually causes its content to become obsolete by the time it reaches the destination. Second, the requirements for resource reservation imposed by flow- and congestion-control methods, together with channel and processor overhead, may result in a decrease of the effective network throughput.

Ideally, achieving perfect control with no overhead results in a linear increase of the throughput as the offered load increases, until the maximum theoretical throughput is reached. In an uncontrolled network, however, the throughput can only reach the ideal maximum when the offered load is low. The network throughput degrades rapidly to a very low value as the offered load continues to increase. Flow and congestion controls must strike a balance between the gain in efficiency that is due to controls and the loss in efficiency caused by overhead and resource sharing.

The complexity of achieving congestion control increases considerably in an internetwork environment consisting of numerous local- and wide-area networks connected by bridges and routers. The difficulty of the problem stems primarily from the dramatic channel-bandwidth mismatch between the links, the sharp contrast between the very high speed of the local-area networks and the packet-processing rate of the routers, and the unbalanced load among different networks. These factors cause congestion of the routers at the boundaries of the network, thereby reducing significantly the effective throughput.

Congestion-control algorithms must include three important components to

- detect the occurrence, or imminent occurrence, of congestion
- notify the sources of traffic about congestion in the network
- respond to the congestion to restore the network to its nominal steady-state operation.

When the queue of buffers at any intermediate node becomes full, the node cannot accept packets for processing and notifies the original source that congestion has occurred. The intermediate node may notify some or all of the sources that follow of a packet: the switching node that accepted the packet from a host at its entry into the network (packet-switch-level congestion control); the source host that originally sent the packet (host-level congestion control); or a higher level entity, such as a transport-layer entity (connection-level congestion control).

Congestion control schemes differ in who is notified of the congestion and how a recipient reacts to the news of the congestion. Several schemes, however, rely only on upper-layer flow control. This approach may prove to be insufficient or infeasible for certain types of traffic. Consequently, some form of control must be implemented at the bridges and router level.

Examples of Network-Layer Protocols

The general issues discussed above were investigated and adopted in the implementation of network layers in public and private networks. Two network layer protocols, X.25 and Internet Protocol (IP), have gained widespread acceptance in the field of computer communications. The purpose of the remainder of this section is to provide a description of the basic functionalities of these network protocols.

X.25

X.25 defines the procedures for the exchange of data between a DTE and a packet-network node DCE. The CCITT-proposed standard encompasses three layers: the physical layer (X.21); the data-link layer (LAP and LAP-B); and the network layer. The network layer, often called the Packet-Layer protocol, provides the primary functions of the network layer, namely, the establishment and clearing of a networkwide connection between two transport-layer protocols.

X.25 operates on the premise of virtual circuit services. Users of the X.25 network perceive a virtual circuit as a dedicated physical circuit. In reality, the physical circuit may be multiplexed among several users. Every connection is uniquely identified by a logical channel number. The standard provides four forms of establishing and maintaining a connection: virtual circuit, permanent virtual call, fast-select call, and fast-select call with intermediate clear. A virtual circuit is similar to a telephone dial-up line. The connection is dynamically set up on one of the available channels, and will exist until one of the DTEs clears it or stops running its communications link. This mode of communication between two DTEs requires a connection setup. The source DTE attempts to establish the required connection by sending a call-request packet to the remote DTE via the DCE. The local DCE forwards the packet to the remote DCE, which in turn passes it to the destination DTE. The destination DTE may accept the connection and respond by a call-accepted packet or may reject the connection. When the source DTE receives the call-accepted packet, the virtual circuit is established. Both DTEs may then use the full-duplex connection to exchange data. Either DTE may terminate the connection by sending a clear-request packet, which is answered by a clear-confirm packet from the other DTE. The communication mode established between a DTE and a DCE under the X.25 protocol is depicted schematically in Fig. 4. Note that the DTE must choose an idle virtual-circuit number, which is either confirmed by the DCE or modified if the virtual circuit chosen is in use at the destination DTE.

In contrast to a virtual-circuit call, a permanent virtual-circuit call does not require call set up prior to data transfer, and the end-to-end connection remains present as long as both DTEs are running. Consequently, prior to establishing a permanent virtual connection, an agreement between the user and the packet-network carrier to allocate a virtual channel permanently to the connection must be reached. Upon allocation, both DTEs may exchange data without the need for call set up or a connection-clearing procedure. Permanent virtual circuits are used occasionally and are usually dedicated to carry network-management traffic.

The main purpose of the fast-select call is to support transaction-oriented types of traffic and short network sessions. These types of traffic cannot use effectively either a switched virtual call, because of the overhead involved in the connection establishment and termination, or a permanent virtual circuit, because its occasional use does not warrant the permanent use of the resource. Fast select provides for two options: fast-select call and fast-select call with intermediate clear. Both modes of transmission allow a DTE to transmit up to 128 octets of data by means of the appropriate request packet. The fast-select facility allows the remote DTE to respond with a call-accept packet, which may also contain data. If desired, the data-exchange session may continue between the two DTEs until completion of the transfer. Fast select with immediate clear, however, requires the remote DTE to transmit a clear-request packet immediately upon receiving a call-request packet.

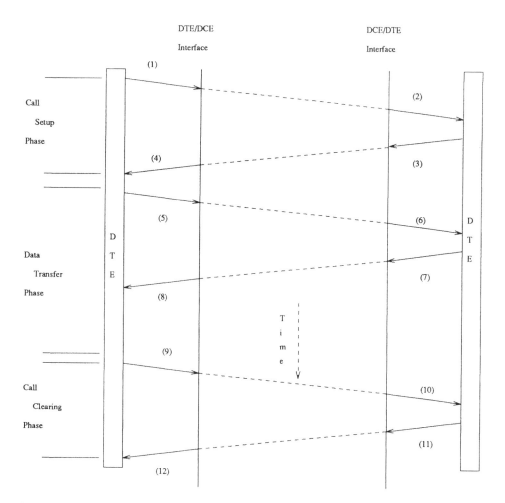

FIG. 4 Phases of an X.25 connection: *1*, call request; *2*, incoming call; *3*, call accept; *4*, call connect; *5*, data packet; *6*, incoming data; *7*, data packet; *8*, incoming data; *9*, clear request; *10*, clear indication; *11*, clear response; *12*, clear confirm.

In addition to the packets described above, the X.25 protocol defines several other packets: the interrupt, receive-ready, receive-not-ready, reject, reset, and restart packets. These packets are used to achieve more control over the connection.

The addressing scheme used by X.25 networks is described in a related standard known as X.121. Based on a telephone number assignment mechanism, the scheme defines a physical address that consists of a 14-digit number, with the last 10 digits assigned to the vendor that supplies the X.25 network service. The standard also provides recommendations for three specifications to support asynchronous terminal interfaces, namely, X.3, X.28, and X.29. The main goal of these standards is to provide a transparent service to the user DTE by providing protocol conversion for a DTE to an X.25 network and a complementary protocol conversion at the receiving end.

Finally, note that the X.25 standard does not contain a routing algorithm. The choice of a routing mechanism is considered to be internal and specific to a particular designer.

Internet Protocol (IP)

Unlike the X.25 protocol, the Internet Protocol exclusively offers an unreliable, best-effort, connectionless, packet-delivery system. The main goal of IP is to build a unified, cooperative interconnection of networks that supports a universal communications service.

The protocol defines the format of the basic units of data transfer, and specifies a set of rules that govern their processing and routing. The basic data unit handled by IP is usually referred to as a *datagram*. Encapsulated into physical network frames, datagrams constitute the data portion of the frame. In order to accommodate several types of networks, IP does not limit datagrams to a specific size, but uses fragmentation and reassembly to carry the original datagram through multiple computers. Fragments of the same datagram contain enough information to allow for their assembly at the source host.

Three fields in the header of the datagram (Ident, flags, and fragment offset) are used to control fragmentation and reassembly. As datagrams arrive, the source host uses the Ident field to identify and assemble fragments of the same datagram. The flag field is used to determine the last fragment of an IP frame. Note that once fragmented, pieces of the same datagram travel as separate datagrams. IP makes an earnest attempt to deliver the fragments, but does not provide guarantees for the delivery of the datagram fragments. Datagrams may be lost or may get out of sequence.

In order to prevent a packet from traveling around the Internet indefinitely, the header of a datagram specifies a time-to-live field, which determines the maximum number of seconds the datagram is allowed to remain in the Internet. Each intermediate node, usually referred to as a *gateway*, checks the remaining time to live of the datagram, and discards the packet if the time has expired. In addition to the basic fixed fields, the header may contain optional fields intended to provide more control over the Internet.

The routing mechanism provided by IP may be categorized as direct or indirect. Direct routing involves two machines within the same network, and does not require the intervention of a gateway. As stated above, the original datagram is encapsulated into a physical frame and sent directly to the destination by means of the underlying transmission system. Indirect routing requires cooperation among gateways of the Internet. The interconnected structure of the gateways uses an Internet routing table that contains destination information used to forward datagrams. In order to keep the size of the routing table to a minimum, IP routing software uses the destination network address as a basis for routing, rather than the host address. More specifically, the routing-table entry specifies the address of the next gateway to receive the datagram destined to the specified host. It may happen that no route is specified in the routing table of the gateway for a particular destination host, in which case the datagram is sent to a default gateway.

Each host in the Internet is assigned a unique 32-bit address. Conceptually, each address (net ID, host ID) defines a network identity and a host identity. Consequently, the addressing scheme does not specify a host but defines a connection to a network. To accommodate different sizes of networks, the addressing mechanism defines three primary classes of networks.

The three high-order bits of the address are used to distinguish among these classes. The first address class is reserved for networks that have more than 2^{16} hosts. In this class, the first high-order bit is set to 0, the seven following bits are used to uniquely identify the network, and the remaining twenty-four bits identify hosts within the network. The second address class is dedicated to networks with an intermediate number of hosts. In this class, the first two high-order bits are set to 10, the next fourteen bits identify the network, and the remaining sixteen bits identify a host within the network. The third class is reserved for networks with a small number of hosts. In this class, the first three high-order bits are set to 110, the next twenty bits are used to identify the network, and the last eight bits identify a host within the network.

In order to allow stations to map their internet addresses into physical addresses, IP defines two protocols: Address Resolution Protocol (ARP) and Reverse Address Resolution Protocol (RARP). ARP allows a host to find the physical address of a target host on the same physical network, given the target's Internet address. This is achieved by allowing the requester to broadcast a special packet that asks the host with the specified Internet address to reply with its physical address. In order to reduce the overhead involved in the resolution of names, hosts maintain caches of recently acquired Internet-to-physical-address bindings. Prior to sending the ARP message, an Internet host first consults its cache.

The RARP protocol is used by diskless machines to inquire for their Internet addresses. As in the ARP protocol, a requesting host broadcasts a name-resolution request packet. The request specifies the host as both the sender and the target machine, and supplies the physical network address of the host. All stations in the network discard the packet, except the RARP server, which replies with a name-resolution packet.

To allow machines to report errors and provide for unexpected circumstances, IP provides a special-purpose protocol called Internet Control Message Protocol (ICMP). This protocol is mainly used to test destination reachability and status. In addition, the protocol is used to control the datagram flow and cope with network congestion. Upon detecting a high rate of datagram arrival, gateways discard the incoming packets and issue source-quench messages to the hosts. There is no ICMP message to reverse the effect of the quench message. Instead, the hosts involved in the network congestion wait until they stop receiving quench messages and gradually increase their traffic.

The Internet protocol is gaining widespread use among the communications community due to its support of heterogeneous host systems and architecture.

Transport-Layer Protocols

The main objective of the transport layer is to provide an interface between the higher application-oriented layers and the underlying network-dependent protocol layers. The interface defines a set of primitives designed to provide the end user with a consistent, reliable, and cost-effective message-transport facility that is independent of the quality of the service provided by the underlying network. *Primitives* are abstract representations of the interactions across the service access points that indicate that information is passed between the service user and the service provider. The set of primitives makes it possible to develop programs for a wide variety of networks without having to deal with different subnet interfaces and the quality of service these networks provide. Transport-layer services are in many ways similar to the services provided by the network layer. Nevertheless, the differences between the two layers are very subtle and often crucial in providing high-quality service to the user of the network.

Features

Transport-layer services are implemented by a transport-layer protocol used between the transport entities. The basic elements of a transport protocol depend on the type of service provided by the underlying network. The spectrum of network services may vary from a flawless, error-free service with no resets to an unreliable service that may cause packet loss and packet duplication. In addition, the network may be subject to resets. Depending on the environment in which it operates, different transport protocols emphasize different features. Nevertheless, the basic features of a transport protocol usually include connection management, addressing, multiplexing, error control, and flow control.

Connection Management

Connection management involves the establishment and termination of a transport-level connection. At first glance, these procedures appear to be easy, inasmuch as they require a connection request from the entity initiating the connection and a connection confirmation from the remote entity. A closer look, however, reveals that problems may occur when the network loses, stores, and duplicates packets. The duplicated packets may be connection-request packets for which acknowledgments were delayed or never received by the sender because of congestion in the network. These packets may surface later, after the original connection is closed, and cause the reestablishment of a closed connection, eventually generating undesired traffic, which may cause negative effects.

Three different methods are commonly used to prevent this situation from occurring. The first method aims at assigning a unique identifier to every transport connection. The identifier allows the transport entities to differentiate between packets generated by different connections. Consequently, stored duplicates of released connections are recognized and discarded. Note that a released connection is listed as obsolete in the connection table. Although simple, this method requires each transport entity to keep track of the history information of different connections over a long period. This requirement may be space prohibitive. In addition, the failure of the machine may cause the loss of the history information.

The second prevention method is based on three-way handshaking. In this method, the connection initiator issues a connection request and includes in it an initial sequence number. The remote transport entity acknowledges the connection request by issuing a connection-confirm packet that includes the remote entity's initial sequence number. Finally, the connection initiator uses the first data packet to piggyback an acknowledgment of the remote transport entity's initial sequence number. The three-way handshaking is, in essence, a reliable way for the transport entities to verify that a connection request is indeed being carried out by a transport entity. The same technique is usually used to terminate a connection.

The third method is based on timers that cause the connection to simply time-out in the absence of activity. In this scheme, the record of the connection is not deleted immediately after the connection is released, but only after an appropriate amount of time has elapsed. Keeping track of previous packets and their acknowledgments makes it possible for the transport layers to detect and discard duplicates. The success of this scheme depends on the careful choice of the time interval. The method requires rather specific knowledge of the worst-case packet delay in the network. The delay may be controlled or bounded tightly in small networks. For large networks, the packet delay may vary considerably, resulting in poor performance of the scheme.

Addressing

The addressing procedure defines a mechanism to discriminate between the identity of the user application process and the location on the network where the user application process resides. The identity of the user is usually represented by a symbolic name, while the location on the network is usually in the form of an address. The address space may be either flat or hierarchical. With flat addresses, every node must maintain the context to map all possible addresses. Consequently, potentially large mapping tables or contexts must be provided. Hierarchical addresses consist of concatenated fields defining subdomains. That is, the address of a process reflects the hierarchical logical geometry of the network.

The functions of allocating unique names to processes and the management of the name database are usually performed by a name server, also referred to as a directory service. To determine the address corresponding to a given service name, the user application process consults the name server. The name server responds by providing a mapping between the symbolic name and the address. In this name management model, a newly created service must be registered with the name server. The name server binds the symbolic name to the address and stores the relationship in the name table.

Multiplexing

Multiplexing-based techniques provide a cost-effective mechanism to carry several transport connections over the same network connection. This form of multiplexing is usually referred to as *upward multiplexing*. Another form of multiplexing, referred to as *downward multiplexing*, allows the transport layer to split the traffic generated by a single transport connection among several network connections in a round-robin fashion. This form of multiplexing increases the throughput and improves the performance of the transport connection.

Error Control

Transport-layer protocols must be designed to handle a variety of contingencies and error conditions when the underlying network cannot be relied upon to provide an error-free, duplicate-free, sequenced delivery of data. These requirements include the capability of detecting errors, duplicate packets, and packets arriving out of sequence. Detection of error conditions and mechanisms to recover from them is a universal problem for transport layers designed to operate in the face of unreliable networks. Note that a number of issues raised at this level are also common to protocols of other layers, such as the data-link layer.

In order to detect duplicates or data packets that arrive out of order, packets are numbered sequentially. A positive acknowledgment with a time-out procedure is also incorporated to ensure that lost packets or those received in error are properly retransmitted.

Flow Control

The flow-control mechanism aims at regulating the amount of traffic between the transport entities involved in the communication, so that such resources as logical channels, buffers, transmission bandwidth, and processor time are efficiently utilized and user performance requirements are met. A common scheme, used to achieve flow control at the transport layer and prevent overutilization of the resources of the receiver, uses an end-to-end window-control scheme. Based on a credit mechanism, the receiver informs the sender of how many packets it is ready to receive. Depending on the window scheme, the receiver may reduce the credits if it senses that the resources are reaching depletion. In some ways, the flow-control problem at the transport layer is similar to the flow-control problem at the data-link layer. However, the number of open connections the transport layer has to deal with usually exceeds the number of host connections the switching node has to manage. This difference makes the buffer-management scheme implemented at the data-link layer impractical at the transport layer.

In the remainder of this section, a few commonly used transport-layer protocols, namely, the OSI transport-layer protocol, the Transmission Control Protocol (TCP), Versatile Message Transmission Protocol (VMTP), and the Network Bulk Transfer Protocol (NETBLT), are described. A discussion of their application to the transpose-layer issues mentioned above is provided.

Transport-Layer Protocols in Public Networks

As stated above, the basic function of the transport layer is to provide reliable, end-to-end data-transport service for the session layer and other higher level entities. The transport layer uses the services of the network layer to provide the functionality mentioned above, but shields the upper layers from the details of the network connections and the types of networks used.

The methodology in the discussion below is based on the ISO transport model depicted in Fig. 5. To request a service from the underlying transport entity, the user of the transport service (TS user) attaches itself to the transport-service access point (TSAP). Communication between the service user and the service provider is accomplished by means of a defined set of primitives.

To cater to different types of networks, the ISO transport layer defines five classes of protocols, labeled Classes 0, 1, 2, 3, and 4. The choice of class is dictated by the underlying network connections and by the quality of service requested by the TS user. The network services are characterized by their error rate and by the rate of failure signaled to the transport layer. The quality of service (QOS), however, is usually characterized by a number of parameters. The first set of parameters deals with connection establishment. These parameters include connection establishment delay and probability of failure. The same parameters are also defined for connection release. The second set of parameters deals with the quality of service provided by the connection. These parameters include the throughput required, the transit delay, the residual error rate, the transfer-failure probability, the level of protection, the priority, and the resiliency of the connection. These parameters are usually specified by the transport-layer user, and negotiated by the end-to-end transport layers.

To handle both the different types of services that may be expected over a transport connection and the wide variety of networks that may be available to provide network services, each class of transport protocols emphasizes specific transport-protocol mechanisms. Class 0 provides the simplest type of transport connection, with a minimum of functions defined. The protocol assumes a network service with a high degree of reliability and an acceptable rate of

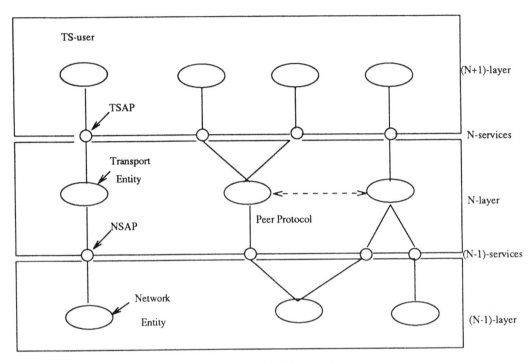

FIG. 5 ISO model of transport layer.

network failure. Consequently, the protocol need be concerned only with setting up simple end-to-end transport connections and segmenting long data messages, if necessary, during the data-transfer phase. The protocol has no provision for recovering from errors and cannot multiplex several transport connections into a single network connection. The protocol relies on the network layer to handle error and flow control. This class is suited for networks with virtual-circuit services at the network layer.

Class 1 protocol is better suited for networks with an acceptable error rate but a high rate of signaled failures. Therefore, the protocol has provision for recovery from such errors as network reset, receipt of packets from an unknown transport connection, and hardware failures. Class 1 protocol numbers the packets in order to recover from a network disconnect and resume the transport service.

Class 2 is similar to the Class 0 protocol, except that it adds support for multiplexing, permits explicit flow control, and makes use of expedited data.

Class 3 is intended to work over network connections characterized by an acceptable error rate, but an unacceptable rate of signaled failures. Consequently, the protocol incorporates error-recovery procedures, in addition to the multiplexing capabilities of the Class 2 protocol.

The Class 4 protocol is designed for low-quality networks characterized by an unacceptable error rate and an unacceptable signaled-failures rate. Therefore, the protocol must handle a variety of such contingencies and error conditions as network failures and detecting and correcting out-of-order packets, duplicates, and misdirected packets.

The services provided by the transport layer in the ISO model fall into two categories:

1. Transport connection management services, which allow the transport user to establish and maintain a logical connection to a correspondent transport user in a remote system
2. Data-transfer services, which provide the means for exchanging both normal and expedited data over an established connection. The protocol also provides a connectionless option. This option allows a transport user to initiate the transfer of an item of user data without establishing a connection. In this mode, every packet carries the full destination address and is transmitted independently of every other packet.

The primitives describing the connection-management services are depicted in Tables 2 and 3. The parameters of the primitives specify the source and destination addresses, the options and QOS requested by the transport-layer user, and an initial data. The QOS requested by the user specifies throughput, transit delay, reliability, and priority of the connection. The transport layer will negotiate these parameters with the peer transport entity at the destination.

The primitives for data-exchange services, together with their parameters, are described in Table 4. The first four primitives allow the transmission of normal and expedited data over an already established connection. The last two primitives associated with the connectionless mode of transmission enable the user to initiate the transfer of an item of user data without first establishing a transport connection.

A state transition diagram for the connection-oriented mode is depicted in

TABLE 2 Connection-Establishment Primitives

Primitives	Parameters
T_CONNECT.request	Callee, Caller, Expedited Data Option, Quality of Services, User Data
T_CONNECT.indication	Callee, Caller, Expedited Data Option, Quality of Services, User Data
T_CONNECT.response	Callee, Caller, Expedited Data Option, Quality of Services, User Data
T_CONNECT.confirm	Callee, Caller, Expedited Data Option, Quality of Services, User Data

TABLE 3 Connection-Termination Primitives

Primitives	Parameters
T_DISCONNECT.request	User Data
T_DISCONNECT.indication	Reason, User Data

TABLE 4 Data-Exchange Primitives

Primitives	Parameters
Connection-Oriented Mode	
T__DATA.request	User Data
T__DATA.indication	User Data
T__EXPEDITED__DATA.request	User Data
T__EXPEDITED__DATA.indication	User Data
Connectionless Mode	
T__UNIT__DATA.request	Callee, Caller, Quality of Services, User Data
T__UNIT__DATA.indication	Callee, Caller, Quality of Services, User Data

Fig. 6. Figure 7 is a time sequence diagram showing the order in which the primitives are used. The sequence of events in the illustration assumes a successful connection establishment phase. The failure of the connection establishment will result in the transmission of T__DISCONNECT.request instead of a T__CONNECT.response.

Transport-Layer Protocols in ARPANET

The Internet uses a protocol hierarchy known as the TCP/IP suite. As discussed above, IP is a datagram-oriented protocol based on the best-effort paradigm. At the transport level, the suite offers both a connectionless transport protocol, (User Datagram Protocol, or UDP) and TCP, a virtually connection-oriented transport protocol.

UDP provides unreliable connectionless delivery service using the underlying IP to transport messages among different machines of the Internet. The protocol provides only peer-to-peer addressing and optional data checksums. The protocol packet headers are extremely simple. They contain the source and destination addresses, the datagram length, and the data checksum. In essence, the protocol adds the ability to distinguish among multiple destinations within a given host to the services offered by IP.

TCP is a general-purpose, reliable, connection-oriented, strewn transport protocol designed with very few assumptions about the underlying network. Consequently, the protocol can be used over a single local-area network such Ethernet, as well as over complex packet-delivery systems (including Internet). The basic features of the protocol include

- Explicit and acknowledged connection initiation and termination
- Reliable, in-order, unduplicated delivery of data
- Flow control
- Out-of-bound indication of urgent data.

The TCP protocol specifies the format of the data and acknowledgments

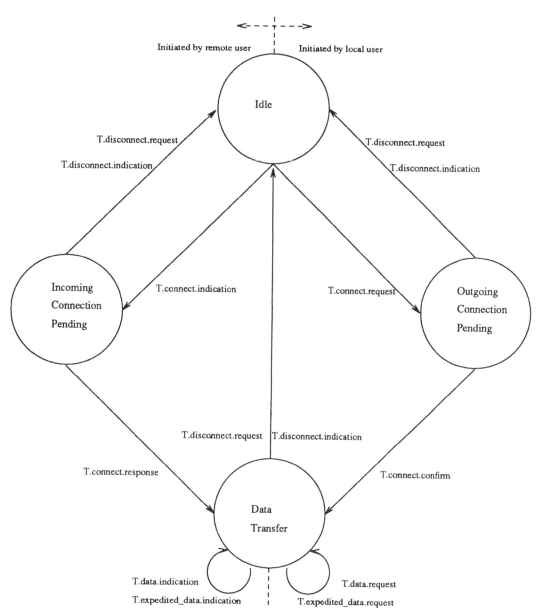

FIG. 6 State transition diagram.

that need to be exchanged between the communicating machines in order to achieve a reliable transfer of data. It defines the procedures that the two communicating machines use to initiate and terminate a stream transfer, and specifies the procedures used by these machines to recover from such errors as lost or duplicate packets. In addition, the protocol specifies the technique used to distinguish among multiple destinations in a given machine.

A TCP connection may be viewed as a bidirectional, sequenced stream of

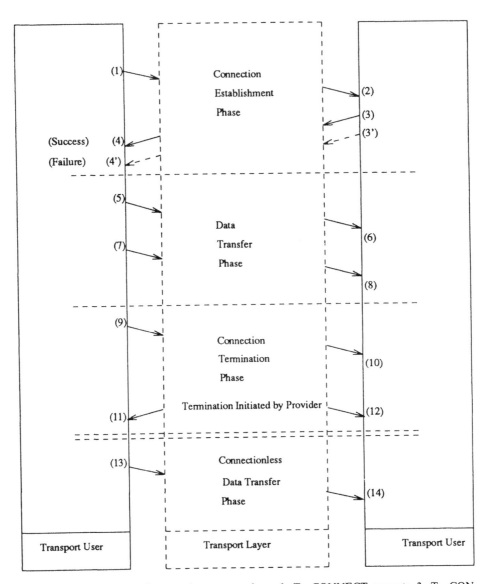

FIG. 7 Time sequence diagram for user services: *1*, T__CONNECT.request; *2*, T__CON-NECT.indication; *3*, T__CONNECT.response; *3'*, T__DISCONNECT.request; *4*, T__CONNECT.confirm (success); *4'*, T__DISCONNECT.indication (failure); *5*, T__DATA.request; *6*, T__EXPEDITED__DATA.request; *7*, T__DATA.indication; *8*, T__EXPEDITED__DATA.indication; *9*, T__DISCONNECT.request; *10*, T__DIS-CONNECT.indication; *11*, T__DISCONNECT.indication; *12*, T__DISCONNECT.in-dication; *13*, T__UNIT__DATA.indication; *14*, T__UNIT__DATA.indication.

data octets transferred between two peers. The data may be sent in packets of varying sizes and at varying intervals. The stream initiation and termination are explicit events at the start and end of the stream. These packets are acknowledged in the same way as data. The initiation packet in each direction carries a sequence number referred to as the initial sequence number. Each data packet of a TCP connection carries the sequence number of its first data and an acknowledgment of all contiguous data received properly.

To prevent data loss, the sender retains the currently transmitted data until it receives an acknowledgment. Consequently, each connection maintains a set of timers to recover from losses or failure of the peer receiver.

In order to allow for multiple application programs on a given machine to communicate concurrently, TCP incorporates abstract objects called *ports* that identify the ultimate destination within a machine. A port is uniquely identified by an integer. These identifiers are assigned locally within the machine. Consequently, the ultimate destination for TCP traffic is uniquely specified by the destination host's network address, and the port number on that host.

To provide for flow and error control, TCP uses a variant of the classical sliding-window protocol. The TCP sliding-window mechanism operates at the byte level. It is based on the usage of three pointers. The first pointer marks the bytes that have been sent and for which an acknowledgment has been received. The second pointer marks the boundary inside the sliding window that separates the bytes that have been sent and not acknowledged from the bytes that have not yet been sent. The third pointer marks the highest byte in the sequence that can be sent. TCP allows the window size to vary over time. This is achieved by including in every acknowledgment additional information containing a window advertisement. This mechanism allows the sender to increase its window size in response to a window-increase advertisement, and to decrease its window size if the acknowledgment carries a window-decrease advertisement.

Congestion control for TCP/IP networks is achieved at the host level. At the detection of congestion, congested nodes send an explicit ICMP quench message to the source hosts that are involved in the congestion. Upon receiving a quench message, a source host proceeds to lower the traffic rate to the specified destination. The rate adjustment of packet transmission is based on the estimates of round-trip delay. The rate reduction remains effective as long as the source host continues to receive source-quench messages. In the absence of the source-quench messages, the source host proceeds gradually to increase its rate until it starts receiving source-quench messages again.

Note that the scheme suffers a few shortcomings. First, the scheme is prone to throughput oscillations. During the traffic-rate reduction period, the throughput decreases considerably. Upon recovery, the throughput increases, resulting in a new congestion state. Second, dropping packets indiscriminately to relieve congestion, although consistent with the "best-effort" delivery philosophy in connectionless delivery, may be counterproductive and generally triggers end-to-end transmission. The third drawback of the scheme is lack of fairness. Upon detection of an imminent congestion, quench messages cause all the source hosts involved to reduce their traffic rate regardless of their contributions to the traffic that affected the intermediate node. In addition, the timers depend heavily on the end-to-end latency. If an acknowledgment for a packet takes longer than the expected time, the transport layer assumes that the network is congested and proceeds to reduce the rate of its packet transmission by adjusting its window size. The difficulty of estimating the actual round-trip delay within the current Internet reduces the efficiency of the congestion-control scheme. Furthermore, the round-trip delay is not a direct measure of congestion, and therefore it may not be an accurate measure of the current network load.

Versatile Message Transmission Protocol (VMTP)

VMTP is basically a reliable transaction-stream protocol. The protocol does not have a concept of a virtual circuit or a connection, but it provides an interface to higher level models to implement a conversation. The protocol uses a packet-group–based flow control. The receiver accepts and acknowledges a group of packets, which constitutes a VMTP message, before further data is exchanged. The protocol uses selective acknowledgment to avoid the problem associated with cumulative acknowledgment. Monitoring of the retransmission request pattern combined with the selective acknowledgment provides a dynamic means of detecting high transmission rates that can cause congestion of the network. Possible congestion can be reduced by adjusting the transmission rate in response to the retransmission requests.

Although the protocol addresses and successfully resolves some of the shortcomings of the TCP protocol, the flow and congestion control in VMTP still presents a few limitations. As stated above, the flow-control scheme used in VMTP is essentially a stop-and-wait protocol at the group packet level that still depends on the end-to-end latency. The approach taken in dealing with congestion control is to recover from congestion after it is detected rather than to avoid it. Congestion avoidance in high-speed, large-sized networks is more effective than congestion detection and recovery. Furthermore, VMTP uses packet retransmission requests as a measure of congestion and a basis for transmission-rate adjustments. This method is not responsive because it might take a long time for the transmitter to make a judgment about packet overruns based on the selective retransmission requests. During this time, the congestion situation may have changed, which causes the transmission-rate adjustment to become obsolete.

Network Bulk Transfer Protocol (NETBLT)

NETBLT is designed for high-throughput, bulk data-transmission applications. The protocol provides a virtual circuit between a client and a server to transfer large blocks of data in series. The protocol implements flow control at two levels. At chart level, the strategy is to confirm through negotiation that both the client and the server have set up matching buffers. Flow control then can be achieved by changing the buffer size or the transmission rate. At the internal level, flow control is the rate-based transmission of packets within the block. Transmission continues at the negotiated rate since the buffer space has been preallocated. Retransmissions carried at the predetermined rate do not change the rate of packet transmission and thus do not induce congestion. The three important features of error recovery in NETBLT include its use of selective acknowledgment, timers based on inter-arrival time, and timers located at the receiving end. Selective acknowledgment eliminates the problems associated with cumulative acknowledgment, while the estimation of retransmission timers based on inter-arrival time is more accurate.

The main limitation of the NETBLT rate-based flow control is that the intermediate nodes are not consulted during the negotiation. The lack of involvement of intermediate nodes in the negotiation may increase the number of retransmissions. Furthermore, a high end-to-end latency may reduce the responsiveness of the scheme to congestion. A condition of overrun may remain unknown to the sender for at least one round-trip delay. Since the transmission rate cannot be negotiated in the middle of transmitting a block, a change in the rate may come too late to relieve congestion. Another drawback of the protocol in dealing with congestion is that the approach taken does not prevent congestion from occurring, but, rather, uses a detection-recovery mechanism. As mentioned above, this approach may cause throughput oscillations.

Application-Oriented Protocols

The three highest layers of the ISO model are the session layer, presentation layer, and application layer. The three layers are referred to collectively as application-oriented layers. They provide the means to exchange information between a set of user application processes that are cooperating to achieve a common distributed-processing goal. These layers rely on the basic, network-independent, reliable, end-to-end communication services of the transport layer to provide a set of useful features to a wide variety of applications. In this section, we discuss some of the issues relevant to the design of protocols for the session, presentation, and application layers. It should be stressed, however, that the layers should be considered collectively as providing a particular application service on behalf of the user application process.

Session-Layer Design Issues

The purpose of the session layer is to upgrade the services provided by the transport connection. More specifically, the session layer provides a way for the session users to establish connections (called *sessions*) and transfer data over those connections in an orderly fashion. In many ways, the session layer is similar to the transport layer. Initially, the existence of such a layer was the subject of considerable debate. The layer, which still does not exist in many networking models, was finally included in the ISO reference model.

The main services provided by the session layer include dialog management, synchronization, activity management, and exception reporting.

Dialog-management services allow a session layer user to establish a logical connection path, transfer data over the established path, and release the connection in an orderly fashion. Dialog management is based on data tokens and release tokens.

A *data token*, mostly useful in a half-duplex mode of communication, gives its holder the exclusive right to transmit data over the logical communication path. Upon completing the transmission of its current data, the data-token holder passes the token to the peer process. The peer process may also request the token from the token holder by issuing an explicit token-transfer message. Depending on its current situation, the token holder may either honor the token-transfer request or reject it. The *release token* allows the two communicating users to negotiate the release of the connection in a controlled way. The session offers an orderly connection release that differs from the abrupt connection-termination mechanism offered by the transport layer. The latter form of connection termination may result in the loss of the data that is in transit.

The types of data defined by the session layer include regular datastream, expedited datastream, typed datastream, and capability datastream. The regular and expedited data types are similar to the datastreams offered by the transport layer. Typed data provide the user of the session service with an out-of-band datastream to implement control information, network-management functions, and system-management routines. Unlike the regular data, typed data do not require the possession of any token. A special arrival indication primitive is used to signal the arrival of such data to the receiver. The capability data are also intended for control purposes. However, these data are used exclusively by the session layer to allow the exchange of session options and parameters. Capability data are fully acknowledged, and are only sent outside of the activity. Prior to transmitting capability data, the sender must hold all three data, synchronization, and activity tokens.

The *synchronization* mechanism allows the user of the session layer to set synchronization points during the dialogue. These points could be used to resume the dialogue at a known state in the event of errors or disagreement. Two different types of synchronization points, major and minor, are defined. Each synchronization point is defined by a serial number that is maintained by the session protocol entity.

Major synchronization points are normally associated with complete units of datastream referred to as *dialogue units*. When inserted into the datastream, major synchronization points are confirmed by the remote peer entity. Minor synchronization points are associated with portions of a dialogue unit and are not confirmed. The session-layer protocol defines primitives of synchronization for each synchronization type and provides two kinds of synchronization tokens for the implementation of the synchronization process. Both synchronization tokens are different from the data token.

Prior to setting any synchronization point, the session-layer entity is required to possess the relevant token. The synchronization process is depicted in Fig. 8. When resynchronization occurs, all the tokens are reset to their previous positions at the instant the synchronization point was set. Resynchronization within a dialog unit may go back as far as the first minor synchronization. In the time interval between major synchronization points, resynchronization is only

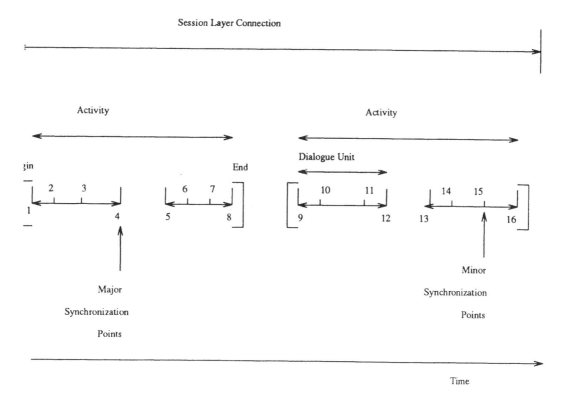

FIG. 8 Session-layer synchronization process.

permitted back to the previous major resynchronization points. In other words, incursion back into a previous dialog unit is not allowed.

The concept of *activity* allows the user of the session layer to distinguish between two independent logical pieces of work. Activities are determined by the user of the session layer, and a session may comprise several activities. Only one activity at a time may be in progress. Consequently, an activity may be interrupted and then resumed, either on the same session connection or on a different connection. This feature is very useful for such applications as the transfer of several files between two sites. Each file transfer may be associated with one activity and accomplished during that activity.

Activity management allows the user to structure a session. It is controlled by a token, the same token used for major synchronization points. Prior to invoking an activity service, the session-layer user must be in possession of the activity token. In addition, the user is also required to hold the minor synchronization token and the data token. This precaution ensures that no minor synchronization is attempted by one user at one end while the other user is attempting to start an activity.

The *exception-reporting feature* provided by the session layer to its users allows for reporting an exception during the session connection. This general mechanism can be utilized either by the user to report the occurrence of unexpected errors or by the service provider to notify the user about internal problems as reported by the transport layer or the layers discussed below. When notified about the nature of the exception, the user may decide if any action is to be taken.

Presentation Layer

The main function of the presentation layer is to provide a mapping of the application-layer service requests into the corresponding session service primitives. These primitives may include dialog control and system-synchronization control. The presentation-layer function must be achieved in an open interconnection environment where machines manufactured by different vendors may engage cooperatively in achieving a common information service. It is likely that the cooperating machines may use different internal formats to represent such different data types as characters or integers. The reliable transfer of these data types between incompatible machines does not guarantee that the meaning of the exchanged information is preserved. The situation becomes even more complicated when more complex data structures, such as records, are involved. Consequently, a mechanism must be provided to build a common environment of shared semantics among the cooperating applications. More specifically, a presentation context that associates an abstract syntax with a compatible transfer syntax is required. The conversion mechanism necessary to preserve the semantics of the transported data defines additional functions of the presentation layer. These functions include providing a way to specify complex data structures, managing the current set of data structures as negotiated and accepted by the cooperating applications, and converting data between external and internal forms.

The negotiation process managed by the presentation layer allows users to define the types of data structure that are required by each context. Note that, since a variety of abstract types may be exchanged during an application session, a number of different presentation contexts may be selected. The content of a context may be altered by either side, and a session may involve more than one context. The presentation contexts to be used during a session are referred to collectively as a *presentation context set*. The context management services allow the users to change the current context.

The negotiated context allows the presentation layer to determine the appropriate transfer syntaxes that are suitable for conveying the type of data message to be exchanged. The role of the presentation layer is to transform the data to be transmitted from its specified abstract syntax into the selected transfer syntax form, and the received data from the transfer syntax form into the specified abstract syntax form for use by the presentation-layer user. The presentation layer receives the data from the application layer in the form of a tagged data element. The tag is used by the presentation layer to identify the presentation context associated with the data, and to determine the corresponding transformation that needs to be applied on each element of the data.

It is clear from the information above that the key issue in the problem of representing, encoding, transmitting, and decoding data structures is to provide a way for an efficient abstract syntax to represent data structure, and an efficient transfer syntax to encode the data structure into a stream of bits. These issues are the subject of the remainder of this section.

As stated above, user applications usually maintain complex data structures that are relevant to their operations and that must be transmitted over the network. In addition, the protocols at different layers manipulate such different data structures as frames and packets. In order to deal with all types of data structures in a global and coherent fashion, the approach adopted within the ISO framework was to define an abstract syntax, known as Abstract Syntax Notation No. 1 (ASN.1), which can be used both as an application syntax and as a means of defining the structure of the packet data units associated with different protocol entities.

The type of identifiers defined by the ASN.1 document are grouped into four classes: universal, application, private, and context specific. The universal type is reserved for the primitive types defined in the ASN.1 standard. The application type is used by the OSI application-layer protocols to denote types used in these protocols. Application types may also be used by users as long as ambiguities are avoided. Private types of identifiers are user defined. The context-specific types are related to the specific context in which they are used.

The universal class contains both the primitive and the constructed identifiers. The primitive type is either a basic data type, or a string of one or more elements, all of the same basic data type. They constitute the building blocks of more complex types. The constructed type is defined by reference to other types, which may be primitive or constructed. The primitive and constructed types of universal identifiers are listed in Table 5. In addition to these universal types, the ASN.1 standard primitive types describe a set of predefined types that are useful in several applications, such as telex, videotex, and graphics.

An additional useful feature of the ASN.1 allows the user to declare fields

TABLE 5 ANS.1 Primitive and Constructed Types of Universal Indentifiers

Primitive Type	Meaning
Primitive	
BOOLEAN	True or false
INTEGER	Integers of arbitrary length
BITSTRING	List of 0 or more bits
OCTETSTRING	List of 0 or more bytes
IDENTIFIER	IA5 string or graph string
NULL	The variable has no type assignment
ANY	The variable can have any type assignment
Constructed	
SEQUENCE	Fixed ordered list of various types
SEQUENCEOF	Fixed ordered list of the same type
SET	Fixed unordered list various types
SETOF	Fixed unordered list of the same type
CHOICE	Fixed unordered list of types selected from a previously specified set of types

OPTIONAL or DEFAULT. OPTIONAL elements may or may not be present in the encoded packet. If the field is omitted, its value is ignored. The DEFAULT feature has a similar meaning, except that when a field is omitted, the receiver assigns a default value to it. It is clear that these features, although useful, may introduce ambiguity at the receiver's side when it comes to determining which fields are actually present in a packet. To resolve the possible ambiguity, ASN.1 supports the concept of assigning a tag to any data type or field. The tag represents a unique identifier of the tagged element. Except for context-specific types, the tag is preceded by the class identifier, namely, UNIVERSAL, APPLICATION, or PRIVATE. Normally, the type of the tagged variable is explicitly sent to allow the receiver to determine what kind of value is being sent. However, since the tag uniquely identifies a type, the receiver can imply the type from the tag field. Variables with implied types may be declared to be IMPLICIT. Notice that variables of the type ANY or CHOICE cannot be declared IMPLICIT, and the explicit type information must be sent to the receiver. An example of ASN.1 type is depicted in Fig. 9.

The SEQUENCE structured type is used to indicate that the data type "reader" is comprised of a number of typed data elements that may be either primitive or structured. All types are tagged, the top level being PRIVATE, and the remaining data types are context specific.

As stated above, ASN.1 is an abstract syntax that defines the type of the data elements to be exchanged but does not determine their concrete structure. Consequently, an encoding method must be associated with the ASN.1 standard to convert each field of the data type that has been defined in ASN.1 form into the corresponding concrete syntax form. The encoding method defines the transfer syntax in order to guarantee that the exchanged data unit has a common meaning to both application entities.

Based on the ASN.1 transfer syntax, the standard representation of the type value transmitted, either primitive or constructed, potentially comprises four fields: an identifier field, a length field, a content field, and an end-of-content flag field. The first field defines the ASN.1 type. The second field determines the number of octets in the data field. The third field determines the contents of the data field. The fourth field, which is optional, is only used when the length of the data type is unknown. Each field comprises one or more bytes. The structure of the identifier byte is depicted in Fig. 10.

In order to encode tags with values larger than 30, a succession of identifier bytes is used. Each identifier byte following the first identifier has the highest-

Reader::=[PRIVATE 10] IMPLICIT SEQUENCE {
name[0] IMPLICIT OCTETSTRING,
birth_date[1] IMPLICIT SEQUENCE{ year INTEGER,
month INTEGER,
day INTEGER},
soc_sec_nbr[2] IMPLICIT OCTETSTRING,
status[3] IMPLICIT BOOLEAN DEFAULT FALSE,
sex[4] IMPLICIT BITSTRING OPTIONAL}

FIG. 9 ASN.1 example.

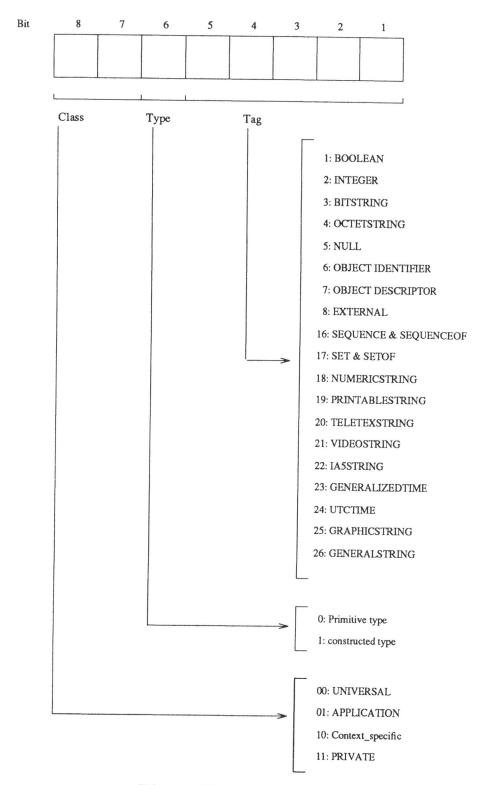

FIG. 10 ASN.1 identifier byte structure.

order bit set to 0, except the last one. The seven remaining bits of each identifier byte may be used to encode tags with high values.

Following the identifier field is the set of bytes that determine the length of the data field. For a multibyte number, the most significant seven bits of the length are transmitted first followed by as many bytes as it takes to encode the total length. Note that all bytes have their high-order bit set to 1. Following the length field is the data field. The encoding of the data field depends on the type of data present.

Application-Layer Protocols

The main function of the application layer is to provide the application protocols that allow the user to access the underlying network environment. The basic concept behind the design approach of the application layer is to offer several utilities and useful mechanisms to support distributed applications and to leave it to the user to select the specific protocols and their associated functions. Consequently, the boundary between the presentation layer and the application layer can be viewed as representing the separation of the protocols imposed by the network designers from those being selected and implemented by the network users.

The application-layer protocols are responsible for transferring information between cooperating user application processes. More specifically, the tasks of the application-layer protocols include

- Allowing users to access network resources and associate them with the right application
- Providing means for the user to transfer files, messages, and documents electronically
- Providing an environment where terminal characteristics are transmitted in a commonly understood format
- Providing standardized command languages for databases and operating systems

The services provided by the application layer form an integral part of the distributed information-service environment and determine how users perceive the networking environment. These services provide increased communication functionality, allowing users to interact with one another as well as with remote machines.

Several of the application-layer services are commonly used. Their widespread use warranted the development of standards for them in an effort to avoid wide proliferation of incompatible protocols supporting these services. The building blocks of these application services, namely, Common-Application Service Elements (CASE) and Specific-Application Service Elements (SASE) are the focus of the first part of this section. A discussion of the design issues involved in designing protocols for such frequently used applications as file transfer, electronic mail, virtual terminals (VTs), job transfer and manipulation

(JTM), manufacturing message service (MMS), and directory service (DS) is provided in the second part of this section.

Application-Layer Design Issues

The general structure of the application layer comprises two classes of basic entities known as *service elements*. The first class, CASE, consists of the elements that are commonly used by a number of applications. More specifically, the elements provide a general framework of information-transfer capabilities to any application-layer entity regardless of its nature. The second class, SASE, defines a set of service elements that allow user applications to access and provide specific services.

CASE is comprised of one or more protocol elements. The more general elements include Association-Control Service Elements (ACSE), Remote-Operations Service Elements (ROSE), and Commitment, Concurrency, and Recovery (CCR) elements. ACSE elements are designed to manage connections commonly referred to as associations. They are concerned solely with the establishment and release of the logical association between communicating SASEs. The set of primitives provided by ACSE includes primitives to establish and release an association and primitives to allow either the user or the provider to abort the current association. ROSE elements are concerned with initiating operations and receiving results from remote SASEs. CCR is a service element that coordinates multiparty interactions and ensures a predictable effect of concurrent accesses to such shared logical resources as files and databases.

To guarantee the integrity and consistency of the shared information, CCR implements the concept of atomic actions. *Atomic actions* can be viewed as a set of operations that is indivisible. Therefore, an action is all or nothing in its effects, and the partial effects of one atomic action remain invisible to other atomic actions. There are two factors that can threaten the atomicity of actions: failure of the processes and computers, and concurrency of access. To implement atomic actions, CCR uses the two-phase commit technique. In this approach, every atomic action has two phases with three associated states: tentative, committed, and aborted. Throughout the first phase, the atomic action is in its tentative state. During this phase, all requested changes are performed on a secondary stable storage. Stable storage is a generic approach designed to ensure that any essential permanent data will be recoverable after any single system failure. Stable operations are implemented using redundant storage for structural information. Upon completion of all required changes, the atomic action enters its second phase and an attempt to commit the changes is made. If the attempt fails, the entire atomic action is aborted and the initiator of the atomic action is informed.

As stated above, SASEs have been developed to allow user applications to access and provide specific services. The application layer supports a range of distributed information-processing services provided by SASEs, such as file transfer, electronic mail, VTs, JTM, MMS, and DS. These SASEs constitute some of the most commonly used protocols. However, the number and the range of these protocols will increase as the applications of the open system environment expand.

The global structure of the application layer and its relationship with the presentation layer is depicted in Fig. 11. In this illustration, the user agent, also referred to as the user element, provides the user interface to the various distributed information services supported by the application layer. Note that the interface primitives between a user application and the user agent need not be the same as the service primitives provided by the application entity being used. This approach was adopted to allow manufacturer-dependent software to be used in the context of the open system environment. The user agent imple-

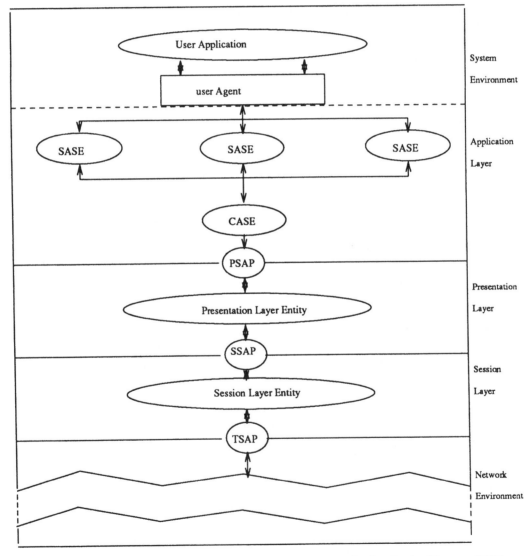

FIG. 11   Application-layer protocols (CASE = Common-Application Service Element; SASE = Specific-Application Service Elements; TSAP = transport-service access point).

ments the required mapping functions between the real system environment and the open system environment. This is achieved by defining a virtual device and an associated set of user primitives for each application service. A software layer then is introduced to map between the service primitives relating to the real device and the corresponding service primitives relating to the virtual device.

File Transfer, Access, and Management (FTAM)

File handling is one of the principal services provided by distributed information systems. File transfer and remote file access constitute the most frequently used file-handling operations. File transfer involves copying an entire file from one machine to another. Remote file access, however, provides the means to interact with parts of a file without the need to transfer the entire file from the remote site to the local host.

File access and management operations are usually defined within the client–server paradigm. This model defines a pattern of interactions among cooperating applications, and is fundamental in understanding the foundation of distributed information systems. The term *server* in this model refers to any program that offers a service to a distributed community of client processes. Servers accept requests that arrive over the network, perform their services, and return the result to the requesting clients. This type of interaction defines a convenient and natural extension of the fairly well-understood mechanism used for inter-process communications on a single machine. The flexibility provided by this model warranted its extensive use as the building block for interaction among cooperating entities in a distributed file system.

Based on the client–server model, a distributed file transfer, access, and management system can be viewed as a set of one or more servers that maintain a potentially large collection of files and provide a distributed community of clients with the facilities to access and update their contents, to create new files, and to delete existing ones.

The content of a file as viewed by clients of the file service is simply a modifiable sequence of data items. The file server, however, may use different conceptual models of what a file is. In its simplest form, a file can be viewed by the server as an unstructured set of data. The absence of structure within the file restricts the operations the server can perform on the file to reading or writing the entire file. The flat-based model provides a more structured view of the file, and considers the file to be consisting of an ordered sequence of records. This view allows the client to address specific records of the file based on their keys or on their positions within the file. In the most general form, a file can be perceived as a tree-structured set of nodes. Each node of the file may have a label and a data record.

File servers maintain a variety of information about each file. These items of information are often referred to as *attributes*. They are usually grouped together into a standard-format descriptor associated with every file in the system. The most commonly used attributes describe the name of the file, the main characteristics of the file (such as length, last access, and last modification), the structure of the file, and the type of access that is permitted to the file.

File servers define a variety of operations that can be performed on parts of the file or on its entire contents. The basic operations allow the client to create, read, write, update, and delete a file. Another operation may include support for directory operations. In a distributed information-service environment, the concurrent execution of these operations by different clients makes the implementation of these access primitives difficult. A widely implemented solution to the problem of concurrency of access is based on the use of shared and exclusive locks. A *shared lock* allows more than one client to perform compatible operations concurrently. An *exclusive lock* restricts the access to the file to the current holder of the lock. Shared locks are typically used for reading, while exclusive locks are often used for writing. Locking-based schemes may create several problems, such as deadlocks, failure detection of clients that are currently holding locks, and recovery from these situations. In addition, the locking-based scheme introduces several design issues that are hard to deal with, such as the granularity of the locking and the maximum lifetime of a lock before it is broken. As an alternative to the locking-based concurrency control mechanism, some file servers support optimistic concurrency control, a form of control based on checks made at the time that a transaction closes, and also support a timestamp-based mechanism that associates with each transaction a unique timestamp value that defines its position in the time sequence.

Several of the design issues have been used in many operating systems. In the section that follows, a description of the specific models produced by ISO and ARPANET are provided. The presentation focuses on the approach these models used to address the issues described above.

OSI File Transfer, Access, and Management.   The OSI FTAM system is based on the virtual-file-store model. The main goal of the model is to provide sufficient flexibility to allow any real file store to be readily accessed and managed without excessive software mapping functions. In this model, each client process is referred to as the *initiator*, and the server process as the *responder*. An arbitrary number of initiators may have an association with the responder file store simultaneously. A set of attributes is associated with each file of the file store. These attributes include the following:

1. file name
2. permitted actions
3. access control
4. account number
5. date and time of the file creation
6. date and time of the last file modification
7. date and time of the last file read
8. date and time of the last attribute modification
9. identity of the owner
10. identity of the last modifier
11. identity of the last attribute modifier

12. the file availability
13. presentation context of file contents
14. encryption key
15. file size
16. maximum future size
17. legal qualifications
18. private use

The legal qualifications attribute refers to the various types of legal information that are private to a specific nation. Transfer of this information is usually restricted to within the boundaries of the nation.

The access structure within the file store is hierarchical and based on an ordered-rooted tree (see Fig. 12). Each node of the structure gives access to a subtree, referred to as a file-access data unit (FADU). The content of the file is stored in data units (DUs).

The virtual file store provides a set of highly connection-oriented service primitives that are grouped into a nested set of regimes. The FTAM regimes and the primitives associated with them are depicted in Fig. 13.

The application-connection regime allows the establishment of an association and the gathering of storage and accounting information. The file-selection regime is concerned with the creation of new FADUs or the unique identification of existing ones. Furthermore, the operations that can be performed on the FADU in subsequent phases are determined. The file-access regime determines the establishment of the file-data-transfer phase. During this phase, a suitable access context is determined and the required capabilities for data transfer are determined. Finally, the data-transfer regime allows the access of the selected FADU. During this regime, such commands as read and write may be issued to access the FADUs of the selected file.

Internet File-Transfer Protocols. The Internet protocol suite includes a File Transfer Protocol (FTP), which allows users to access remote machines, list the contents of remote directories, and copy files between machines. In addition, the protocol allows the user to execute a few simple commands remotely. Security is handled by requiring the user to specify a user name and password to the other computer.

Unlike FTAM, FTP does not use the virtual-store concept as a building block for its design. The main idea behind the design of the protocol is to allow file transfer, taking into consideration the possible differences between the involved machines. Consequently, the main characteristic of FTP resides in its ability to understand the basic file formats and convert among frequently used character-representation sets, such as ASCII or EBCDIC (Extended Binary Coded Decimal Interchange Code). FTP recognizes four types of files: ASCII, EBCDIC, image, and logical byte files. File transfer can occur in stream mode, compressed mode, or block mode. The stream mode is used to transfer regular files. The compressed mode allows the compression of the file prior to its transfer. Compression is achieved by encoding runs of similar consecutive characters into a substantially shorter string of characters. The block mode allows

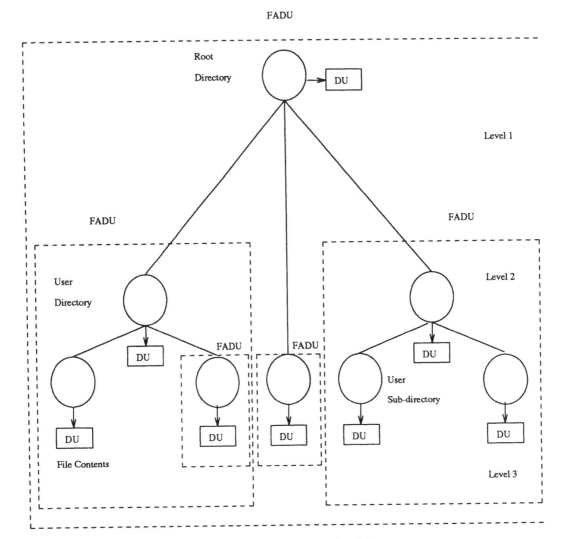

FIG. 12   File transfer, access, and management virtual-file-store access structure.

the transfer of highly structured files consisting of blocks of data, each with a header giving its size, position, and type.

FTP can also handle third-party transfers. In this mode, the client first establishes connections to the servers on two remote machines, then issues the command to transfer the files. The two servers establish a transport connection between themselves and carry out the transfer of data. The client retains control of the transfer but does not participate in moving data. The FTP uses TCP, a reliable stream service, to transport the data between the two machines. It allows a user to access multiple machines in a single transfer session. In addition, FTP maintains separate transport connections for control and data transfer.

The Internet protocol suite also offers an unsophisticated service for file transfer known as Trivial File Transfer Protocol (TFTP). Unlike FTP, TFTP

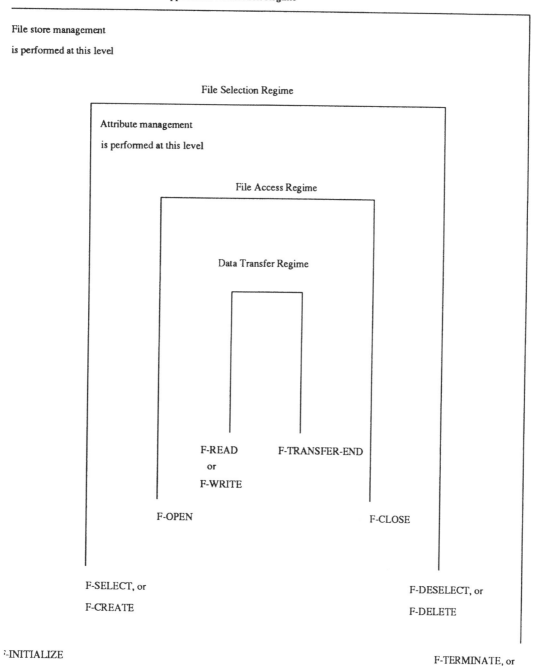

FIG. 13 File transfer, access, and management regimes and primitives.

uses UDP, an unreliable packet-delivery service, as a transport mechanism. The protocol is intended mainly for applications that do not require complex interactions among clients. The protocol has been used in such applications as bootstrapping for diskless workstations and electronic mail systems. In the former case, the protocol was encoded into read-only-memory chips, and used for system bootstrap. In the latter case, the protocol was fully integrated with the electronic mail system.

Electronic Mail Systems

Electronic mail systems provide a generalized facility for two application entities to exchange electronic messages. In many ways, electronic mail systems are similar to file-transfer systems. Their resemblance stems from the fact that they both involve the transfer of a file from a local to a remote machine. However, electronic mail systems usually involve the exchange of highly structured documents intended to be used mostly by human beings rather than by machines. Consequently, electronic mail systems must provide for a friendly user interface to compose, edit, and read a message, as well as a transport mechanism to deliver the message to its destination and report a notification to the sender. A conversion mechanism may also be necessary to convert a message into a format suitable for display at the recipient terminal. Furthermore, the mail system must provide its users with facilities to create and manipulate mailboxes, as well as such advanced features as creation of carbon copies, high-priority mail, and alternative recipient distribution lists.

Several models have been proposed for the implementation of electronic mail systems. Most of these models define the basic structure of the messages as composed of an envelope and the contents. The *envelope* represents the information used for transferring the message within the electronic system, and the *contents* represent the actual message to be delivered to the recipient. In the remainder of this section, a description of the Message Handling System (MHS) proposed by OSI and the Simple Mail Transfer Protocol (SMTP) is provided.

Message Handling System (MHS). The MHS provides a set of facilities for electronic message exchange in an open system interconnect. The service provided by MHS is based on ISO X.400 recommendations. The conceptual model of MHS is depicted in Fig. 14. The basic entities of the MHS model include a message-transfer agent (MTA) and the user agent (UA). The MTA can be viewed as a virtual mail server. The basic function of the MTA is to relay the message from the sender to the receiver. Note that this operation is fundamentally different from a typical file-transfer operation since the recipient UA may not be currently present in the system. Collectively, the MTAs form the message-transfer system. The message-transfer system is divided into administrative domains usually operated by different authorities. The interaction between MTAs is described by a set of defined services and protocol standards.

The UA is concerned with managing the dialogue with the message originator, negotiating the delivery and receipt of messages with the message-transfer system, and managing the message store (MS).

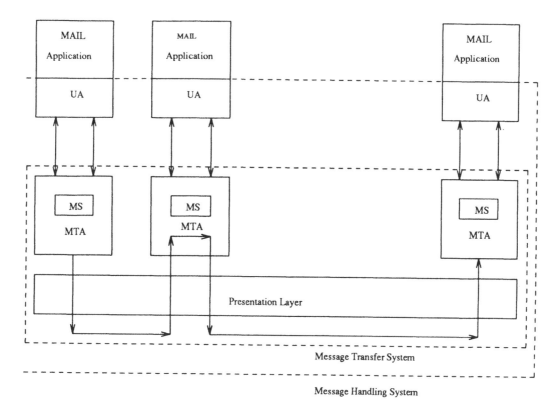

FIG. 14 Message Handling System model: MS = message storage; MTA = message-transfer agent; UA = user agent.

MHS provides a range of different service primitives. Most of the services provided by these primitives are confirmed, although the confirmation is generated by the local MTA protocol rather than by the remote entity. The basic primitives include LOGON.request, LOGON.confirm, SUBMIT.request, SUBMIT.confirm, LOGON.indication, LOGON.response, DELIVER.indication, LOGOFF.request, and LOGOFF.confirm. A set of specific parameters is associated with each of these primitives. The parameters specify the originator application-entity address, the content and its type, and the priority of the message.

Internet Simple Mail Transfer Protocol (SMTP). The Internet specifies the SMTP standard for the exchange of mail between machines connected to the internetwork. The protocol is concerned mainly with the underlying mail delivery system. The protocol specifies the mechanism used by a given machine to pass the message to the next machine across the link. More specifically, the protocol does not specify the interface to the user or the method used to store and forward messages and the frequency at which the server attempts to deliver the message.

The protocol used between the client and the server is modeled after human interaction. Prior to sending a message, the client establishes a connection to the server and waits for the server to send the READY FOR MAIL message. Upon receipt of this message, the client greets the server by sending a HELO command, to which the server responds by identifying itself. The mail transaction begins with a MAIL command, which includes the sender identification and a FROM field to which error status must be reported. The sender then proceeds to issue the RCPT command, which contains the identity of the recipient. Note that both the identity of the sender and the recipient must be acknowledged by the server before the data-transfer phase is entered. This last phase starts when the client issues the command DATA and terminates with the termination sequence.

The protocol supports only the transfer of regular ASCII text, and does not provide any mechanism for transfer of facsimile, digitized voice, or graphics. SMTP does not make any distinction between the envelope and the message, and considers the electronic message as consisting of a file containing a specific header at the top. The headers usually determine the envelope functions, the address headers, the message contents, and other useful information, such as the date, message identity, and encryption field.

Other Application Services

As the application of OSI has expanded, several application protocols have been defined. Some of the services these protocols define may apply to a general distributed information-processing environment. Others are more tailored to specific industrial environments. The most common services provided include JTM, VT, MMS, and DS.

Job Transfer and Manipulation. JTM service allows a user to submit a job for execution by a remote-application entity. The main function of JTM is to guarantee that the input files of the submitted job are collected and submitted for execution at the appropriate locations, and that the files containing the results are deposited at the user requested locations. Note that the JTM is not required to know about the content of the files, the job control language used, or the type of processing required. Several models have been proposed to carry the above tasks, including the ISO JTM model described below.

The OSI JTM service can be viewed as a collection of SASEs located on different systems that form the JTM service provider. The services provided by JTM allow a user to submit a work specification to a remote-application-layer entity and monitor the execution of the submitted job. The work specification specifies the type of job requested, the location at which the job is to be executed, the location at which the involved files reside, and the location at which the output is to be stored. The application entity that submits the job specification is referred to as the *initiating agency*. The application entity that actually executes the job is known as the *executing agency*. The executing agency contacts the source agency in order to correct information and data relating to the job. An application known as *job monitor* may also be specified by the initiating agency to report on the current status of the job that is being executed. The job monitor gathers its information from the report document created by the JTM service, which is provided and used to store the main events as they occur during the lifetime of the executing job. Upon completion of job execution, the results are collected in a document that is submitted to a nominated-application entity known as a *sink agency*.

In order to carry the task above, ISO JTM provides a set of primitives including

J__INITIATE to submit a job specification
J__CIVE to allow the JTM service provider to request a document
J__DISPOSE to pass a document to an execution or sink agency
J__TASKEND to signal the end of execution
J__STATUS to report the current progress of the job execution
J__KILL to terminate all the activities associated with the job
J__STOP to stop the execution of a job temporarily.

Virtual Terminal. The VT provides a facility for a local-application entity to communicate with a remote-application entity in a standard fashion, independently of the characteristics of the local terminal currently being managed by the local application.

The most frequently used types of terminals may be grouped into three classes: scroll mode, page mode, and form mode. Scroll-mode and page-mode terminals incorporate a basic Universal Asynchronous Receiver/Transmitter (UART) and utilize a simple asynchronous character mode communication protocol with an echo-checking error control. Consequently, these types of terminals are incapable of executing any protocol to communicate with the network. To overcome this shortcoming, a packet assembler/disassembler (PAD) is required. The PAD implements the protocol required by the network and uses the RS-232 specifications to interface to the terminal.

Form-mode terminals have built-in microprocessors. These terminals have the capability to download a form from a remote computer, allow the local editing and execution of macroinstructions, and finally upload the modified document across the network to the remote computer.

The model proposed by OSI Virtual Terminal Service (VTS) is based on the concept of conceptual communication area (CCA). The CCA holds five different types of objects: conceptual data store, control object, device object, token-handling information object, and the ASN.1 parameters definition object. The conceptual data store object holds the required information for a commonly accepted international alphabet. In this alphabet, support for diacritical marks has been provided. The control objects relate to terminal-handling functions other than the ones concerned with text processing. This function includes bells, modeling function keys, and interaction with mouse buttons. The device object handles the interface to each real device, such as a keyboard or printer, attached to the terminal. The final two types of objects relate to the ownership of the token in a synchronous-mode communication, and to the current ASN.1 parameters negotiated and agreed upon. In essence, the model implemented by OSI VTS is similar to the form mode's virtual terminal.

Manufacturing Message Service.   MMS provides a facility for an application entity to communicate with remote-application entities controlling an automated manufacturing environment. In essence, MMS is similar to a regular mailing system with the exception that the system is intended for communication between machines and robots, for instance, rather than for communication between human beings. The services offered by MMS allow the establishment of associations with specific controllers, the reading of data files, the remote control of the operations of a controller, and the identification of the services provided by a specific controller.

Directory Services.   A DS offers a "White-Page" type of service that allows a user to look up people in order to learn information about them. In its simplest form, a White Pages service provides what the White Pages of a telephone book provide. Based on a set of attributes, one can find the name, address, and telephone number of users of the environment. In a network environment, there may be many other kinds of location information, such as electronic mailbox, electronic calendar, or file server, where one might leave a file for the recipient. In addition, the electronic White Pages may support a much more sophisticated set of mechanisms for lookup. One might match on a more complex set of attributes than first and last name. In addition, the searching might span more than one local White Page service.

Several models of naming and directory-service specifications have been proposed and implemented in the field. They differ by the type of functionality they provide and the mechanisms they use to address that functionality. Within the world of networking today, there are a number of partial solutions to the directory service problem. Examples of these are the Internet Domain Naming Service (DNS), Clearinghouse, DECnet Network Architecture Naming Service (DNANS), and X.500.

The Domain Naming Service provides a directory service most commonly used for host naming and mail delivery. Clearinghouse and DNANS are, respectively, the Xerox and Digital Equipment Corporation corporate naming services, originally for mail delivery. The CCITT recommendation for directory services is the X.500 (ISO DIS 9594) document. One of the primary goals of the standard is to provide the naming service needed for message handling (X.400). X.500 is still developing, and needs further evolution to cover all the requirements of different types of public and private networks. Nevertheless, the standard is expected to form the basis of commercial products, and become the directory service in many types of networks.

Summary and Conclusions

In this article, we examine some of the fundamental aspects of communications networks. In the first part of the article, we deal with the fundamental issues involved in data communication between two devices that are directly connected by some form of point-to-point transmission medium. Although simple, the model raises several questions concerning data transmission, data encoding, data-link control, and multiplexing. A more sophisticated model, based on a communications network, is introduced.

Two concepts are paramount in the discussion of computer communications and computer networks. The first concept is concerned with the rules and regulations that govern the access to the network. These rules are commonly referred to as protocols. A protocol is characterized by its syntax, semantics, and timing. The second concept deals with the appropriate architecture, which defines the structure and framework for the development and implementation of the required protocols.

The ISO/OSI seven-layer model is adopted as the backbone architecture for the discussions of the protocols relevant to each layer of the model. For each layer, the design issues relating to the functionalities of the layer are introduced, and the solutions provided for their implementation in the real world are described.

It is important to note that the world of data communications is changing rapidly. The most recent advances in computer technology and VLSI design have created new opportunities for new applications. These applications include digitized voice, still and motion video, bulk data transfer, computer imaging, distributed scientific computation and visualization, multimedia conferencing, remote interactive processing, and data acquisition. These applications have a wide range of performance characteristics and exchange information. The design of adequate communications protocols and the development of a new set of quality services to support these applications bring about new challenges in the world of computer communications. These challenges and the solutions to the problems they entail will be the focus of current and future research in computer communications.

Glossary

ACCESS CONTROL. Field used in token ring MAC to control access to the channel according to the priority and reservation algorithm.

ACCESS METHOD. A set of rules for transferring data between hosts.

ACKNOWLEDGMENT. Message returned from receiver to the sender to indicate the successful reception of transmission.

ACK0, ACK1. BSC acknowledgments for even and odd numbered messages, respectively.

ADCCP (*Advanced Data Communication Control Procedures*). ANSI standard version of the bit-oriented data-link-control protocol.

ADDRESS. Data structure used to identify the destination of a message.

ALGORITHM. Method for solving a problem that is suited to computer implementation.

AMPLIFIER. Hardware device that may be inserted at various points in the communication system to impart a gain in signal strength.

AMPLITUDE MODULATION. Modulation technique that varies the amplitude of the carrier signal to convey information.

AMPLITUDE SHIFT KEYING. Modulation technique in which the two binary values are represented by two different amplitudes of the carrier frequency.

ANALOG SIGNAL. Continuously varying electromagnetic wave that may be propagated over a variety of media.

ANALOG TRANSMISSION. The transmission of analog signals regardless of their content.

APPLICATION LAYER. The seventh layer of the OSI model.

ARPANET (*Advanced Research Projects Agency Computer Network*). One of the earliest and most influential computer networks in the United States.

ASCII (*American Standard Code for Information Interchange*).

ASYNCHRONOUS BALANCED MODE. HDLC mode for balanced configuration used only in point-to-point operation. The configuration consists of two combined stations and supports both full-duplex and half-duplex transmission.

ASYNCHRONOUS RESPONSE MODE. HDLC mode for unbalanced configuration used in point-to-point and multipoint operation. The configuration consists of one primary station and one or more secondary stations and supports both full-duplex and half-duplex transmission.

ASYNCHRONOUS TRANSMISSION. A type of transmission in which each character is transmitted independently of other characters. Each character is preceded by a start bit and ended by a stop bit. This type of transmission is often referred to as start–stop transmission.

ATTENUATION. The loss of power a signal suffers as it travels from the transmitting device to the receiving device.

ARQ (*Automatic Repeat Request*). Error-control technique in which receiver detects packets with errors and requests their retransmission.

BACKOFF. Random delay before another attempt at transmitting is made by a colliding station in random-access protocols.

BALANCED CONFIGURATION. This configuration consists of two combined stations and supports both full-duplex and half-duplex transmission. The configuration is used only in point-to-point operation.

BALANCED TRANSMISSION. Mode of transmission in which the information is conveyed by comparing voltages in two circuits. For digital signals, this technique is known as differential signaling. It aims at minimizing the effect of induced voltages.

BANDWIDTH. The difference between the two limiting frequencies of the spectrum. It is expressed in hertz.

BASEBAND. Transmission of signals without modulation. Since the entire scheme is occupied by one signal, this technique does not allow frequency division multiplexing.

BAUD RATE. Number of signal changes per second, used to convey information. Often misused for bit rate.

BIT-ORIENTED PROTOCOLS. A class of data-link-layer protocols based on the use of a flag character to delimit the frame.

BIT RATE. Number of bits per second transmitted or received.

BIT STUFFING. The insertion of extra bits into a datastream to avoid the occurrence of unintended control fields.

BLOCK CHECK CHARACTERS. Extra characters added to frames for error protection in the BSC data-link-layer protocol.

BROADBAND. The use of coaxial cable for providing data transfer by means of analog signals. Modems are used to transmit digital signals over one of the frequency bands of the signal.

BROADCAST COMMUNICATION. A communication network in which transmission is received by all stations at the same time.

BSC (*Binary Synchronous Communication*). A character-oriented data-link-layer protocol for half-duplex applications.

BUS. A conductor used for transmitting signals or power.

BYTE-COUNT-ORIENTED PROTOCOLS. A class of data-link-layer protocols that rely on byte-count fields in the frame to determine the length of the frame.

CARRIER. A continuous frequency that is impressed with a second signal to convey information.

CHARACTER-ORIENTED PROTOCOLS. A class of data-link-layer protocols that uses standard alphanumeric control characters for communication.

CHECKSUM. An error-detection code based on a summation performed on the bits to be checked.

CIRCUIT SWITCHING. Mode of operation usually used by telephone networks. This mode requires the establishment of a communication path prior to the exchange of information.

CLASS 0 TRANSPORT SERVICE. OSI transport protocol for Type A networks with error control, sequencing, and flow control. The transport service provides minimal connection establishment, data transfer, and connection–termination facilities.

CLASS 1 TRANSPORT SERVICE. OSI transport protocol for Type A networks with error control, sequencing, and flow control. The transport service provides minimal connection establishment, data transfer, and connection-termination facilities. In addition, it provides capability to recover from network-layer resets or disconnects, and offers expedited data capabilities.

CLASS 2 TRANSPORT SERVICE. OSI transport protocol for Type A networks with error control, sequencing, and flow control. The transport service provides connection establishment, data transfer, and connection facilities for multiplexing multiple transport-layer connections over a network-layer virtual circuit. The service uses credits to provide for transport-layer flow control.

CLASS 3 TRANSPORT SERVICE. OSI transport protocol for Type B networks. The protocol adds the multiplexing and control functions of Class 2 to the error-handling functions of Class 1.

CLASS 4 TRANSPORT SERVICE. OSI transport protocol for Type B networks. The protocol is designed to handle worst-case networks. It provides comprehensive error detection and recovery mechanisms.

COAXIAL CABLE. Cable used for large bandwidth electrical signals. It consists of an inner conductor, usually a small copper wire, a braided outer conductor, and shield.

COLLISION. Result of simultaneous transmission of packets over the same medium.

COMBINED STATION. Combines the features of both primary and secondary HDLC stations. This type of station may issue both commands and responses. They are used in an asynchronous balanced mode.

CONNECTIONLESS DATA TRANSFER. Data transfer without prior establishment of a connection between the sender and the receiver. The exchanged units of information are usually referred to as datagrams.

CONNECTION-ORIENTED DATA TRANSFER. This mode requires the establishment of a connection before any exchange of data, and the termination of the connection upon completion.

CONTENTION. Attempt by two stations to use the same channel at the same time.

CONTINUOUS ARQ (*continuous Automatic Repeat Request*). Standard approach to ARQ for data-link-control protocols that allows frames and acknowledgments to occur simultaneously. Depending on the implementation, the sender may be required either to selectively retransmit packets with errors, or retransmit the packet that was in error and all packets sent after it.

CONTROL CHARACTER. A character in a standard character set used to convey control information to a device.

CONTROL CIRCUIT. X.21 circuit used to send control information from a DTE to a DCE.

CONTROL FIELD. Field in a frame containing control information.

CREDIT. Permit to transmit a number of messages as prescribed by the permit message. Used to allow variable-sized windows.

CSMA (*Carrier Sense Multiple Access*). A control method used to access a shared transmission medium.

CSMA/CD (*Carrier Sense Multiple Access with Collision Detection*). A control method used to regulate the access to a shared transmission medium. A station wishing to transmit must first sense the channel and transmits its message only if the channel is idle (no carrier). The transmitting station must monitor the actual signal for a duration of an end-to-end propagation delay to detect a possible collision with simultaneous transmissions from other stations. In the event of a collision, the colliding stations reschedule their attempts to transmit randomly in the future.

CYCLIC REDUNDANCY CHECK. An error detection code in which the code is the remainder resulting from the division of the bits checked by a predetermined binary number.

DATAGRAM. A self-contained packet of information that is sent over the network with minimum protocol overhead. The transmission of datagrams do not require the setting of a connection. Datagrams are handled independently of each other.

DATA-LINK LAYER. This layer corresponds to the second layer of the OSI reference model and most other network architectures. The main functions implemented by this layer include the reliable transfer of data across the data link and the control of data flow between the sender and the receiver.

DATA SET READY (*DSR*). An RS-232C interface circuit activated by the data set (modem) to indicate that it is powered up and ready to use.

DCE (*data circuit terminating equipment*). Standard committee (CCITT, EIA) terminology for the attachment of user devices to the network. The data equipment takes on different forms for different network types.

DDCMP (*Digital Data Communication Message Protocol*). A byte-count-oriented data-link-control protocol developed by Digital Equipment Corporation.

DEMAND ASSIGNMENT MULTIPLE ACCESS. A technique used to allocate the in-demand part of the access-medium capacity to users.

DLE (*Data Link Escape*). A communication control character used by BSC. The purpose of this character is to change the semantics of a limited number of contiguous characters that follow it.

DS (*directory service*). A protocol entity forming part of the application layer. The basic function of this protocol is to translate symbolic names as used by the application processes into fully quantified network addresses as used with the open system interconnect environment.

DTE (*data termination equipment*). A generic name for data-processing equipment connected to a data network, such as visual display units, computers, and workstations.

EBCDIC (*Extended Binary Coded Decimal Interchange Code*). Eight-bit alphanumeric code used in many IBM products.

ENCAPSULATION. The addition of control information to the data obtained from the user.

ENQ (*Enquiry*). A control character used in standard alphanumeric codes and by some data-link-layer protocols, such as BSC, for polling and selection or for assertion of the receiver status.

EOT (*End of Transmission*). A communication control character used to conclude a BSC transmission that may have contained one or more text messages and headings.

ETHERNET. A CSMA/CD-based local-area network.

ETX (*End of Text*). A communication control character used in BSC to terminate the transmission of an entity.

EXPEDITED DATA. A short packet with high priority for treatment. It is usually used in emergency cases.

FAST SELECT. An option of the X.25 virtual call that allows the inclusion of data in the call-setup and call-clearing packets.

FIBER OPTICS. A communications medium used for high-speed networks. It is usually composed of glass or similar fiber and associated circuits to transmit optical signals.

FLAG. A special configuration of bits used by data-link-control protocols to achieve frame synchronization.

FLOW CONTROL. A technique used to regulate the traffic between the sender and the receiver. It aims at preventing the sender from overflowing the buffers of a slow receiver.

FRAGMENTATION. The need to break up data into a smaller-bounded size to accommodate a network with short maximum packet length. Other motivations for fragmentation include more efficient error control, more equitable access to shared communications media, and better management of small-sized buffers. The method may, however, incur more processing overhead and generate more interrupts to be served by the communications devices.

FRAME. The unit of information transferred over a data link. It is composed of control information and user data.

FREQUENCY. Rate of signal oscillation (in hertz).

FREQUENCY MODULATION. Modulation in which the frequency of an alternating current is varied.

FREQUENCY SHIFT KEYING. Modulation in which the two binary values are represented by two different frequencies near the carrier frequency.

FTAM (*file transfer, access, and management*). A protocol entity forming part of the application layer that allows the user application to access a distributed file system.

FULL-DUPLEX TRANSMISSION. Data transmission in both directions at the same time.

GATEWAY. A device that connects two systems and allows data to be passed from one system to the other. When the two communicating systems operate different protocols, the gateway performs protocol conversion.

GO-BACK-N ARQ. A variant of continuous ARQ. Upon detection of an erroneous frame, the receiver drops the frame in error and all succeeding frames. The receiver waits for the arrival of the correct version of the frame that was in error to resume receiving the succeeding frames.

HALF DUPLEX. An information exchange strategy in which data transmission may take place in either direction, one direction at a time.

HANDSHAKING. Sequence of message exchange to convey information and verify its receipt.

HDLC (*High-Level Data-Link Control*). Standard data-link-layer protocol used to regulate the data exchange between point-to-point data links or multidrop data links.

HEADER. Protocol-defined control information that precedes user data.

ICMP (*Internet Control-Message Protocol*). An integral part of the Internet Protocol (IP) that handles errors and control messages. It allows hosts to interact with gateways, and gateways to report problems about datagrams to the original source that sent the datagram. It also includes an echo request/reply used to test host reachability.

IEEE 802. A set of standards for local-area networks being developed by the IEEE 802 committee.

IEEE 802.2. Logical link-control protocol developed by the IEEE 802 committee.

IEEE 802.3. CSMA/CD MAC protocol developed by the IEEE 802 committee.

IEEE 802.4. Token bus MAC protocol developed by the IEEE 802 committee.

IEEE 802.5. Token ring MAC protocol developed by the IEEE 802 committee.

IMPULSE NOISE. A high-amplitude, short-duration noise pulse.

INTERFACE. Boundary between equipment or protocol layers.

INTERNET PROTOCOL. An internetworking protocol that provides connectionless service across multiple packet-switched networks.

INTERNETWORKING. Communication among devices across several networks.

ISO (*International Organization for Standardization*). International organization responsible for telecommunications networking standards including the OSI reference model.

JTM (*job transfer and manipulation*). A protocol entity forming part of the application layer. It is used by user application processes to transfer and manipulate documents describing job processing tasks.

LAN (*local-area network*). A general-purpose network that spans a limited geographical area.

LAP-B. CCITT X.25 version of bit-oriented data-link-control protocol.

LINK. Communications facility connecting nodes in a communications network.

LOOPING. Routing that sends packets back to the source node. It is used for testing and checking.

MAC (*Medium access control*). Lower part of the data-link layer in the IEEE 802 architecture.

MANCHESTER ENCODING. A digital encoding technique in which there is a transition in the middle of each bit time.

MESSAGE SWITCHING. Technique used to transfer messages between nodes of a network.

MHS (*message-handling system*). A protocol entity forming part of the application layer. It provides a generalized facility for exchanging electronic messages between systems.

MODEM (*modulator-demodulator*). Translates a digital signal into a form suitable for transmission over an analog communications facility, and translates back an analog signal into a digital form suitable for processing by a digital computer.

MULTIPLEXING. A technique used for sharing a resource among several contenders.

MULTIPOINT. A configuration in which more than two stations share a communication path.

NACK (*Negative Acknowledgment*). A message or a control character indicating to the sender a negative acknowledgment.

NETWORK LAYER. Third layer of the OSI reference model.

NODE. Data-processing equipment that is part of a telecommunications network.

NRM (*normal response mode*). Unbalanced configuration in the HDLC transfer modes of operation. In this mode, the primary may initiate a data transfer to a secondary, but a secondary may only transmit data in response to a poll from the primary.

NRZ (*Non-Return-to-Zero*). A digital signaling technique in which the signal is at a constant level for the duration of a bit time.

NULL MODEM. An appropriate wiring of the RS-232-C circuits that allows two devices to communicate directly with each other.

OPTICAL FIBER. A transmission medium used for transmission of data in the form of light waves of pulses.

OSI REFERENCE MODEL. An architectural model of networking developed by ISO and adopted as an international standard.

PACKET SWITCH. A mode of operation of a data communications network in which each message to be transmitted is first divided into packets. As they travel across the network, packets may be stored and later forwarded to the next node in the path.

PARITY BIT. Extra bit added to transmitted data to enhance error control.

PCM (*Pulse Code Modulation*). A process in which the signal is sampled, and the amplitude of each sample with respect to a fixed reference is quantized and converted by coding to a digital signal.

PHASE MODULATION. A modulation technique that varies the characteristics of the carrier signal phase to convey information.

PHYSICAL LAYER. The first layer of the OSI reference model and most other network architectures.

PIGGYBACKING. A technique used to return acknowledgment information across a full-duplex data link without the use of special acknowledgment messages.

POINT-TO-POINT. A configuration in which two stations share a communication path.

PRESENTATION LAYER. The sixth layer of the OSI reference model.

PRIMARY STATION. The master station that is in charge of the HDLC link.

PROTOCOL. A set of rules and conventions that regulates the exchange of data between two communicating devices.

QOS (*quality of service*). Parameters that describe a quality of service as perceived by the user.

RANDOM ACCESS. Unscheduled access to the communications medium.

RING. Configuration of stations in shape of a ring, and in which communication normally flows in one direction.

ROSE (*remote-operations service element*). A protocol entity forming part of the application layer that provides facility for initiating and controlling operations remotely.

ROUTING. A technique used to determine the path of a message from the source to the destination.

RS-232-C, RS-449, RS-422-A, RS-423-A. Standards developed by the Electronic Industries Association for a physical-layer interface between communications devices.

RTS (*Request To Send*). The DTE activates this circuit in the RS-232-C interface when it is ready to send data.

SASE (*specific-application service element*). A collection of protocol entities forming part of the application layer responsible for providing various specific application services.

SDLC (*Synchronous Data Link Control*). A data-link-layer protocol developed by IBM. It is the IBM version of HDLC.

SECONDARY STATION. A station in an HDLC configuration that is not primary.

SEGMENTATION. Fragmentation of a message or stream of data into small fragments.

SERVICE ACCESS POINT. Used in the OSI reference model to define a point at which a service is provided.

SESSION CONTROL LAYER. Fifth layer of the OSI reference model.

SIMPLEX TRANSMISSION. A type of information-exchange strategy between two communications devices whereby transmission can only occur in one direction.

SLIDING-WINDOW FLOW CONTROL. A method of flow control in which transmitting stations may send numbered packets within a dynamically changing window of numbers.

SOH (*Start of Heading*). Communication control character used in BSC at the beginning of a sequence of characters that contain address or routing information.

SPECTRUM. Refers to an absolute range of frequencies.

STOP-AND-WAIT ARQ. A flow-control strategy in which the sender sends a block of data and waits for an acknowledgment before proceeding with the following block.

STX (*Start of Text*). Communication control character used in BSC to indicate that the sequence of characters that follows, usually referred to as text, is to be treated as an entity.

SUPERVISORY FRAME. Frame used by data-link-control protocols for supervision of the link.

SYNCHRONOUS TRANSMISSION. Data-transmission mode in which the time of occurrence of each signal representing a bit is related to a fixed time-frame.

TCP (*Transmission Control Protocol*). The TCP/IP standard transport-level protocol that provides a reliable, full-duplex stream service. It is a connection-oriented transmission protocol, and relies on the IP protocol to transmit data across the physical network.

TIME DIVISION MULTIPLEXING. A technique used to share the channel capacity of a communications channel by carrying different signals at different time slots.

TOKEN BUS. Local-area network architecture using a coaxial cable as a physical transmission medium, and a token to guarantee a regulated access to the medium.

TOKEN RING. Local-area network architecture in which all stations are connected in the form of a physical ring, and messages are transmitted by allowing them to circulate around the ring. A regulated access to the medium is guaranteed by a token.

TRANSPORT LAYER. Fourth layer of the OSI reference model.

TWISTED PAIR. A transmission medium consisting of two insulated wires arranged in a regular spiral pattern.

UDP (*User Datagram Protocol*). The TCP/IP standard transport-level protocol that allows an application layer on one machine to send a datagram in another application layer to another machine. It is a connectionless protocol, and relies on the IP protocol to deliver the datagrams to the destination.

UNBALANCED CONFIGURATION. An HDLC link configuration used in point-to-point and multipoint operations. The configuration consists of one primary and one or more secondary stations. It supports both full- and half-duplex transmission.

UNNUMBERED FRAME. HDLC frames to handle such operations as link startup and shutdown, specifying modes, and other maintenance-related operations.

VIRTUAL CIRCUIT. A transmission path set up by the connection protocol from the source to the destination.

WACK (*Wait before Transmit Positive Acknowledgment*). Response that contains an acknowledgment and requests the sender to pause for a time duration.

X.21. Physical layer interface included in the OSI reference model.

X.25. Standard defining the user interface to public data networks.

Bibliography

The field of distributed computing and information-processing systems has been rapidly gaining importance among computer and information scientists. The development of these systems and the design of the communications protocols that support their interaction has also become a major focus among the scientists and engineers. The result is a large volume of publications. The list that follows, arranged by general topic, provides a selection of some of the publications that have made major contributions in the computer communications field.

Computer Communications

Bertsekas, D., and Gallager, R., *Data Networks,* Prentice-Hall, Englewood Cliffs, NJ, 1987.

Black, U. D., *Data Communications and Distributed Networks*, 2d ed., Reston Publishers, Reston, VA, 1987.

Day, J. D., and Zimmermann, H., The OSI Reference Model, *Proc. IEEE*, 71:1334–1340 (December 1983).

Folts, H. C., *Compilation of Data Communications Standards*, 2d rev. ed., McGraw-Hill, New York, 1983.

Halsall, F., *Data Communications, Computer Networks, and OSI*, 2d ed., Addison-Wesley, Reading, MA, 1988.

Henshall, J., and Shaw, A., *OSI Explained: End to End Communication Standards*, Ellis Horwood, Chichester, England, 1988.

Linnington, P. F., Fundamentals of the Layer Service Definitions and Protocol Specifications, *Proc. IEEE*, 71:1341–1345 (December 1983).

McNamara, J. E., *Technical Aspects of Data Communication*, 3rd ed., Digital Press, 1988.

Martin, J., *Computer Networks and Distributed Processings: Software Techniques and Architecture*, Prentice-Hall, Englewood Cliffs, NJ, 1981.

Martin, J., *Introduction to Teleprocessing*, Prentice-Hall, Englewood Cliffs, NJ, 1972.

Quarterman, J. S., and Hoskins, J. C., Notable Computer Networks, *Commun. ACM*, 29(10):932, 971 (October 1986).

Schwartz, M., *Telecommunication Networks, Protocols, Modeling, and Analysis*, Addison-Wesley, Reading, MA, 1987.

Seyer, M. D., *RS-232 Made Easy: Connecting Computers, Printers, Terminals, and Modems*, Prentice-Hall, Englewood Cliffs, NJ, 1982.

Sherman, K., *Data Communications, A User's Guide*, 3rd ed., Prentice-Hall, Englewood Cliffs, NJ, 1990.

Stallings, W., *Data and Computer Communications*, 2d ed., Macmillan, New York, 1985.

Tanenbaum, A. S., *Computer Networks*, 2d ed., Prentice-Hall, Englewood Cliffs, NJ, 1988.

Zimmermann, H., OSI Reference Model—The ISO Model of Architecture for Open Systems Interconnection, *IEEE Trans. Commun.*, COM-28:425–432 (April 1980).

Physical Layer

Bell Telephone Laboratories, *Transmission Systems for Communications*, Bell Telephone Laboratories, Murray Hill, NJ, 1982.

Bertine, H. U., Physical Level Protocols, *IEEE Trans. Commun.*, 28(4):443–444 (1980).

Bleazard, G. B, *Hand Book of Data Communications*, NCC Publications, 1982.

Chou, W., *Computer Communications*. Vol. 1, *Principles*, Prentice-Hall, Englewood Cliffs, NJ, 1983.

Cooper, E., *Network Technology*, Prentice-Hall, Englewood Cliffs, NJ, 1986.

Davies, D. W., *Communication Networks for Computers*, Wiley, New York, 1973.

Freeman, R, *Telecommunication Transmission Handbook*, Wiley, New York, 1981.

Held, G., *Data Communication Components*, Hayden, Rochelle Park, NJ, 1979.

Intel Incorporated, *The Intel 8251A/S2657 Programmable Communications Interface*, Intel Incorporated, 1978.

Luetchford, I. C, CCITT Recommendations—Network Aspects of the ISDN, *Jrnl. Sel. Areas Commun.*, SAC-4:334–342 (May 1986).

McClelland, F. M., Services and Protocols of the Physical Layer, *Proc. IEEE*, 71:1372–1377 (December 1983).

Peterson, W. W., *Error Correcting Codes*, MIT Press, Cambridge, MA, 1981.

Yanoschak, V., Implementing X.21 Interface, *Data Commun.* (February 1981).

Data-Link Layer

Conard, J. W, Service and Protocols for the Data Link Layer, *Proc. IEEE*, 71:1378–1383 (December 1983).

Doll, D. R., *Data Communications: Facilities, Networks, and System Design*, Wiley, New York, 1980.

Field, J. A., Efficient Computer-Computer Communication, *Proc. IEEE*, 23:756–760 (August 1976).

Fraser, A. G., Delay and Error Control in a Packet Switched Network, *Proc. ICC.*, (22.4):121, 125 (1977).

Fraser, A. G., Towards a Universal Data Transport System. In: *Advances in Local Area Networks* (K. Kummerle, F. Tobagi, and J. O. Limb, eds.), IEEE Press, New York, 1987.

Rudin, H., An Informal Overview of Protocol Specifications, *IEEE Commun.*, 23:46–52 (March 1985b).

Media-Access-Control Sublayer

Burg, F. M., Chen, C. T., and Folts, H. C., Of Local Networks, Protocols, and the OSI Reference Model, *Data Commun.*, 129–150 (November 1984).

Bux, W., et al., A Reliable Token Ring System for Local Area Communication. In: *Proceedings of the National Telecommunications Conference*, New Orleans, A.2.2.1–A.2.2.6 (1981).

Carpenter, R., A Comparison of Two Guaranteed Local Network Access Methods, *Data Commun.*, (February 1984).

Clark, D., Program, K., and Reed, D., An Introduction to Local Area Networks, *Proc. IEEE*, 66:1497–1516 (November 1978).

Dixon, R. C., Strole, N. C., and Markov, J. D., A Token-Ring Network for Local Data Communications, *IBM Sys. J.*, 22:47–62 (January–February 1983).

Eswaran, K. P., Hamacher, V. C., and Shedler, G. S., Collision Free Access Control for Computer Communication Bus Network, *IEEE Trans. Software Eng.*, SE-7: 574–582 (November 1981).

Fine, M., and Tobagi, F. A., Demand Assignment Multiple Access Schemes in Broadcast Bus Local Area Networks, *IEEE Trans. Computers*, C-33(12):1130, 1159 (December 1984).

Fratta, L., Borgonovo, F., and Tobagi, F., The EXPRESS-NET: A Local Area Communication Network Integrating Voice and Data. In: *Performance of Data Communication Systems* (G. Pujolle, ed.), Elsevier North-Holland, New York, 1981, pp. 77–88.

IEEE, *Carrier Sense Multiple Access with Collision Detection (CSMA/CD) Access Method and Physical Layer Specifications*, American National Standard ANSI/ IEEE Std. 802.3-1985, New York.

IEEE, *Logical Link Control*, American National Standard ANSI/IEEE Std. 802.2-1985, New York.

IEEE, *Token-Passing Bus Access Method and Physical Layer Specifications*, American National Standard ANSI/IEEE Std. 802.4-1985, New York.

IEEE, *Token Ring Access Method and Physical Layer Specifications*, American National Standard ANSI/IEEE Std. 802.5-1985, New York.

Network Layer

Bell, P. R., and Jabbour, K., Review of Point-to-Point Routing Algorithms, *IEEE Commun.*, 24:34–38 (January 1986).

Bertsekas, D., and Gallager, R., *Data Networks*, Prentice-Hall, Englewood Cliffs, NJ, 1987.

Chapin, A. L., Connections and Connectionless Data Transmission, *Proc. IEEE*, 71: 1365–1371 (December 1983).

Davies, D. W., et al., *Computer Networks and Their Protocols*, Wiley, New York, 1979.

Deasington, R. J., *X.25 Explained: Protocols for Packet Switched Networks*, 2d ed., Ellis Horwood, Chichester, England, 1986.

Dhas, C. R., and Konangi, V. K., X.25: An Interface to Public Packet Networks, *IEEE Commun.*, 10:74–84 (June 1977).

Green, P. E., Protocol Conversion, *IEEE Trans. Commun.*, COM-29:726–735 (May 1981).

Hawe, B., Kirby, A., and Stewart, B., Transparent Interconnection of Local Area Networks with Bridges, *J. Telecommun. Netw.*, 3:116–130 (1984).

Meijer, A., and Peters, P. *Computer Network Architectures*, Computer Science Press, Rockville, MD, 1982.

Network Service Using X.25 and X.21, ISO TC97/SC5/N2743, ISO, 1981.

Transport Layer

Chong, H. Y., Software Development and Implementation of NBS Class-4 Transport Protocol, *Computer Networks and ISDN Systems*, 11:353–365 (May 1986).

Comer, D., *Internetworking with TCP/IP: Principles, Protocols, and Architecture*, Prentice-Hall, Englewood Cliffs, NJ, 1988.

Groenbaek, I., Conversion between the TCP and the ISO Transport Protocols as a Method of Achieving Interoperability between Data Communications Systems, *IEEE Jrnl. Sel. Areas Commun.*, SAC-4:228–296 (March 1986).

ISO, Transport Service/Protocol, *Information Processing Systems*, ISO 8072/3, 1983.

Nagle, J., Congestion Control in TCP/IP, *Computer Commun. Rev.*, 14:11–17 (October 1984).

Open System Interconnection, Basic Reference Model, *Information Processing Systems*, ISO 7498, 1984.

Pouzin, L., and Zimmerman, II., A Tutorial on Protocols, *Proc. IEEE*, (November 1978).

Rose, M. T., and Cass, D. E., OSI Transport Services on Top of the TCP, *Computer Networks and ISDN Systems*, 12:159–173 (1987).

Watson, R. W., and Fletcher, J. G., An Architecture for Support of Network Operating System Services, *Computer Networks*, 4:33–49 (February 1980).

Watson, R. W., and Mamrak, S. A, Gaining Efficiency in Transport Services by Appropriate Design and Implementation Choices, *ACM Trans. Computer Systems*, 5:97–120 (May 1987).

Application-Oriented Protocols

Albert, A. F., *Videotex/Telex—Principles and Practices*, McGraw-Hill, New York, 1985.

Huffman, A. J., E-Mail—The Glue to Office Automation, *IEEE Network Magazine*, 1:4–10 (October 1987).

Linington, P. F., The Virtual Filestore Concept, *Computer Networks*, 8:13–16 (1984).

McLeod-Reisig, S. E., and Huber, K., ISO Virtual Terminal Protocol and Its Relationship to Mil-Std TELENET, *Proc. Computer Networking Symp.*, 110–119 (1986).

MMS: Service and Protocol Definition, Draft 5, EIA RS-511, 1986.

OSI: ACSE Service/Protocol Specification, ISO 8649(2)/50(2), ISO, 1986.

OSI: CCR Service/Protocol Specification, ISO 8649(3)/50(3), ISO, 1986.

OSI: FFAM Service/Protocol Specification, ISO 8571/1-4, ISO, 1985.

OSI: JTAM Service/Protocol Specification, ISO 8831/2, ISO, 1985.

OSI: ASN.1/Encoding Rules, ISO 8824/5, ISO, 1985.

OSI: Presentation Service/Protocol Specification, ISO 8822/3, ISO, 1985.

OSI: Session Service/Protocol Specification, ISO 8326/7, ISO, 1985.

OSI: VT Service/Protocol Specification, ISO 9040/1, ISO, 1985.

Reference

1. Brandt, R. P., Binary Serial Data Interchange, EIA-232-D Standard. In: *The Froehlich/Kent Encyclopedia of Telecommunications*, Vol. 2 (F. E. Froehlich and A. Kent, eds.), Marcel Dekker, New York, 1991, pp. 113–123.

TAIEB F. ZNATI

Communication Terminals*

Introduction

A communication terminal is an instrument by which a user communicates, through a network, with another user or with some service. It is a recent concept, an evolution, generalization, and synergism of the telephone set, video display terminal (VDT), and the television receiver. Communication terminals are differentiated by the services provided to the user through the terminal, the physical equipment, the user interface, the network interface, and the distribution of intelligence.

A *terminal* is the end of something. A bus terminal is a building at one end of a transportation route and a battery terminal is an electrical contact at one end of a battery. A *transmission terminal* is the equipment at one end of a carrier link where the multiplexing equipment resides. A *communication terminal* is an instrument at one end of a telecommunications channel.

Telecommunication occurs over physical and logical channels that reside in systems of channels called networks. Independent networks have evolved, optimized for each of the three major telecommunications media: voice telephony, data, and broadcast video. Independent terminals have evolved similarly, also optimized for each of these media. These familiar special-purpose terminals are: the telephone set, the VDT, and the television receiver.

The advantages of digital electronics technology have caused many different types of networks to be made from this same cloth, which suggests (to some people) the integration into one network. During the last 15 years, the study of integrated digital networks included the study of integrated terminals and services; this led to the development of the Integrated Services Digital Network (ISDN), which is discussed in two other sections of this article.

The purposes of this article are to categorize and review the state of the art in communication terminals and to present a view of the future. Some material is tutorial and some philosophical. Numerous examples are provided.

Telecommunications

Telecommunications is a means of sharing remote information. More than moving bits of information from one machine to another, telecommunications is moving concepts from one brain to another. The telecommunications environment is discussed in this section. The three major telecommunications media are described and two perspectives on telecommunications are compared.

Three Major Media

The three major telecommunications media are voice telephony, data, and broadcast video. Voice telephony is user-to-user, point-to-point, two-way communication. Data communication is also usually point to point and two way, but it is usually user to service, while telephony is usually user to user. Video communication is usually user to service also, but it is usually a broadcast, one-way form of communication. Some typical characteristics of these three media are shown in the table below.

Medium	Type of Service	Fan-Out	Direction	Regularity	Rate (bps)
Telephony	User to user	Point to point	Balanced	Synchronous	10^5
Data	User to service	Point to point	Unbalanced	Bursty	10^3–10^7
Video	User to service	Broadcast	One way	Synchronous	10^8

While voice telephony and data communication are both usually two way, telephony is usually balanced, while data communication is usually unbalanced, with more data moving from service to user than from user to service. Telephony and video are typically synchronous, while data communication is typically bursty. The bandwidth required for telephony is specified by its digital equivalent of 64 kbps. Full-motion video requires about 1000 times this bandwidth. Keyboard-to-computer data can be 100 times slower than voice telephony, while disk-to-processor data is typically 100 times faster. While radio is an alternate voice medium with higher rate and fidelity than telephony, it more closely resembles the video medium because it is user-to-service, broadcast, and one-way. Telex and electronic mail have some characteristics of data communication and some of voice telephony. Facsimile transmission has some characteristics of data and some of video. Video conferencing has some characteristics of voice and some of video.

The general telecommunications environment, illustrated by Fig. 1, looks so simple that it would seem to be well understood and predictable. But the telephone and the television were much more successful than their inventors could have ever imagined, each spawning huge industries. Both met market needs that were not even known to exist at the time of invention. Conversely, many examples of communications systems, apparently addressing obvious needs, have been commercial failures. The apparently simple telecommunications environment shown in Fig. 1 is subtly complex and poorly understood.

Two different perspectives for analyzing the elements shown in Fig. 1 are proposed: the window perspective and the interface perspective.

FIG. 1   The telecommunications environment.

The Window Perspective

From the window perspective, the purpose of the terminal and the network is to enable the user to "see" the service (or another user). If the user and the service were co-located, then the terminal and the network would not be needed. From this perspective, the terminal and the network are intermediaries and, as such, should be transparent. The terminal and the network represent a simple two-way window through which the user perceives the service.

Example 1. The Emergency Call System (ECS) (AT&T) is a communication terminal that provides a specialized service. ECS is a special-purpose, originate-only, programmable telephone that adds its singular service to the residence. It is a small modular system that connects to the network like a conventional telephone. ECS is available in two commercial forms: Medical Alert and Smoke/Fire Alert. ECS links locally to a natural trigger mechanism, a manually activated trigger in the Medical Alert model and activation by the sound of the user's commercial smoke/fire detector in the Smoke/Fire Alert model. When triggered, the Emergency Call System places a telephone call to some party programmed by the user, typically a rescue squad for Medical Alert or a fire department for Smoke/Fire Alert. When the called party answers, ECS recites an appropriate prerecorded message. The output and user feedback are computer-synthesized voice, so there is no visual output display. User program input is from a console with fixed-feature buttons.

If the user owned a private-line connection to a rescue squad or fire department, he or she could connect a trigger to the private line at the user's end, and a bell or light at the service's end. Using the public switched telephone network instead of a private line requires additional equipment at the user's end between the trigger and the telephone line. From the window perspective, ECS provides a natural and transparent way to make the switched connection resemble the private-line connection.

While the service is possible with a conventional telephone, which the user probably owns anyway, the general utility of the telephone makes it slower and less transparent, which justifies the ECS for many users. In comparing the conventional telephone with ECS, a general rule is observed: special-purpose communication terminals can be designed for improved interaction with specific services, but as a terminal is made more general in purpose, it becomes less efficient and less transparent.

The Interface Perspective

The interface perspective is the second perspective used to analyze the elements of the telecommunications environment portrayed in Fig. 1. From the interface perspective, each box in Fig. 1 represents a stage of transduction and each line represents an interface between successive transducers. From this perspective, a communication terminal is a transducer between the user's senses and the network's mechanisms. If the user and the service were co-located, certainly the

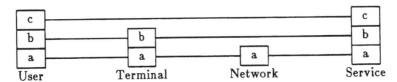

FIG. 2   Layered processes in the communications environment.

network would not be needed; but, the user and the service (or other user) might be so incompatible that a terminal would be needed as a translator. For example, it is virtually impossible for a human and a computer, even in the same room, to communicate without a VDT.

All the components shown in Fig. 1 are dynamic, including the user. One cause of the chaos in this field today, and a reason that standards are important for the future, is that each component in Fig. 1 must be designed to interface with its adjacent components, but they may change.

Figure 2 is a schematic representation of a simplified layered protocol with only three layers. We call the layers the interface layer (a), the interaction layer (b), and the perceived layer (c). If all four components of Fig. 1 are present in this protocol, then they must be physically and logically interconnected as indicated by the a layer in each component in Fig. 2. Once the network connection has been established, it is transparent. The user interacts with his or her terminal and the terminal interacts with the service, as indicated by the b layer of these three components in Fig. 2. By the window perspective, if the terminal is well designed, the user should barely perceive its presence. Instead, the user would perceive a direct communication with the service, as indicated by the c layer of these two components in Fig. 2.

In circuit-switched networks, the user must first set up the call. This act is different from that pictured in Fig. 2, which is relevant only in the steady state. Again, as shown in Fig. 3, all four components from Fig. 1 must be physically interconnected (indicated by the a layer of each component in the figure). Since the network does not know which service the user will request, the connection is incomplete through the network during call setup. But the user must be physically connected to the network through his or her terminal and every service must be connected to some port on the network. While the user still interacts with his or her terminal during call setup, the terminal interacts directly with the network, as indicated by the b layer of these three components in Fig. 3. At a well-designed terminal, the user perceives a direct communication with some connection-request service on some processor in the network, indicated

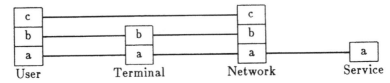

FIG. 3   Call setup in the communications environment.

by the c layer in these two components in Fig. 3. When the user sets options in his or her terminal, an event even more primitive than call setup, he or she both interacts with and directly perceives the terminal. Thus, the user is connected to it at all three layers.

Both the window perspective and the interface perspective are correct. Any new communication terminal must be both a transparent window and a two-way transducer between the user and the network.

Services

The design of a communication terminal should be specified primarily by the set of services intended to be provided through it. Complicating these designs is the recent notion of integrated services.

An Integrated Voice/Data Service

The integration of service is different from the union of sets of services. An example of a true integrated service is the coupling of directories with autodialing. The reader is, of course, familiar with the scenario in which a telephone user wishes to call someone whose number is not known. In the typical telephone environment, we think of this as using two distinct services.

For the first service, the user looks up the name of the desired party in some directory and retrieves the telephone number. If the user does not have the correct directory, he or she must call an information operator. This operator has the correct directory, looks up the name for the user, and gives the telephone number to the user, who typically writes it on a piece of scrap paper. Once the user has the called party's telephone number, for the second service, the call is placed conventionally from his or her telephone. Would it not be convenient if telephones were intelligent enough that the user could point to the entry in the directory or hold up the scrap of paper and the telephone would read the number and place the call? Most of us would pay a couple of dollars more for a telephone that could do this.

Many companies have recently computerized their corporate directory. The main advantage for the company is that those employees with VDTs do not need a printed directory. The main advantage for the employee with a VDT is that this directory is current. In this case, to call someone whose number is not known, for the first service, the user types into his or her computer the command for the program that accesses this directory, types the called party's name to that program, and then the program prints the telephone number on the screen of the VDT. The user, acting as a simple transducer for the second service, reads the number from the VDT screen and pushes the buttons on the telephone keypad. Would it not be convenient if the telephone were intelligent enough to read the number directly off the VDT screen and place the call? Most of us would pay a couple of dollars more for a telephone that could do this.

In this scenario, each of the two distinct services illustrates the window perspective. If the called party were co-located, the user would not need the telephone or the network, at least for this call. If the directory were co-located with the user, he or she would not need the VDT or the network connecting it to the computer, at least for finding this number.

Another reason for discussing this scenario is to illustrate an integrated service. The suggested integrated service is "Ring the person named _____ (and do not bother telling me the phone number)." The implementation suggested is a technology by which a greatly modified telephone reads numbers off a cathode-ray tube (CRT), but alternate implementations are available in more conventional technologies. In a second implementation, the computer on which the directory resides and the computer that controls the telephone center communicate about the user's intentions. In a third implementation, a slightly modified telephone acquires the telephone number because it has tapped the line between the VDT and the computer. As long as they cost the same, these three implementations would be virtually equivalent for the user. Providing this new service through existing familiar terminals and networks requires some modification to either or both. In a fourth implementation, these two separate terminals are merged into one integrated communication terminal. It connects to the computer like a VDT, to the voice network like a telephone, and it seems designed especially for this integrated service.

Example 2. Displayphone (Northern Telecom, Inc.) is such an integrated communication terminal. It has a telephone handset, a full-sized alphanumeric keyboard, and a 9-in CRT display, all integrated into an attractive package with a reasonably small footprint. For many years, a Displayphone appeared on J. R. Ewing's desk in the television show "Dallas." Displayphone has not been a large commercial success, mainly because it is much more expensive than a telephone.

Service Sets

While terminals may be specialized to one service like ECS, they are usually designed for a set of related services. Even the telephone is used for more and more services every year, and every year a new telephone comes out with another new button on it because the classical telephone was inadequate for some new service that somebody has added to the service set. A terminal is integrated if it interfaces to a union of historically independent service sets or to more than one telecommunications medium. An integrated terminal may be required because previously independent services have been integrated to form a new service set.

For example, most people who work in an office environment expect their office telephone to provide a familiar set of telephone services. Sometimes the office telephone has six buttons on the bottom or an adjunct speaker to accommodate some services that are typically not available from their residence telephone. Similarly, these people expect a familiar set of text services to be available from their word processor or VDT. The union of these two historically independent service sets, the presentation of this combined service set through an integrated terminal, clever ways to integrate old services from each subset, and the invention of new integrated services has come to be called *office automation* (OA). An OA terminal resembles a VDT, but with an attached telephone handset, or resembles a telephone, but with an attached CRT and keyboard. In recent years, entire technical conferences and journals have been dedicated to OA. While OA has not been a large commercial success, it probably will be once the OA terminal becomes universal and the service set becomes fixed, understood, and familiar. This topic is discussed further in the section below concerning the user interface.

This relationship between the service set and the communication terminal has two parts. First, as already discussed, the terminal must have the necessary features so that the user may interface to every service in the service set. Second, the set of services to which a user may interface from a given terminal must be complete. With OA, for example, if there is some telephone service or text service that some user wants but that the OA terminal cannot provide, then this user would need a telephone or word processor anyway. It is safe to say that no desk will have a telephone, a word processor, and an OA terminal on it; probably the OA terminal will be discarded. For example, television receivers that answer the telephone can be criticized because the user needs a telephone in the room anyway to originate calls. A more important issue is that any new service set, and corresponding integrated communication terminal, must meet a real need. But the single most important issue is cost.

The Golden Service

The total commerce of so-called enhanced services is considered by many to be a potentially lucrative business, if it could ever get started. Worse than a chicken-and-egg problem, this commerce is a three-way, you-go-first scenario because the total commerce of providing new integrated services has three separate elements: the terminal, the network, and the service, as depicted in Fig. 1. The positions taken by providers of each of the three elements are as follows:

- The terminal manufacturer cannot produce a low-priced terminal until it can be mass produced. High sales volume requires a mass-market demand, which will not happen until service software is available and networks are integrated, standardized, and real.
- The network providers cannot optimize their networks (to ISDN, perhaps) or offer such networks at low cost until many users own communication terminals and use existing services.
- The service vendors cannot afford to invest in the development of service software until integrated networks are real and many users own terminals.

A *golden service* is some wonderful service that consumers want so much that they not only will pay for this service, but also will buy the terminal and subscribe to an expensive network. If one such golden service exists, it is likely to be in the areas of entertainment, education, or security. But, so far, it has been elusive. The promise of many "brass services" has been insufficient to get the total business started.

With neither a ubiquitous terminal nor a golden service, the expected future sales of enhanced services must somehow subsidize the initial price of the terminal and the network interface. In France, where the government controls the telephone network, communication terminals are available at a very low price because of such subsidizing and the redirection of money saved by not printing directories. In the post-divestiture United States, price subsidizing of terminals will require creative regulation, and that is unlikely. The impact of the network on this scenario is discussed further in the section below concerning network influence.

Further complicating an already complex market, some services that have received a little publicity may be resisted by vested interests. ISDN and digital out-of-band signaling provide the ability to display the telephone number, and even the identity, of the calling party on the screen of the called party's communication terminal during ringing. The subscriber to this service has the option of not answering undesired calls. This "calling party identification service" has been tariffed recently in some states and banned in others. Field-trial users have typically liked this service, but enterprises that depend on mass calling and people with unlisted directory numbers have been vocal opponents. A host of customized call-blocking services can be imagined, depending on the legal and regulatory outcome of calling party identification. A host of countermanding and overriding services can also be imagined. The users on both sides of this issue will feel like they are dealing with an armaments vendor that alternately sells increasingly better armor to one side and increasingly better armor-piercing shells to the other, at increasing prices, of course.

Integration with Image and with Video

The integration of voice and data is only beginning to be understood, at least in terms of the service set and the integrated terminal (certainly not in the network, as discussed below). Complicating this progress, however, is the imminence of networks and terminals that provide image services. Attempted integration of these poorly understood services and terminals with the only slightly better understood voice/data services and terminals has added to the confusion. But the payoff is potentially huge.

The most primitive image terminal is the facsimile (FAX) machine, which enables the remote copying of a piece of paper. The information copied includes written and printed text, sketches, and script, such as a signature. Gray-scale photographs did not work in early facsimile versions, but work reasonably well in later versions. The calling party inserts the original into the FAX transmitter and dials the called party's FAX machine. The information on the page is scanned, interpreted as a stream of black and white dots, converted into modem-like electronic signals, and carried over conventional telephone lines to the called party's FAX machine, which produces a copy of the original.

But FAX is conventionally a paper-in/paper-out process. One obvious extension of OA and integration of data with image is to deliver FAX images to the screen of a communication terminal. Some recent personal computers (PCs) and word processors have optional boards that interface to FAX, but the media are not integrated. It is relatively easy to convert an ASCII (American Standard Code for Information Interchange) file into a formatted document and to transmit that document by FAX. Receiving a text document by FAX and storing it as an ASCII file, however, requires equipment for optical character recognition (OCR) that is relatively expensive and not completely reliable.

Example 3. Wang Computers offers a program called Freestyle that runs on any PC that has a high-resolution, bit-mapped display screen. The resulting system allows the user to examine a text document and to assign notes to selected points in the document. The notes may be voice messages or they may be handwritten in cursive script. The system also handles signatures for authorizations and embeds pictures in the document.

Special-purpose terminals have been proposed and/or built for remote examination of x-ray photographs, drafting and reading blueprints, very-large-scale-integrated (VLSI) design, and a host of computer-aided design/computer-aided manufacturing (CAD/CAM) applications. Some of these include gray-scale and even color applications (discussed below). The next step in this integration progression is to extend this interaction to full-motion video.

Example 4. Integrated voice/data/video services are presented on the Aerial Terminal. Aerial is the terminal portion of the voice/data/video system that was custom built for the EPCOT® Center at Walt Disney World®. It is a component of a complex system of terminals, host computers, and video disk players. System components are scattered among a centralized hub and kiosks containing subsystems with several terminals. The visual display uses color televisions, and data is overlaid on conventional, full-motion color video. The user interface is menu driven and tree structured, and service selection uses a touch-sensitive screen. The presentation is simultaneous voice/data/video.

Terminal Equipment

The hardware equipment of a communication terminal includes its internal components, the attached input and output devices, and its physical packaging. Special attention is given to the styling and quality of image in terminals with visual displays. This equipment affects the terminal's capabilities, size, cost, and quality. These characteristics, in turn, determine the usable service set and contribute to the user's perception of the terminal and the services.

Components

Communication terminals are controlled by a variety of internal components. The most powerful internal architecture resembles a general-purpose computer, with a processor, a large amount of random-access memory (RAM) and read-only memory (ROM), and fixed and optional special-purpose components, such as processor peripherals. Processor software may even run in an operating system environment. More economical but less powerful architectures may be built around one-chip microcontrollers that typically do not have internal operating systems. Fixed, special-purpose components typically include chip sets that interface to a keyboard, at least one data port, an analog or digital telephone line, and CRT electronics.

Commercially available one-chip keyboard controllers sense key operations on a coordinate grid pattern and transmit a serial code over the keyboard-to-terminal cable (or infrared link) to a Universal Asynchronous Receiver/Transmitter (UART) that interfaces with the microprocessor. A typical chip set for an asynchronous data port includes a UART, RS-232-C interface chip, and glue chips. Analog and digital telephone interface circuits also are commercially available.

Several manufacturers make chip sets that act as VDT video controllers for ASCII characters and mosaic pixels. This discussion uses typical sizes and numbers for the description that follows. The CRT screen in the system with this chip set is partitioned into a 24-by-80 area of picture elements (pixels). The corresponding pixel RAM, with 1920 words ($24 \times 80 = 1920$), is written by the terminal's main microprocessor and is read by the chip set. Each pixel word identifies which one of 256 patterns is displayed in the corresponding pixel, along with such characteristics as intensity, blink rate, and the like. This set of 256 patterns includes the traditional alphanumeric characters, punctuation symbols, the two-letter symbols used to display ASCII control characters, a cursor symbol, and mosaic patterns for simple graphics. Each pattern is a 10-by-8 array of dots and, at one bit per dot, a pattern ROM requires 2.56K bytes ($256 \times 80/8 = 2.56K$). The chip set delivers the entire displayed screen, consisting of 153,600 dots ($240 \times 640 = 153,600$), in row-sequential, bit-serial format to the CRT electronics 30 times per second (about 4.6 million dots per second). The chip set also provides for input from a light pen and translates to X-Y position for output to the microprocessor.

Physical design and shielding of internal circuits are important to reduce electromagnetic interference to and from the outside and among internal CRT, microprocessor, and telephone. Standards are carefully upheld by the Federal Communications Commission (FCC) and the Underwriter's Laboratory (UL). Optional components include special-purpose hardware on cards, cartridges, and modules that are externally attached or internally installed. Optional software is conveniently sold on diskettes or in the familiar videogame packaging—ROM or battery-backed RAM in plug-in cartridges. Software download is discussed but seldom implemented.

The terminal's external devices provide the network interface and provide for user input and output. Network interface devices and modems are discussed below. Two categories of user input devices are data input devices and pointing devices. Data input devices include a microphone for voice, switches and knobs for control functions, and an alphanumeric keyboard or numeric keypad for text or numbers. Some terminals could even have a camera for video input or a FAX scanner or OCR system to read written or printed paper. While keystrokes may be used to move a cursor and select from a menu, the mouse is a much better cursor-moving device. Cursor-free pointing devices include light pens and touch screens. These devices detect when the CRT's electron beam is under the pen or finger and the time of this event is translated into X-Y coordinates.

User output devices include a speaker for voice, individual lamps or light-emitting diodes (LEDs) for status indications, a linear array of bit-mapped or seven-segment displays for short text messages, and a display screen for both text and image. While various flat-panel display technologies have been investigated, the CRT, while bulky, is still most common. A terminal optionally may include a printer, but the object frequently is to avoid or replace printed paper.

Packaging

A terminal's packaging includes its size, degree of modularity, styling, and name. Size is an important characteristic. If a terminal overpowers an end table or occupies too much counter or desk footprint, it will not be tolerated in the living room, kitchen, or office. A terminal that is so large that it must be considered a piece of furniture must provide a wonderful, highly golden service. The other size extreme for a terminal is the Dick Tracy-type wrist radio. *Integrated packaging* is illustrated by the example that follows.

Example 5. The Genesis Telesystem (AT&T) is a telephone in function and size, but it has more generalized function and local intelligence than the traditional voice terminal. It has many extra features and a small visual display, but its outstanding characteristic is the ability to accept cartridges and modules that augment its functionality. Modules are appended to the body and provide additional hardware, such as a keyboard. Cartridges are inserted internally and provide additional software capability, such as an electronic date book, custom calling convenience, autodialing, and electronic padlock. A highly augmented Genesis is still a single, small package.

Modular packaging is illustrated by the example that follows.

Example 6. Videotex is a prototype system that integrates voice, data, and static cartoon-quality images. Systems have been tested in Great Britain, France, West Germany, Japan, and the United States. The prototypes have been modular, with a control unit and keyboard tied to the user's color television receiver. A videotex color display features text and geometric and mosaic graphics. The human interface is by typed entry at the keyboard and uses tree traversal and menus for access and selection, as discussed above.

A discussion of the variety of communications interfaces in the different prototypes of videotex is a worthwhile digression. Potential communications interfaces are the user's video cable and the user's telephone line; one or both have been used. In alternate architectures

- the remote centralized video database is dial accessed, screen requests are transmitted upstream, and the requested image is transmitted downstream, all through the telephone network and without using video cable
- similar communication occurs over two-way video cable and the telephone network is not used
- screen requests are transmitted upstream over the telephone line and the requested image is transmitted downstream over one-way community antenna television (CATV) cable
- the entire database is repeatedly broadcast over one-way CATV cable and the controller grabs the desired image

The styling of most commercial and prototype terminals, like that of most VDTs, has been simple and linear but not boxy. Besides being economical, molded plastic provides the desired look. While beige and light blue VDTs have been commercially available, the preferred color seems to be off-white with charcoal trim. This typical styling and color have a high-tech, futuristic effect that may actually contribute to some users' resistance.

Communication terminals, like VDTs, have been given different kinds of names:

- functional names like Displayphone and Teleterminal
- high-tech names like Sceptre (AT&T) and Aerial
- license-plate names like 510 BCT (AT&T) and HP 2621 (Hewlett-Packard)

The Getset, described in the section on the user interface below, has a refreshing name and the original model (Fig. 4) has been described as "cute."

Display Format and Quality

The format and quality of graphical displays lie in a complex spectrum from static light-on-dark picture-element displays to full-motion, color, high-definition, real-image video. The point on this spectrum selected for a communication terminal will strongly determine the services provided through the terminal, its cost, its internal architecture, and its network interface. Five overlapping neighborhoods on this spectrum are described below.

1. Static light-on-dark (or dark-on-light) displays of picture elements were used in early VDTs and currently are used in inexpensive VDTs. The screen is partitioned into a rectangular array, typically 24 by 80, of pixel locations, each providing the space for a rectangular array of dots. The pixels are predefined, encoded, and stored internally, and the control is pixel mapped, as described above. Pixel definition may vary; a pixel defined on a 7-by-5 array of large dots has much lower quality for font production and lower granularity than one defined on a 12-by-9 array of smaller dots. But the CRT electronics of the latter requires more than triple the bandwidth of

the former, and has higher cost. While the predefined set of pixels includes the usual alphanumeric characters, depending on the terminal and its intended application, this set can also include patterns for mosaic graphics. A more elite pixel display may expand the simple light-on-dark patterns to include gray scale and color. The communication format at the terminal's network interface is a simple message stream, such as the ASCII/RS-232-C format, discussed in the section concerning network influence below. Such formats are extendable to cover expanded pixel sets, gray scale, or color.

2. Static displays of sketch-quality images, used in modern and expensive VDTs and PCs, allow arbitrary shapes and lines equivalent to those used in FAX technology. The control is bit mapped directly from an internal frame store whose capacity is one bit for every dot on the display. The bandwidth required of the CRT electronics is the same in pixel mapping and bit mapping. But while pixel mapping requires only 15.36K bits of display memory ($24 \times 80 = 1920$ bytes $= 15.36$K), bit mapping requires at least 10 times this amount, depending on the dot density. Pixel mapping is incorporated indirectly into a bit-mapped display by writing pixel bit patterns into the frame store. More advanced and more expensive displays provide gray scale and color. Highly advanced PCs can store information about a three-dimensional image, and sophisticated software can process hidden lines and surfaces, scaling, rotations, and perspective. While such terminals must be compatible with ASCII-like communications, they must also allow for entire bit maps to be up- or downloaded. While full-motion video (cartoons) can be generated internally, particularly for real-time rotations of three-dimensional images, downloading an entire bit map 30 times per second requires a communications link of 6 Mbps, at least, with neither gray scale nor color, depending on the dot density.

3. Static displays of real images, equivalent to analog pictures such as photographs, require a terminal similar to the terminal above, but with higher resolution and finer granularity. The CRT is more expensive and the frame store is much larger. Most applications require at least the resolution of that of commercial television. Since gray scale is necessary and color is common, increased quality (and cost) comes from raising the resolution. With gray scale and color, downloading even one static image can take an extremely long time over conventional voice/data communications systems. Alternatives are to use a high-rate local-area network (LAN) or conventional analog television networks and convert one frame to digital in the terminal with a device called a *frame grabber*.

4. Full-motion displays of real images require a television receiver and all its analog electronics. Color is expected and high-definition standards are under investigation.

5. Windows are logical displays on the same physical display. The common implementation is a rectangular area on a screen that directly corresponds to some communications link. These areas may overlap, and some may be entirely hidden by others. Windows are common on VDTs, where each window communicates with a different active process on the host. Incorporation of sketch-quality images in windows is becoming common on bit-mapped terminals. Transmission of television with windows has been com-

mon for years—for example, in viewing baseball games. Windowing several channels on the same television receiver is only just beginning to be commercially available. Coexistence of full-motion video windows and typical data windows on a terminal's screen is only restricted by the network interface.

User Interface

The user's cognitive and emotional perception of a communication terminal superficially comes from its packaging. But, at a deeper level, this perception also comes from the software and the services that are perceived through the terminal. This perception is related to the friendliness of the services themselves, but also to the procedures and command language involved in using the terminal and the services. In this section, the terminal's ability to provide access to services and to allow the user to select the desired service are described. The terminal can act as an agent or translator between the user, assumed to be naive, and the system of network and services, assumed to be difficult to use. The user's level of sophistication is also discussed.

The User

From either the window perspective or the interface perspective, Fig. 1 implies that the users and services are the end points of some system of terminals and networks. We usually view such end points as boundary conditions or fixed parameters of a system's specifications. Realistically, however, the services are part of the total system and must be designed and optimized to conform to the networks, terminals, and users. Less obviously, viewing users as fixed boundary conditions is also a mistake (1).

While users cannot be designed and built, they can be selected and trained, within limits. A given system of services, terminals, and a network can be designed for a set of users who are assumed or required to have a certain minimum level of training, intelligence, and sophistication. Some, but certainly not all, of these qualities can be acquired. While this is a large, frequently underestimated problem, there is an even larger one.

Users are different from other system components, because they are human; their performance varies daily, they get tired and bored, they have other things on their minds, and they have feelings.

One of the early experiments with OA, performed in a bank in the mid-1970s, illustrated an interesting phenomenon by being ahead of its time. A strong subliminal message was that bank executives at that time, and particularly if they were women, did not want to be seen performing any activity that others might think was typing. Unfamiliar VDTs were confused by the user's peers with the typewriter, their use was demeaning, and women were especially sensitive. Only five years after this experiment, the VDT not only became familiar, but people with VDTs in their offices were seen by their peers and bosses as rising stars and on the fast track. People placed terminals and PCs on their desks, even if they did not know how to use them, because of the image that it gave them.

The User's Agent

When dealing with anything complex, we have the choice of learning the complexity with which we must deal or of having some intermediary, or agent, act on our behalf. For instance, when you bring your automobile to a service station for maintenance, you are not expected to understand the complexity of the modern automobile, to diagnose the problem yourself, and to tell the mechanic exactly what you want him to do. Instead, you explain the symptoms to a service manager, who acts as your agent, and he writes the shop order. Even the most naive people can bring their automobiles to a service station for maintenance.

A similar situation arises when a new user sits at an unfamiliar terminal connected to an unfamiliar computer that runs an unfamiliar operating system with an unfamiliar program for servicing electronic mail. This new user knows he or she has mail and wants to read it. The user is expected to deal with all the complexity and is given volumes of poorly written documentation to help. Typically, the user is given little human assistance and is made to feel stupid when he or she requests this assistance. Most users are smart enough to learn it all, but mastering this complexity is not their primary job. The user who takes the time to master it all has probably shirked on his or her primary job. Many users of computer-based services could use the assistance of an agent. This agent can assist the user in each direction of the transmission between the user and the service.

The transmission of data from the user includes typed commands that connect the terminal to the required host computer, log-in and password protocols to networks and hosts, setting transmission and terminal options, traversing a file directory, typed commands in the operating system's command language to invoke the desired service, character deletions and line deletions to recover from typing errors, and exit commands for graceful and nongraceful termination of a requested service. This list does not include the commands used while dealing with the requested service itself; it is merely the surrounding complexity of the operation. Two ways that an agent can simplify these activities for the user are by (1) allowing the user to select a service by pointing to an entry on a menu and (2) using icons instead of text to represent activities. Some people believe that spoken input to computers will be the panacea for these ills, but the user will still need to know the complexity of the system. Speaking the "computerese" jargon, instead of typing it, may even make it worse.

The transmission of data to the user includes prompts for various kinds of user input, error messages, and garbage (or nothing at all) whenever terminal and host are mismatched in some way. The repertoire of error messages from most operating systems is classic for providing little assistance and for insulting the user. Since operating systems seem to be designed to be obnoxious, an agent can, and should, be used to intervene.

This agent is some kind of program that interfaces between the user and the host operating system. It could reside on the host and be run as a kind of command interpreter and terminal driver. If it resides in the terminal, it can emulate any terminal for the host and it can also assist the user in connecting to the host.

Getset

Getset is a research prototype of an integrated voice/data terminal (2,3). It is a combination of a small VDT and a telephone with an autodialer. It may be recognized as having been part of the Bell System exhibit at the 1981 World's Fair in Knoxville, Tennessee. Two generations of the system are reported.

Example 7.   Getset-32 (shown in Fig. 4) has the footprint of a telephone, much smaller than even Displayphone. It has a 5-in CRT screen that displays a field of 16 × 32

FIG. 4   Getset-32.

characters. There are six function buttons on each side of this screen and a small alphanumeric keyboard that is adequate for only two-finger typing.

Getset-80 (shown in Fig. 5) has a 50% larger footprint, a low-profile, 24 × 80 character display screen, a total of 25 function buttons on three edges of the screen, a touch-typing keyboard, and an interface that simulates a conventional six-button telephone. While Getset appears integrated to the user, it does not have an integrated network interface. Two separate connections are required: an analog telephone line for voice communications and an RS-232-C port for data communications. While both versions have internal microprocessors, the resident software only controls local operations and messaging with a host. Connection to a host computer is required during all operations.

Getset was the hardware instrument for the research of integrated voice/data services and the human–machine interface to them. This project was part of a 10-year experiment involving digital switching, terminals, and services (4). Services include calling by name, organization chart access, access to a variety of public and personal directories, mail transactions including call memos, data access, calendar, reminders, games, and all the services typically found in a host computer operating system environment. Getset-80 is more appropriate for OA services than Getset-32 and was used for several years in a field trial of voice/data services among executives and their secretaries.

FIG. 5 Getset-80.

Service Access and Selection

The user's interaction with a service has four stages:

1. The user determines what services are available (*access*).
2. The desired service is selected.
3. The user is served.
4. The user terminates the service.

The first two stages are an important part of the user interface to the service system and are discussed briefly next. With computer services, the traditional approach is to provide the user a thick manual that describes all the commands; the user selects a desired service by typing the command. Such manuals are backward in the same way that we are told to check the spelling of a word by looking it up in the dictionary. The user must know, or suspect or guess, the name of the service before he or she can select it.

The access of services is another of those issues that is apparently obvious but deceptively complex. A service access mechanism using menu presentation is a good solution to this complexity. If hundreds of services are available, the menu must be organized into a tree-like structure so the user is not presented with too many choices at the same time. The menu entries can be short descriptions of the service, such as "text editor," instead of typical operating system commands, such as "emacs." Tree-structured menus provide the user the ability to browse for services.

Selection of the desired service is easiest if the user is allowed to point to the desired service on the menu. Many pointing mechanisms are available. In the Getset, when a menu page is presented to the user, the entries are the labels for the buttons on the sides of the display screen. The user selects a service by traversing the tree-structured menu to the page on which this service is displayed and then pushing the corresponding button. Many modern workstations use a mouse-directed arrow to point to the desired service on a menu and then a click of a mouse button to start it. Touch-sensitive screens, as in Aerial, and light pens are also effective pointers.

The ease of access and selection is illustrated by the "ring the person named _____" service, described above, as it is implemented on a Getset. Suppose a user, a professor at the University of Pittsburgh, decides to call a colleague in the English department of the university. One of the entries on the Getset screen at the root of the tree-structured menu is "directories." This user pushes the corresponding button and a new menu page is displayed. One of the entries on this page is "Pitt," for the organizational chart directory of the university. After selecting "Pitt," one of the entries on the subsequent page is "A&S," for the School of Arts and Sciences. After selecting "A&S," one of the entries on the subsequent page is "English." After selecting "English," the subsequent page displays the names of the faculty and staff in this department. The user pushes the button that is labeled for the name of the person to be called, lifts the receiver of the Getset, and the call is placed. The user never sees the person's telephone number because it was never needed.

Network Influence

The relationship between the network and the communication terminal is discussed in this section. Network interface equipment is described, and an example of the problem created when this equipment is incorporated in the terminal is given. The difference between integrated networks and integrated access to parallel networks is also discussed.

Network Interfaces

The properties of several networks are summarized and the equipment for interfacing to these networks is described.

The traditional *analog telephone network* appears as a two-wire, AC (alternating current) -coupled transmission line for audio-grade signals and, simultaneously, a direct current (DC) loop for call supervision and dial pulsing. The usable signal bandwidth is between 100 Hz and 4 kHz. The protocol and network control is traditional circuit switching, requiring call setup when a connection is first established, but the resulting channel is dedicated and contention free. The signal format is analog and usually continuous, with small amplitude and high noise susceptibility. The line is two way and susceptible to reverse-channel cross talk (echo). The end-to-end circuit-switched connection may be digitized and multiplexed over some spans, but these digital spans frequently outperform the traditional copper spans. The interface between the telephone and the network requires a hybrid transformer to separate the two directions of signal flow, impedance-matching circuitry, a DC switch for supervision signaling, and a dual-tone multiple frequency (DTMF) or dial-pulse signaling device for call placement. Mobile and cellular networks are emerging that give the user mobility.

Currently, digital data is transmitted over analog lines using a *modem*. Modems convert binary data, such as asynchronous ASCII/RS-232 signals, into some analog format, such as frequency shift keying (FSK). FSK performs better over an AC-coupled transmission line than binary pulses would and solves the problem of reverse-channel cross talk. Modems that operate at 300 baud transmit about 30 characters per second, which is sufficient for keystrokes but inadequate for downloading data to a terminal's screen. Host-to-user data, downloading to a terminal's screen, is more acceptable at higher data rates, such as 1200 or 9600 baud. The protocol and control is to use a conventional telephone to establish the circuit-switched connection and then to bridge the modem onto the line either by a local switch or by acoustic coupling. Modems with autodialers are available as peripherals for most PCs.

Ethernet is a common packet-switched local-area network (LAN). The medium is coaxial cable, which can support more than 1 Mbps, but the architecture is a bus instead of a star. Users compete and contend for common resources. Since control is packet switched, all transmissions must conform to a packet format in which the destination address is included in the packet. The interface between the terminal and the network was initially quite large and expensive, but has been reduced in size and cost significantly in the past several years.

The standard interface to the broadcast television networks has evolved for many people from a home antenna to a *CATV connection*. Each channel's signal uses about 5 MHz of analog bandwidth and each channel is frequency modulated into allocations of the radio spectrum, in the case of reception by a home antenna, or into frequency bands in the spectrum of the CATV coaxial cable. The signaling is one way only. The receiver requires a circuit for selecting the desired channel and demodulating it. The required equipment is discussed in detail in the section on inconsistent equipment. Digitized video, high definition, compression, narrowcast, and video jukebox are variations in stages of development.

The *ISDN* will provide digital access over the same facilities now used for analog telephony. The basic ISDN format provides 8000 words per second, consistent with the traditional sample rate for telephone-quality audio. Each word has 18 bits, two 8-bit subwords, and a 2-bit subword. Each 8-bit subword provides a 64-kbps digital "bearer" channel, consistent with traditional pulse-code modulated digital telephony. The two-bit subword provides a 16-kbps "data" channel intended for signaling in a digital packet-switched format. One ISDN line provides two digital voice channels and a signaling channel, all at 144 Kbps, net. It is intended to replace the classical analog telephone termination. Higher order ISDN formats are defined, but are not intended for universal application. ISDN is discussed in the next section.

Many proposals have appeared for so-called *broadband networking*. Part of the confusion among the research community is the definition of "broadband," varying from 45 Mbps to 2.4 Gbps. Proposals at the low end use existing digital transmission facilities, while proposals at the high end require optical fiber and photonics technology.

Network Integration

The networks that have been optimized for the various media are moving away from their separate analog implementations, some faster than others. They are all moving toward digital implementations, even broadcast video. This suggests the possibility that one integrated network could eventually deliver every medium and provide every service. The advent of optical fiber and photonics technology strengthens this suggestion because the bandwidth requirements of the different media become less significant. Figure 1 becomes a picture of the actual architecture if we hypothesize an integrated communication terminal and an integrated network. It is very appealing.

However, such an integrated network could never be optimized for every medium that it delivered. It would be optimized for one medium and be highly nonoptimal for the others or it would be optimized for some compromise specification. Examples of nonoptimal use of networks are packet-switched voice and circuit-switched keystrokes. The question that must be answered is whether the net cost of parallel but optimal networks exceeds the net cost of a single but nonoptimal network. Related questions are whether a single network is desirable, or should competitive networks, in a free market, be encouraged. A sensible compromise intermediate position exists.

The recent divestiture and deregulation of the telephone industry in the United States suggests an interesting hypothesis. We have observed many enterprises competing in the long-distance market, but no one tries to compete with the local telephone company by offering alternative residential telephone service. We hypothesize from this observation that integration is more important in the local portion of the network than in the long-distance portion. If the hypothesis is true, then the future may require an integrated local subnetwork that provides access to many parallel, separately optimized, long-distance networks.

If Fig. 1 illustrates an integrated network, then Fig. 6 illustrates integrated access to separate networks. Figure 6 is an expansion of Fig. 1, in which the block labeled "Network" from Fig. 1 is expanded. While the difference between these two kinds of integration is extremely important to network planners, there is no difference to the designers of communication terminals. Ten years ago, we would have all agreed that the ideal format for this integrated local subnetwork would be ISDN. But the network community has stalled, debated, and quibbled over details for such a long time that we must ask:

- Is it still a good idea?
- Is it too late?

Many experts now believe that optical fiber will replace economically the classical copper telephone line into everyone's home and office. If fiber to the home really happens, ISDN will be out of date. The potential for fiber to the home has further delayed ISDN deployment.

Today's terminal designers have shown the propensity to find unusual ways to interface to and control the existing network. The mechanisms would not need to be so unusual or the existing network bypassed, if ISDN were currently in place. Once enough equipment like this is installed, there will be a natural opposition to changing the network, even if the change is to a network that may be easier with which to deal. The first ISDN symposium was held in 1976— a short time ago in the minds of network planners. Two men named Steve started Apple Computer in 1976, a long time ago in the minds of the users,

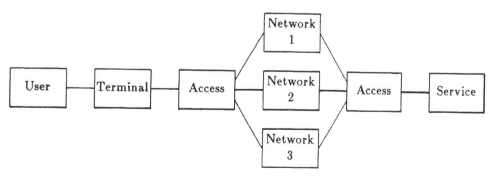

FIG. 6 Integrated network access.

service vendors, and terminal manufacturers. The users, terminal manufacturers, and service providers alike do not seem to care if the network is ISDN or step by step as long as they get useful, economic, friendly, reliable services. And if the existing network is perceived as cumbersome or expensive, they will build their own networks (5). It has been said that ISDN stands for "It Still Does Nothing."

Inconsistent Equipment

Example 8.   The configuration of commercial equipment found in many homes illustrates how such equipment can be difficult to manage and can even be inconsistent. Figure 7 shows a conventional television receiver, a videocassette recorder (VCR), and the interface "box" for cable TV. Each piece of equipment comes with its own infrared (IR) remote controller. In the only wiring configuration that works, we connect the box's input to the cable and its video output to the antenna-lead input on the VCR, and set the VCR to Channel 3 or 4 permanently. We connect the VCR video output to the antenna-lead input on the TV and set the TV to Channel 3 or 4 permanently. We use the TV's IR remote to turn the TV on and off and adjust sound level, the VCR's IR remote to control tapes or to view from cable, and the box's IR remote to change channels; this is a nuisance, but acceptable.

So-called cable-ready television receivers work well with many CATV systems, eliminating the need for an interface box. But many people purchase them and bring them home, only to find that they do not work out; the motto here is *caveat emptor* (let the buyer beware). The configuration in Fig. 7 requires a cable-ready VCR, not a cable-ready TV. Furthermore, if the CATV system scrambles its signals, the user must own a box for descrambling anyway.

Many VCRs allow the user to record a program from one channel while viewing a program on a different channel. Many VCRs also may be preprogrammed to record, unattended, several programs on different channels at preset times. These features are possible, without reconfiguring, only when the VCR input receives all channels, as when connected to a home antenna. When

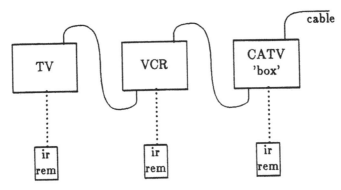

FIG. 7   Television with videocassette recorder (VCR) and community antenna television (CATV).

the channel is selected by the CATV interface box, the VCR cannot automatically record multiple formats. Switches are available, but the nuisance has increased. Furthermore, some security-conscious CATV systems require that the user's box upload a code and receive a downloaded descrambling key for each channel selection. In these systems, unattended recording by the VCR of cable-only programs is impossible, regardless of configuration.

The very important lessons pointed out by the discussion above are that

- reliance on market evolution may lead to deadlocks
- an architecture should have been planned (but by whom?)
- modular packaging does not always work
- privacy and security are difficult in broadcast systems
- interface specifications and standards are not enough
- services depend on the architecture of the network and the terminal equipment

Architecting Intelligence

In designing telecommunications systems (as shown in Fig. 1) the principal architectural issue is the distribution of processing intelligence between the service and the communication terminal. The economically optimal architecture depends on the relative costs of computing and communicating and this is subject to change. The impact of optical fiber and photonics technology is discussed in this section, particularly with respect to the design of wideband terminals.

Communicating versus Computing

Over the years, the steadily declining price of digital electronics has dramatically reduced the cost of memory and computing. Although systems for digital transmission and digital switching also use this technology, the ratio of the cost of storing and processing bits to the cost of transmitting bits decreased steadily in the interval from 1955 to 1985. This trend was exacerbated because the price of computing was competitive in an open market, while the price of communicating had been largely regulated.

In a price/performance optimization, it follows that, over this 30-year interval, the distribution of intelligence between the terminal and the host (service) would reflect the state of this cost ratio. Gradually more intelligence was built directly into terminals and we saw the evolution of the VDT into the PC. Simply put, communication terminals became gradually more complex and self-contained because it became gradually more economical and convenient to provide local intelligence than it was to communicate with shared intelligence.

Besides a price/performance optimization, this architectural trend was also driven by the bandwidth demanded by users and services. The conventional 300/1200-baud connection between terminal and host became gradually more inadequate. As LANs came on the scene and provided greater bandwidth at the burst rate, the architectural trend took a slight turn. With a PC, the user became responsible for administrative tasks (notably, disk back-up) that conventionally had been performed by staff at centralized computing centers. High burst data rates over LANs allowed the mass memory, previously in the PC, to migrate back to a centralized facility, called the *file server*, where it could be administered. A new kind of terminal, called a workstation, has evolved for this environment. A *workstation* is a VDT with an internal processor or a PC without a floppy-disk drive.

Four Terminal–Host Architectures

A generalized terminal-to-host architecture is presented in Fig. 8-*a*. Each of the four interfaces (A–D in the illustration) is a candidate for the boundary between the terminal on the user's premises and the host in some central location. Figures 8-*b* through 8-*e* represent architectures in which the boundary between the terminal and the host has been selected at each of these candidate interfaces. The first three (Figs. 8-*b* through 8-*d*) are classical configurations. The fourth (Fig. 8-*e*) is proposed for the future.

Figure 8-*b* illustrates the architecture for connecting a classical VDT to a classical computing center. The VDT on the user's premises contains a CRT and electronics, a keyboard, a byte-mapped memory, digital controller, and network interface. The computing center in a centralized location contains a general-purpose, time-shared, mainframe host computer with terminal interface peripherals and a cluster of disk drives. The network connection between the classical VDT and the central host system is typically circuit switched and uses a low data rate. If a multiple-access LAN is used for this connection, the communication is still only point to point.

Figure 8-*c* illustrates the architecture for interconnecting many personal computers over a LAN. The user has, on his or her premises, a complete stand-alone PC, including a floppy-disk drive and a network interface. The multiple-access LAN need only support low-rate data transmission.

Figure 8-*d* illustrates the architecture for connecting a modern workstation to a file server over a LAN. The workstation is equivalent to a VDT with an internal processor or to a PC without a disk drive. The file server in a centralized location contains the disk drives for all the workstations on the LAN. The LAN must support more data traffic and at a higher rate than the LAN in Fig. 8-*c*. The connection between the user's workstation and the centralized file server could be circuit switched and point to point. A multiple-access network is preferred if a workstation is served by more than one file server or if there is considerable terminal-to-terminal traffic.

Example 9. An experimental multimedia teleconferencing system, called Rapport (6), has been designed, prototyped, and reported. Rapport presents voice, data, and video

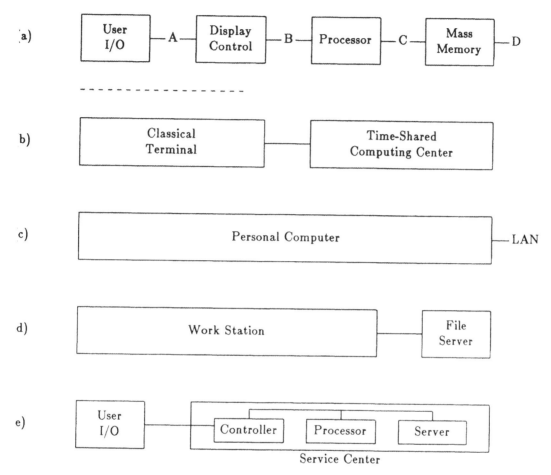

FIG. 8 Terminal-to-host interconnection alternatives: *a*, generalized terminal-to-host architecture; *b*, connection between a classical video display terminal (VDT) and a classical computing center; *c*, connection between a personal computer (PC) and a local-area network (LAN); *d*, connection between a workstation and a file server over a LAN; *e*, proposed future connection.

media to the user at an integrated terminal. The service set includes many services that provide for teleconferencing in all three media. It performs most of the functions described in the section above concerning integration of voice, data, and images. The implementation is represented by Fig. 8-*c*. The terminal is very complex and expensive and the network is a simple Ethernet.

The Impact of Photonics

The advent of optical fiber and the accompanying photonics technology should dramatically reverse this migration of electronic complexity into the terminal. While some people view photonics technology as an evolutionary replacement for the physical point-to-point channels in existing architectures, others believe it will affect the architectures of communications networks and even computing systems in revolutionary ways. Photonics technology will change telecommunications to the same degree that the transistor changed computing—improving throughput, cost, availability, and human capability by orders of magnitude. We are on the leading edge of the next major revolution.

The virtually unlimited bandwidth of photonics technology can provide a channel between the terminal and the services that should dramatically affect the kinds of services that might be made available and even redefine the boundary between terminal and network.

Figure 8-*e* illustrates a fourth alternative architecture. The equipment on the user's premises is even simpler than the classical VDT illustrated in Fig. 8-*b*. In the architecture in Fig. 8-*e*, the digital controller and possibly even the bit-mapped memory are removed from the terminal and pooled in a centralized service center. This service center also includes the equipment found in the classical computing center in Figure 8-*b*. Communications requiring multiple access for multiple file serving and for terminal-to-terminal data uses the LAN internal to the service center. The network that interconnects the simple user's instrument with the terminal controller can be circuit switched and point to point, but it must support a very high data rate. Communications is squandered, but the communication terminal may be affordable and universal. Perhaps a future implementation of the Rapport system could be represented by Fig. 8-*e*.

Services by or through the Terminal

Services may be provided directly by the terminal, as in ECS and the Genesis Telesystem. Software can be internally resident by inserting floppy disks or cartridges or it can be downloaded over the network from some file server. Alternatively, services may be delivered through the terminal from a host computer, as in Getset, videotex, and the Aerial Terminal.

Whether services reside locally in the user's terminal or remotely at the other end of some network can affect cost, performance, administration, reliability, security, and user interface. But they should be perceptually equivalent. If these performance differences are small, the services will be perceived to be local. The user should not care whether services are actually local or remote. With an appropriate terminal and a flexible network, services could be offered either way. The best procedure is to allow part of the service to be local and part to be remote and to let performance issues and market forces dictate how much is done where.

Summary

The communication terminal is defined as an instrument by which a user communicates with another user or with some service through a network. It is described as an evolution, generalization, and synergism of three primordial special-purpose terminals: the telephone set, the VDT, and the television receiver. The communication terminal is described as both a transparent member and as a two-way transducer in a generalized telecommunications environment. We show how a communication terminal should be specified primarily by the set of services intended to be provided through it. Integrated networks, terminals, and services are all described, as well as various components, packaging, and display options. The user interface to service access, selection, and to services themselves are described. The terminal's interface to various networks is discussed. Four alternative architectures are presented that define how equipment is partitioned between the user's location and a central hub and how the intervening network is affected. In addition, broadband terminals are discussed.

Acknowledgments: Epcot Center and Walt Disney World are registered trademarks of Walt Disney Productions.

References

1. Smith, M., A Model of Human Communication, *IEEE Commun.*, 5–14 (February 1988).
2. Hagelbarger, D. W., et al., "An Experimental Teleterminal," seven papers presented in a dedicated session at the National Telecommunications Conference, New Orleans, LA, 1981, and special issue of *IEEE J. Sel. Areas Commun.*, February 1983.
3. Hagelbarger, D. W., and Thompson, R. A., Experimenting with Teleterminals, *IEEE Spectrum*, 40–45 (October 1983).
4. Bergland, G. D., Experiments in Telecommunications Technology, *IEEE Commun.*, 4–14 (November 1982).
5. Terminals and the Emerging ISDN, complete session at the Workshop on the Integrated Services Digital Network, Tarpon Springs, FL, October 1983; summary paper presented at International Communications Conference, Amsterdam, The Netherlands, May 1984.
6. Ahuja, S. R., Ensor, J., and Horn, D., The Rapport Multimedia Conferencing System, paper presented at the Conference on Office Information Systems, Palo Alto, CA, March 1988.

RICHARD A. THOMPSON

Communication with Intelligent Modems

Introduction

Modems provide a means of converting serial or parallel signals from data termination equipment (DTE) to a form suitable for transmission over some medium (channel). The data to be transmitted by a modem may be analog or digital (usually binary if digital). Analog signals may be used for facsimile data, for example. Digital data is most common and, where intelligent modems are involved, an asynchronous DTE-to-DCE (data circuit terminating equipment) interface is often used. "Intelligent" devices abound. The meaning of *intelligent* is vague, but it seems to mean a device that will do more than its main function; it is one with which a person or program can also communicate.

An *intelligent modem* is one that can recognize and respond to commands from its DTE over the same DTE interface that is used for data transmission. That is, an intelligent modem can interact with its DTE as well as send and receive data signals (1). The modem user can communicate with the modem as well as send data through it. This implies that the DTE interface is digital, usually an asynchronous interface operating at some convenient speed such as 1200 bps (bits per second). Intelligent modems are usually designed to use a serial or computer bus DTE interface and communicate over telephone lines, using the Public Switched Telephone Network (PSTN). These modems may provide data transmission at speeds from near 0 bps to over 19,000 bps. The advent of low-cost microprocessors made intelligent modems practical. The "intelligence" resides in the program of a microprocessor.

This article discusses the history of intelligent modems, what they do in addition to sending and receiving data signals, how they are controlled (control languages), and details of selected control languages.

History

Modems, in the sense of modulators and demodulators for digital signals, have been in use since the 1920s (2). Early digital modems used amplitude modulation and, later, frequency-shift-keying (FSK) modulation. High-speed modems now use more exotic modulation schemes (3).

In the United States, the first modems that used the PSTN for data transmission appeared in the 1950s with the introduction of modems by AT&T and others. These modems were what may now be called dumb modems. That is, they were controlled by binary control signals; each control wire affected one function. Such modems could go off hook and establish connection with a distant modem and could answer incoming calls automatically. The usual practice was to accomplish the dialing function of call setup by means of a separate device—a telephone or an automatic-calling unit—which used a separate cable to the DTE for calling control.

The first intelligent modem was introduced by Bizcomp in 1980 (4). D. C. Hayes Associates (now Hayes Microcomputer Products, Inc.) introduced the second, a 300 bps modem, in 1981. Since that time, dozens of manufacturers have introduced intelligent modems. Figure 1 shows the Hayes Smartmodem 9600. The Bizcomp modem provided a basic command set of 12 commands (4). Hayes added automatic DTE speed recognition by the modem and an improved command set. The Hayes Standard AT Command Set, with some variations, is the most widely used command means in use today.

The first intelligent modems used asynchronous communication between DTE and modem. Synchronous interfaces were developed a few years later.

The International Telegraph and Telephone Consultative Committee (CCITT) Recommendation V.25bis was approved in 1984. V.25bis has a limited command and response set, and provision for asynchronous DTE interface and two synchronous DTE interface protocols, Binary Synchronous Communication (BSC) and High-Level Data-Link Control (HDLC). Also in 1984, Racal-Vadic introduced a synchronous DTE–modem command language, called Synchronous Auto Dial Language (SADL). SADL provides for synchronous control of a modem using BSC, Synchronous Data Link Control (SDLC), and

FIG. 1 V-series Hayes Smartmodem 9600. (Photo courtesy Hayes Microcomputer Products, Inc.)

HDLC protocols. While similar to V.25bis, SADL has more commands and features. The Hayes, V.25bis, and SADL command languages are discussed below.

Functions

Intelligent modems provide functions that facilitate control of communications. Typical functions are shown in the table below.

Function	*Comments*
Line control	The modem contains a relay to open or close its connection to a telephone line.
Dialing	The modem is able to send dual-tone multiple frequency (DTMF) tones or dial pulses to establish a telephone call.
Dial tone	Most modems can detect dial tone before dialing. Many countries require dial-tone detection and do not allow blind dialing.
Busy tone	A call attempt can be aborted if busy tone is detected.
Answer calls	The modem contains a ringing-signal detector and can answer a call after a predetermined number of rings or a command.
Number-storage options	Telephone numbers may be stored in a modem. This facilitates the use of frequently called numbers.
Options	Such user-changeable options as timing limits, speaker volume, DTMF versus dial-pulse dialing, and the like can be changed by commands.
Test modes	Test functions can be turned on and off by commands.
Error control	Some modems provide error-free data transmission, usually by error detection and retransmission.
Compression	Data compression enhances the apparent transmission speed of a modem. Many modern modems provide this function, which is especially useful for asynchronous terminals.
Retry count	A modem may keep track of the number of retries to a phone number and blacklist numbers that may not be dialed again without manual intervention. This function is required in some countries.

Error correction received much attention by the CCITT in 1987 and 1988, culminating in a new recommendation, V.42 (5). This recommendation describes error-correcting protocols for use with duplex data circuit terminating equipment (DCEs) to accept start–stop (asynchronous) data from a DTE and transmit in synchronous mode. The recommendation is an HDLC-based protocol called Link Access Procedure for Modems (LAPM).

Modem Control

An intelligent modem designed for use on the PSTN normally has several modes of operation or states. Four states may be used to describe the main functions of an asynchronous intelligent modem. These states are listed in the table below.

Modem State	Comment
Command	Signals from the DTE are treated as commands. Ringing signals may cause the modem to answer the call.
Call setup	The modem is setting up a call: detecting dial tone, dialing, waiting for answer tone, or handshaking. The modem may be returned to the command state by a busy tone, an abort instruction from the DTE, or a time out.
On line	The modem treats signals from the DTE as data. The data is transmitted to the distant end. The modem examines data from the DTE and may recognize an "escape" sequence, which is a command to go to the on-line command state.
On-line command	The modem maintains communication to the distant end but treats signals from the DTE as commands. This state allows on-line commanding of the modem. This state can be reached by an asynchronous modem. The synchronous command sets discussed below do not have any provision for this state.

The scenario that follows illustrates the operation of an intelligent modem. See Fig. 2 for an indication of the flow of data, commands, and responses.

When the modem is turned on, it is in the command state, ready to communicate with its DTE. After dialing or other instructions are given to the modem, it may go off hook, detect dial tone, and dial a number. After the number is dialed, the modem will wait for an answer for a specified maximum time. If a busy tone is detected before the other end answers the call, the modem will report the busy tone to the DTE and abort the call, returning to the command state.

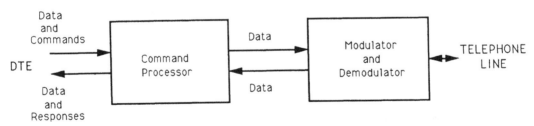

FIG. 2 Block diagram of generic intelligent modem. While the modem is in the command state, signals from the data termination equipment (DTE) are treated as commands and responses are sent from the modem to the DTE. While in the data state, signals from the DTE are passed on to the modulator and signals from the demodulator are passed to the DTE. Some intelligent modems also provide data-error-control protocols, data compression, and other processing of the data.

If the call is answered by another compatible modem, the two modems execute a procedure known as handshaking. This procedure includes sending signals to disable echo suppressors that may be in the PSTN channel (on long-distance trunks), signals to determine the line speed (bit rate) at which to connect, and signals to facilitate adaptive equalizer training (if required). After handshaking, the modem informs its DTE of the connection and connection speed, and goes to the on-line state.

If the DTE needs to give some instructions to the modem while it is in the on-line state, the DTE may send a unique data sequence that the modem can recognize. The modem then stops sending data from its DTE to the distant end, maintains the connection, goes to the on-line command state, and asks the DTE for instructions. The DTE then may communicate with the modem and, when done, instruct the modem to return to the on-line state.

When the call is completed, the modem detects a loss of carrier or long space signal from the distant end and goes on hook and to the command state. A call may also be discontinued by a command from the DTE (while in the on-line command state).

Test routines and modes can be activated by the DTE while the modem is in the command or on-line command states.

Note that V.25bis and SADL do not have any provision for an escape from the on-line state to an on-line command state. Thus, the statements above about commands to a modem while on line apply only to modems with an asynchronous DTE interface.

Control Commands

Commands, issued while the modem is in one of the two command states, cause the modem to take immediate action or determine an action that will be triggered by later events. Commands are also used to establish certain modem parameters, such as how many ringing signals must be received before the modem will answer a call.

There are many command sets, some for asynchronous DTE interfaces and some for synchronous ones. The three described in this article are the commands specified in CCITT V.25bis (6), the Hayes Standard AT Command Set (7), and Racal-Vadic's SADL (8). There are many variations of the Hayes commands and several other command sets; however, space does not allow a description of every command set.

International Telegraph and Telephone Consultative Committee V.25bis Commands

CCITT Recommendation V.25bis specifies commands (signals from DTE to DCE), indications (from DCE to DTE), and procedures for automatic calling and answering using serial communication between a DTE and its DCE. The recommendation provides commands to make a call request, to program the modem, to request information from the modem, and responses to indicate various types of call failures. V.25bis has no mechanism to go to the on-line command state.

V.25bis provides for asynchronous communication between the DTE and modem (DCE), and two types of synchronous communication: synchronous character-oriented operation and HDLC operation. In asynchronous operation, each message (command or indication) is followed by a carriage return (CR) and line feed (LF) character. The 7-bit code specified by CCITT Recommendation T.50, which is similar to the well-known ASCII code described in American National Standards Institute (ANSI) Standard X3.4-1977, is required by V.25bis. In the two synchronous modes, the CR and LF characters are replaced by suitable block delimiter codes.

The V.25bis commands, except those that are under study, are given in Table 1.

The V.25bis indications are listed in Table 2.

Hayes-Compatible Modems

The Hayes Standard AT Command Set provides a wide range of modem control functions and responses. A complete description of the Hayes commands is beyond the scope of this encyclopedia. The important commands, sufficient to indicate the variety of the command set, are described below.

All commands and responses (result codes or information text) are sent asynchronously using ASCII code; there is no provision for synchronous DTE-DCE communication. Result codes, unsolicited messages from the modem to the DTE, are used to indicate an incoming call, successful connection to a distant modem, and so forth. Information text is sent to the DTE in response to certain information-request commands. Result codes and information text are not preceded by prefix letters; they may be preceded by CR or LF.

Hayes-compatible modems allow the user to specify which ASCII characters a modem will interpret as CR, LF, and BS (backspace). This provides flexibility for communications software.

Hayes-compatible modems will recognize a unique bit sequence while in the on-line state and go to the on-line command state. The unique bit sequence contains a user-selected quiet time (during which no characters are sent from the DTE) and a selected ASCII character. This is selected by register S2 (see below).

Hayes Standard AT Command Set

All commands, except the repeat command, are prefixed with AT (or the lower-cased at) and ended with a CR character. The modem determines the DTE speed from the A and the parity from the A and the T. The selection of commands in Table 3 may be found in Ref. 7. A/ is used to repeat the preceding command string. Most commands may be strung together on one line; separate command

TABLE 1   CCITT V.25bis Commands

Command	Meaning
CRN XXXXXX	Go off hook and dial number XXXXXX.
CRI XXXXX YYYYY	Dial number XXXXX with identification (ID) YYYYY.
CRS MMMMM	Dial the number whose memory location is MMMMM.
PRN MMMMM XXXXX	Store number XXXXX in location MMMMM.
PRI YYYYY	Store the ID number of the station.
RLN	List the stored phone numbers.
RLF	List the forbidden phone numbers.
RLD	List the delayed phone numbers.
RLI	List the ID number of the station.
DIC	Disregard incoming call.
CIC	Answer incoming call.

TABLE 2   CCITT V.25bis Indications

Indication	Meaning
DLC	Number will be delayed SSSSS seconds
INC	Incoming call
VAL	Valid command
INV	Invalid command
LSN	Stored numbers
LSF	Forbidden numbers
LSD	Delayed call numbers
LSI	Identification (ID) number

TABLE 3   Selection of Hayes Standard AT Commands

Command	Meaning
A	Go off hook and answer a call
D	Dial the following phone number
DS $=n$	Dial phone number in location n
H	Hang up (go on hook)
H1	Go off hook
Ln	Set speaker volume to n
Mn	Turn on speaker according to n
O	Go on line
P	Use pulse dialing
T	Use tone dialing (DTMF dialing)
Qn	Use result code display option n
Sn	Select S register n
Sn?	Display contents of S register n
Vn	Use result code format n
Yn	Use long space disconnect option n
Z	Reset the modem (soft reset)
&Dn	Respond to DTR change by method n
&Qn	Use communication mode n
&Z$n=x$	Store phone number x in location n
&Wn	Store current modem parameters in location n

lines are not required. Responses are called result codes by Hayes. Some of the result codes for a Hayes-compatible modem are given in the Table 4. Memory locations, called *S registers*, are used to store modem parameters. Table 5 shows a selection of S registers to illustrate the types of parameters.

Synchronous Auto Dial Language (SADL)

SADL defines commands, responses, call progress messages, protocols, and "rules" for communication between a DTE and an intelligent modem. The protocols follow the standard protocols for BSC, SDLC, and HDLC, and are not described here. The commands and call progress messages (see Table 6) define the functions of a modem; see Ref. 8 for more details.

TABLE 4 Some Result Codes for a Hayes-Compatible Modem

Result Code	Meaning
OK	Previous command acknowledged or executed
CONNECT	Modem is connected to distant modem
RING	Ringing signal on and off was detected
NO CARRIER	Failure to connect or loss of carrier
ERROR	Command error
CONNECT 1200	Modem is connected and DTE speed is 1200 bps
NO DIALTONE	Dial tone was not detected within time limit
BUSY	Busy tone was detected
NO ANSWER	Answer tone was not detected within the time limit
CONNECT 2400	Modem is connected and DTE speed is 2400 bps

TABLE 5 Selection of S Registers for a Hayes-Compatible Modem

Register	Meaning of Contents
S0	Number of rings on which to answer a call
S1	Number of rings detected so far
S2	Decimal representation of escape character
S3	Decimal representation of CR character
S4	Decimal representation of LF character
S5	Decimal representation of BS character
S6	Wait time before blind dialing
S7	Time limit on wait for answer
S8	Pause time for comma command
S9	Carrier detect delay
S10	Time carrier can be off without hangup
S11	DTMF dialing on and off time

TABLE 6 Commands and Call Progress Messages in Synchronous Auto Dial Language

Commands	Comments
Enquiry	Precedes every message from DTE to modem
Option	Sets modem options
Dial	Gives modem number to dial
Second dial tone	Instructs modem to wait for dial tone again
Call Progress	
Dialing	Tells DTE modem is ready to dial
Ringing	Modem has detected distant ringing (ringback)
Answer tone	Modem has detected answer tone
On line	Modem has completed handshaking; going on line
Busy	Modem has detected busy signal; will abort call
Failed call	Timer expired without answer tone; will abort call
No dial tone	Modem did not detect dial tone in time
Invalid	Modem received invalid command or message

Conclusions

Intelligent modems have revolutionized computer communications. Virtually all desktop computer communications use intelligent modems and most other computer communications do also. Communication programs are available from many sources to facilitate the control and use of intelligent modems, making them very easy to use with a personal computer.

The command and control paradigm devised for modems is being extended to Integrated Services Digital Network (ISDN) terminal adapters, making them respond to the same commands that are used for modems. This allows modem communication software to be used with some ISDN terminal adapters also.

References

1. Dunlap, E. M., and Rife, D. C., Personal Computer Communications via Telephone Facilities, *IEEE J. Sel. Areas Commun.*, SAC-3(3): 399–407 (May 1985).
2. Bennett, W. R., and Davey, J. R., *Data Transmission*, McGraw-Hill, New York, 1965.
3. Bingham, J. A., *The Theory and Practice of Modem Design,* Wiley, New York, 1988.
4. Bizcomp, U.S. Patent 4,387,440, Modem Control Device Code Multiplexing, issued June 7, 1983.
5. CCITT, *Error-Correcting Procedures for DCEs Using Asynchronous-to-Synchronous Conversion*, CCITT Recommendation V.42, ITU, Geneva, 1988.
6. CCITT, *Automatic Calling and/or Answering Equipment on the General Switched*

Telephone Network (GSTN) Using the 100-Series Interchange Circuits, CCITT Recommendation V.25bis, ITU, Geneva, 1984.

7. Hayes Microcomputer Products, *V-Series Smartmodem 9600 User's Guide,* Hayes Microcomputer Products, Norcross, GA, 1987.
8. Racal-Vadic, *SADL Designer's Guide,* 2d ed., Racal-Vadic, Milpitas, CA, 1987.

DAVID C. RIFE

Communications, Acoustic

(see Acoustics in Communications)

Communications and Information Network Service Assurance

Purpose and Background

Purpose

This article has been prepared to introduce the reader to a pragmatic approach to the design of quality network performance. It is not intended to be a technical treatise, but instead a means of organizing one's thoughts toward designing for network reliability. The technologies that can be used to design network solutions change so rapidly that focusing on a particular approach would be obsolete as soon as it is written. As an alternative, quality principles are emphasized as the foundation for a rational approach to reliable network design.

Background

Communications and information networks serve an increasingly important role in the migration of the United States to an information-intensive society. Through the capabilities of information networking, we have enabled access to and usage of information on an unprecedented scale. In many businesses, both large and small, electronic access to information is essential to the work itself, and without such access the business process stops. Coincidentally, productivity and profitability deteriorate. As a consequence, communications network reliability has assumed an increased level of importance in the business infrastructure. The reliability of these networks is often tested through the unfortunate adversity of failure. It is not unusual that, under the stress of a network failure, a company finds its normal way of doing business substantially disrupted. Depending on the length and severity of the communications outage, the functions of a business can be impaired, with accompanying significant financial impact.

Real Failures

Current examples of significant communications disruptions have included the devastating local central-office fire at Hinsdale, Illinois, and the recent signaling network failure of a national interexchange carrier. In both cases, the impact on customers was significant and, as a result, many communications companies and their customers began a systematic review of the vulnerability of their networks to single-point failures as well as system failures.

The Classic Approach

The classic approach to managing network failures has been to restore the failed network element rapidly through patching and other techniques to reroute

433

around the failure condition. While this approach continues to be necessary, customers still experience a less than acceptable level of disruption and aggravation. For certain critical business functions, this level of disruption could be commercially unacceptable and therefore must be prevented wherever economically feasible.

Integration Issues

Most end-user network needs are provisioned in a multisupplier environment. For large firms, although a single communications manager may be utilized, the manager integrates the functions of a number of network suppliers. A typical mix of suppliers could include

- Local telephone company access, switching, and distribution networks
- Communications software and equipment
- Long-distance switching and transport
- Computer software and equipment
- Local and emergency power for on-site systems

A cursory examination of this partial list of suppliers suggests that an organized approach is essential to achieving reliable, integrated network performance. Integration of individual supplier efforts to create a reliable end-user network requires a quality-centered design technique that meets the user's performance requirements.

Quality: A Preventive Approach

Quality Concepts

Quality concepts have been applied by the communications industry for many years. During the decade of the 1970s, substantial energy was directed toward individual network-component reliability, and attempts were made to design network components as a system with overall reliability objectives. Design objectives were established for component-level mean time between failures (MTBF), as well as understanding failure probabilities for a network of components.

Communications networks throughout the 1970s were mostly analog (as compared to today's growing digital environment) with individual network components carrying relatively limited volumes of information. As a result of the distributed nature of the networks and technologies employed, single-point failures usually did not create major concerns for customers unless a particularly critical element was out of service for a very long period of time. Examples of critical elements include entire copper cables or switching systems that were simply too expensive to duplicate.

Impact of Digital Technologies

As such newer technologies as fiber-optic transmission systems and very large digital switching systems were constructed during the 1980s, the level of risk associated with single-point failures was increased. For example, an older 300-pair copper cable typically carried a comparable number of voice circuits operating at 64 kilobits per second (Kbps) for a bandwidth of 19,200 Kbps. Replacement of this same cable with a 10-pair fiber-optic cable and associated electronics could now provide 10 optical channels capable of carrying 500 megabits per second (Mbps) or higher for a total bandwidth of at least 5 gigabits (Gb). The impact of a total cable outage now imposed risk levels several orders of magnitude higher. Improved network design considerations are required, even though the system failure rates are generally lower than the older technology, because the amount of information at risk is so great.

Given the higher risks of the newer digital technology, a new design approach is emerging. This new approach to quality service emphasizes the prevention of disruption from a customer's viewpoint. Simply stated, the design goal must move from a "not to exceed" outage level to a network that does not appear to break down at all. While the concept is superficially simple, making it actually happen in practice can be very costly to both the customer and the communications carrier. Consequently, a rational approach must be employed to discern where added preventive investment can have the greatest impact on the desired result. Such an approach is encompassed in a concept called *designing for service assurance*. This approach uses the Pareto principle, which suggests that a substantial part of a system's results can be controlled by a smaller subset of performance levels or decisions.

Defining Service Value to a Business

Most businesses today are made up of a large number of functional components or processes. While each of the components is necessary to support business performance over time, each is not necessarily equal in terms of the impact on the firm should it experience a short-term communications disruption. The obvious questions are: How much of a disruption and for how long? How does one set a value on avoiding disruption?

As an example, one could envision a firm involved with continuous commodity or other trading operations. In the short term, the communications capability for the trading operation is likely to be far more important to the firm than that of the executive or administrative offices. The potential revenues or profitability of the firm are readily linked to continuing performance of the trading department. The trading department's performance is, in turn, inextricably linked to communications and access to data and other trading firms. An analysis of the revenue impact of each of a hypothetical company's departments or functions is shown in Table 1.

Although these numbers are an illustrative example, it is quite clear that the

TABLE 1   Departmental Financial Analysis, ABC Traders, Inc:
Potential Losses for Communications Outages (In
Millions of Dollars)

Department	Length of Outage		
	1 Hour	4 Hours	24 Hours
Executive	$ 0	$0.1	$0.3
Personnel/Administration	0	0	0
Accounting	0	0	0.5
Trading	1.0	2.0	6.0
Sales	0.2	0.5	1.0

firm should be willing to expend additional funds to protect the trading department from catastrophic communications failure. Conversely, if the personnel
department were disrupted for a day, while inconvenient to many, it would not
have a major financial impact on the firm.

Such an approach is possible with most businesses and provides an organized
way of establishing a value to the firm of providing additional service assurance.
This is the first step toward determining which communications service is most
valuable to the firm and which can be given a lower priority if a disruption
occurs.

Disaster Prevention

In the example above, we isolated the trading department as the firm's first
priority for protection from disruption. The next step in the analysis requires
an examination of the various factors that could cause a communications disruption to this department's operations.

Let us first list some of the physical factors that must be examined concurrently because of their impact on communications requirements.

- Is this the only location where trading operations occur?
- How reliable is the electrical service for the location?
- How reliable is the physical security for the location?
- What level of fire protection is provided?
- Does the firm have an alternate location to which calls could be routed?
- Could personnel be relocated easily to an alternate site if a disruption oc-
  curred?
- Can the communications and data systems handle a redirection to an alter-
  nate site?

This list is certainly not all inclusive but is a sample of questions to be asked to determine the firm's vulnerability to disruption. Let us assume that these and other questions have been answered satisfactorily and focus next on generic approaches to examining communications reliability at a specific location.

Public Network Solutions

Vulnerability Analysis

The concern to be addressed is how vulnerable is the design of the total communications system? The response to this question must be viewed in the context of a system consisting of more than just the equipment in a single location.

A step-by-step analysis utilizing a schematic diagram is helpful; Fig. 1 depicts a typical arrangement for a communications service. For simplicity, only eight network components are shown. Greater or lesser complexity is possible in any specific location. A prolonged failure of any of the eight network elements shown in the illustration would have a severe impact on the sample firm's principal revenue source—the trading floor. The repair time in the event of

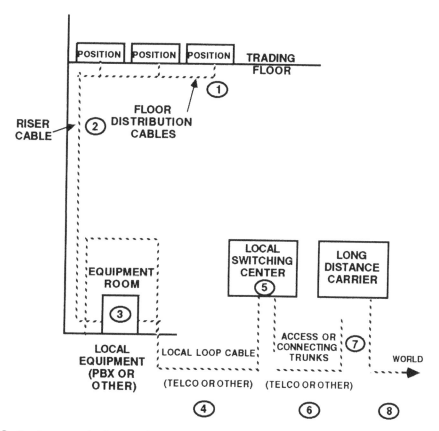

FIG. 1 A communications service arrangement (numbers 1 through 8 refer to items discussed in Table 2).

single-component outage is likely to exceed 2 hours. Without a preplanned service-backup capability, such an outage time would be costly to the firm. The need for prevention of the outage is obvious, and communications companies have solutions to offer, if the need is preplanned and the capability is provided in advance.

Preplanning Service Assurance

Preplanning is the key to communications service assurance, and the cost-versus-benefit relationship establishes the extent of protection employed. Table 2 lists some possible solutions for the potential failure of the network elements shown in Fig. 1.

From Table 2, one can observe that solutions are available from network suppliers; however, each business situation is unique and may require different approaches. How far an individual firm proceeds will depend on the specific relationship between costs and benefits.

The fast-paced evolution of technologies that are available to communications suppliers is constantly improving the options available and the flexibility of response to individual needs. Once again, the approach recommended is to

TABLE 2   Service-Assurance Solutions

Communications Network Element	Possible Solutions for Failure Prevention or Mitigation
1. Floor distribution	Duplicate floor distribution cable Serve alternate trading positions from alternate cables
2. Riser cable	Duplicate riser cables with physical separation Serve alternate trading positions across two cables
3. Local equipment [private branch exchange (PBX) or other]	Fail-safe fire protection and security Two equipment rooms, physically separated with alternate positions served by equipment in different rooms
4. Local loop	Duplicate local-loop cables with physical diversity in different duct runs
5. Local switching center	Diversify local service arrangements to two switching centers
6. Connecting trunks	Request route diversification of connecting trunks to long-distance carrier Diversify arrangements to two or more long-distance carriers
7. Long-distance carrier	Have connecting trunks diversified to two separate long-distance-carrier locations, if possible Diversify arrangements to two or more long-distance carriers
8. Long-distance carrier connecting network	Request route diversity for connections in the long-distance network

Numbers 1 through 8 refer to network elements in Fig. 1.

preplan with communications carriers so that continuity of the most important services is assured. Once a disruption occurs, it will be too late to design an effective response that prevents significant losses.

Private Network Solutions

Private Network Considerations

Many firms use both public and private network alternatives to meet their communications needs. While the section above focused on a typical public network arrangement, most private networks are generically similar. As for the public network, a diagram comparable to that shown in Fig. 1 should be prepared for the specific layout and routing of the private network and similar questions posed and answered.

The design alternatives are similar for the two types of networks, but the private network designs do impose some additional questions, such as

- Do I have a single communications company handling my network design?
- How do I work with multiple providers to ensure the appropriate solution?
- Are my private network circuits truly diversified from my public-network circuits such that protection is a reality?
- Are the public and private arrangements coordinated with each other?

Information and Data Systems

The increasing levels of integration of communications and information within the business mainstream are discussed above, along with an examination of a typical serving arrangement for communications and a discussion of several service-assurance approaches. In many instances, the communications serving arrangement may also be an integral part of the data/information architecture of a firm. Therefore, a discussion of communications service assurance is equally applicable to data networks.

In other instances, the distribution of data throughout the firm is accomplished through a separate distribution network, such as coaxial cable or fiber-optic local-area networks. Again, the communications network analysis methodology can be employed, but the specific technologies employed may be different.

Large data-processing systems and their unique communications needs require a special level of preplanning for service assurance. Each application has its own unique requirements due to the highly concentrated data centers often employed.

Service assurance planning for large data systems requires examination of such questions as

- How important is the data-processing function to the short-term performance of the business?
- Are these functions essential to the revenue performance of the firm's most critical business units (e.g., the trading department in the example above may require on-line computer applications to function effectively)?
- Is this the only data-processing center serving this business center?
- Are the applications on line and serving other business segments or are they batch processes?
- What are the emergency power arrangements for this center?
- Is the emergency power arrangement on line continually (sometimes called uninterruptable power) or must it be switched over and restarted?
- How long would a restart take?
- Are the data lines duplicated and diversified?
- Does the data center require special environmental conditioning and fire protection?
- What arrangements do I have for transferring to a backup data center in the event of failure?
- When was the last time the backup arrangements were tested?

The preceding list is by no means exhaustive but can serve as a starting point for examining the service-assurance aspects of the data system in a business context. As with communications, well-trained professional help (either in-house staff or consultants) is important to a thorough review. With the continuing confluence of both data and communications within business functions, each must be examined within the context of the other.

Developing Contingency Plans

Recognizing the Need

The most important step in developing a communications and information system contingency plan is the recognition of the need to have one. Most information systems and networks are highly sophisticated, information intensive, and integrated into the business flow. Consequently, both need to be examined concurrently. In many cases, neither is useful without the other.

Joint Planning with Communications Suppliers

In this article, the need for preplanning for contingencies is emphasized. An effective plan must be developed jointly with the firm's service suppliers. The following are some preliminary checklist items.

1. Determine who is responsible for planning service assurance within the firm.

2. Prepare a list of companies that supply communications and data services to the firm.
3. Establish contacts with these companies.
4. Has a detailed plan been prepared for service assurance for various situations? Are the suppliers included in those plans?
5. Has the appropriate level of service protection for the business been designed?
6. Have the firm's needs changed since the plans were prepared?
7. When was the last time the emergency plans were tested?
8. What arrangements are available for off-hours emergency support?
9. How secure are the facilities of the firm's communications suppliers?
10. Is your firm prepared for a total loss of capability at its location despite its best efforts to preplan and design protection into the system (e.g., insurance for loss of business continuity)?

Preplanning with Communications Suppliers

Service-assurance planning is not an easy task and requires a carefully planned dialog with communications suppliers and vendors. Depending on the level of network vulnerability, the detail required will vary from situation to situation.

Most local telephone companies as well as long-distance carriers are continually improving their networks to provide a high level of performance. There are, however, many instances (such as sparsely populated or highly congested areas) where costs prohibit high levels of redundancy as a standard offering.

Communications suppliers have access to a wide and growing array of technologies to meet individual service-assurance needs. In some instances, these arrangements are standard tariff offerings, particularly in urban centers. In other cases, custom design arrangements can be offered. In either case, communications suppliers are becoming increasingly aware of the need to meet service-assurance requirements of customers.

Typical Service-Assurance Solutions within Communications Supplier Networks

Each communications supplier may use different equipment to achieve degrees of service assurance appropriate to the markets they serve. Regardless of the detailed methodologies, they fall into two generic categories: diversity (physical separation of paths) and redundancy (duplication of capabilities).

The *diversity protection method* provides network protection through automatic or manual rerouting of communications paths around a failure point. Figure 2 shows a typical high-bit-rate fiber-transport system. Two levels of protection are possible in the design shown. In the first, an electronics failure

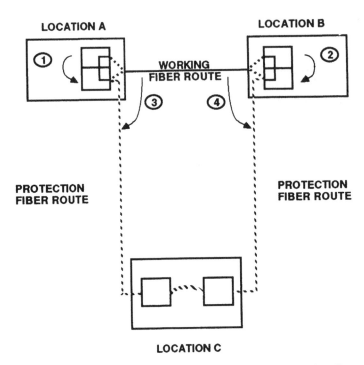

FIG. 2 Diversity protection for transport facilities. Refer to the text for discussion of the elements in this illustration.

at either Location A or B is protected by switching to standby electronics at both locations (depicted by Arrows 1 and 2). In the second level of protection, a fiber cut between Locations A and B is protected by switching to an alternate route through Location C (depicted by Arrows 3 and 4).

Figure 3 shows a typical public-switched-network connection with several possible levels of diversity. In the first level, the customer location is served diversely by two local telephone company wire centers, Locations A and B. The technologies required to make this possible are just beginning to emerge (e.g., Signaling System No. 7). In the second level of diversity, the local telephone company wire center at Location A has two diverse routes to reach Long-distance Carrier 1 at Location D. Traffic can be routed alternately through an access tandem at Location C. In the third level, the two separate long-distance carriers (Arrows 1 and 2) at Locations D and E can be reached by several diverse routing arrangements.

Many of the arrangements shown in Fig. 3 typically are available in major metropolitan areas, but should be confirmed through discussions with the local telephone company serving a specific location. If not readily available, custom designs can usually be offered to meet specific customer needs.

Full redundancy, which is essentially the duplication of all facilities, is generally not an economically attractive alternative except in the most demanding situations. Generally, combinations of diversity and partial redundancy can meet most needs economically.

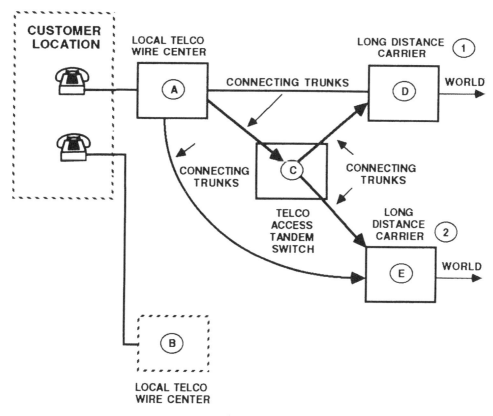

FIG. 3 Redundancy and diversity protection for switched-network facilities. Refer to the text for discussion of the elements in this illustration.

Emerging Technologies with a Potential Impact on Service Assurance

Several new and rapidly emerging technologies hold significant promise for increasing the range of options for service assurance. *High-capacity digital cross-connection systems* provide the capability to switch and reroute high-capacity digital transport facilities around failure points quickly. *Intelligent network platforms* will provide very flexible and dynamic capabilities to translate and reroute traffic in the public network. The implications of this technology to assist in rerouting customer traffic under unusual failure conditions could be significant.

The emerging capabilities of *Integrated Services Digital Networks* (ISDN) and *Signaling System No. 7* will offer substantial flexibility to reroute integrated voice and data traffic to alternate locations. While multisupplier equipment compatibility issues remain to be resolved, this technology provides the capability to handle business transactions at multiple locations while appearing to be at a single location from a network perspective. The opportunities to simplify voice and data arrangements will be significant, as well as the ability to provide service-assurance capability at multiple locations.

The discussion of emerging technologies is not intended to be exhaustive, but instead to provide a forward-looking view of potential planning techniques for service assurance.

Summary

This article introduces the reader to the basic concepts for a pragmatic approach to providing communications and information network service assurance. The subject is explored from several points of view, including those of the end-user, the public-network provider, and the private-network provider.

A common theme is presented regardless of the planning perspective adopted. According to this theme, preplanning for service assurance using preventive quality principles is an appropriate way to minimize the impact of network disruptions on business operations.

Extensive bibliographies for this important and growing area of expertise are not available; however, industry journals frequently discuss new service-assurance approaches. Readers desiring a more thorough discussion of current alternatives should consider contacting their local exchange telephone company to determine the options that are possible in their area.

CARL V. RIPA

Communications Frequency Standards

The Role of Frequency Standards in Communications

Stable oscillators control the frequencies of communications systems. As communications technology evolved, improvements in oscillator technology allowed improved spectrum utilization. (The evolution of spectrum utilization in commercial two-way communications systems is reviewed elsewhere in this encyclopedia [1].) In modern systems, stable oscillators are used not only for frequency control but also for timing.

In digital telecommunications (2,3) and spread-spectrum (4) systems, synchronization plays critically important roles. In digital telecommunications systems, stable oscillators in the clocks of the transmitters and receivers maintain synchronization, and thereby ensure that information transfer is performed with minimal buffer overflow or underflow events (i.e., with an acceptable level of slips). Slips cause such problems as missing lines in facsimile (FAX) transmission, clicks in voice transmission, loss of encryption key in secure voice transmission, and the need for data retransmission. In AT&T's network, for example, timing is distributed down a hierarchy of nodes (2). A timing source-receiver relationship is established between pairs of nodes containing clocks, which are of four types in four "stratum levels." The long-term accuracy requirements of the oscillators range from 1×10^{-11} at Stratum 1 to 3.2×10^{-5} at Stratum 4.

The phase noise of the oscillators can lead to erroneous detection of phase transitions (i.e., to bit errors) when phase-shift-keyed digital modulation is used. For example, assuming a normal distribution of phase deviations and a root mean squared (rms) phase deviation of $\pm 4.5°$, the probability of exceeding a $\pm 22.5°$ phase deviation is 6×10^{-7}. Spread-spectrum techniques are being used increasingly in both military and civilian communications systems. The advantages of spread-spectrum use can include (1) rejection of intentional and unintentional jamming, (2) low probability of interception, (3) selective addressing, (4) multiple access, and (5) more efficient use of the frequency spectrum. As an illustration of the importance of accurate clocks in such systems, consider one type of spread-spectrum modulation—frequency hopping. In a frequency-hopping system, accurate clocks must insure that the transmitter and receiver hop to the same frequency at the same time. The faster the hopping rate, the higher the jamming resistance, and the more accurate the clocks must be. For example, for a hopping rate of 1000 hops per second, the clocks must be synchronized to about 100 microseconds (μs). Such system parameters as the autonomy period (radio silence interval) and the time required for signal acquisition (net entry) are also closely dependent on clock accuracy.

Frequency Standards

Frequency standards can be divided into two major types: quartz crystal oscillators and atomic frequency standards. All atomic frequency standards contain a crystal oscillator.

Generalized Crystal Oscillator

Description

Figure 1 is a greatly simplified circuit diagram that shows the basic elements of a crystal oscillator (XO) (5,6). The amplifier of an XO consists of at least one active device, the necessary biasing networks, and may include other elements for band limiting, impedance matching, and gain control. The feedback network consists of the crystal resonator and may contain other elements, such as a variable capacitor for tuning.

The frequency of oscillation is determined by the requirement that the closed-loop phase shift $= 2n\pi$, where n is an integer, usually 0 or 1. When the oscillator is initially energized, the only signal in the circuit is noise. That component of noise, the frequency of which satisfies the phase condition for oscillation, is propagated around the loop with increasing amplitude. The rate of increase depends on the excess loop gain and on the bandwidth of the crystal network. The amplitude continues to increase until the amplifier gain is reduced, either by the nonlinearities of the active elements (in which case it is *self limiting*) or by an external level-control method.

At steady state, the closed-loop gain $= 1$. If a phase perturbation $\Delta\phi$ occurs, the frequency of oscillation must shift by a Δf in order to maintain the $2n\pi$ phase condition. It can be shown that for a series-resonance oscillator

$$\frac{\Delta f}{f} = -\frac{\Delta\phi}{2Q_L}$$

where Q_L is the loaded Q of the crystal in the network (5). ("Crystal" and "resonator" are often used interchangeably with "crystal unit," although "crystal unit" is the official name. See Refs. 6 and 7, and Chapter 3 of Ref. 8 for further information about crystal units.)

A quartz crystal unit is a quartz wafer to which electrodes have been applied,

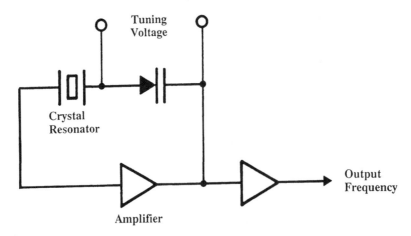

FIG. 1 Crystal oscillator (XO) (simplified circuit diagram).

and which is hermetically sealed in a holder structure. Although the design and fabrication of crystal units comprise a complex subject, the oscillator designer can treat the crystal unit as a circuit component and just deal with the crystal unit's equivalent circuit. Figure 2 shows a simplified equivalent circuit, together with the circuit symbol for a crystal unit. A load capacitor C_L is shown in series with the crystal. The mechanical resonance of the crystal is represented by the motional parameters L_1, C_1, and R_1. Because the crystal is a dielectric with electrodes, it also displays a static capacitance C_0. Figure 3 shows the reactance versus frequency characteristic of the crystal unit. When the load capacitor is connected in series with the crystal, the frequency of operation of the oscillator is increased by a $\Delta f'$, where $\Delta f'$ is given by

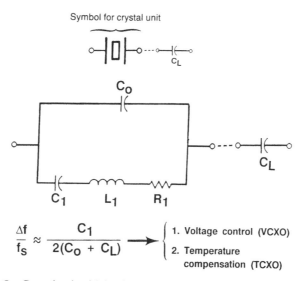

Symbol for crystal unit

$$\frac{\Delta f}{f_S} \approx \frac{C_1}{2(C_0 + C_L)} \longrightarrow \begin{cases} \text{1. Voltage control (VCXO)} \\ \text{2. Temperature} \\ \quad \text{compensation (TCXO)} \end{cases}$$

FIG. 2 Crystal unit with load capacitor (simplified equivalent circuit).

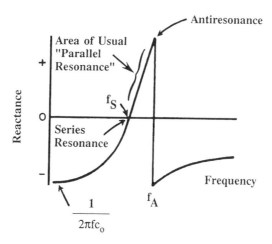

FIG. 3 Reactance versus frequency of a crystal unit.

$$\frac{\Delta f'}{f} \approx \frac{C_1}{2(C_0 + C_L)}$$

When an inductor is connected in series with the crystal, the frequency of operation is decreased.

Stability versus Tunability

In most crystal oscillator types, a variable-load capacitor is used to adjust the frequency of oscillation to the desired value. Such oscillators operate at the parallel resonance region shown in Fig. 3, where the reactance versus frequency slope (i.e., the "stiffness") is inversely proportional to C_1. For maximum frequency stability with respect to reactance (or phase) perturbations in the oscillator circuit, the reactance slope (or phase slope) must be maximum. This requires that the C_1 be minimum, and that the Q_L be maximum. The smaller the C_1, however, the more difficult it is to tune the oscillator (i.e., the smaller is $\Delta f'$ for a given change in C_L). The highest stability oscillators use crystal units that have a small C_1. Since C_1 decreases rapidly with overtone number, high-stability oscillators generally use third- or fifth-overtone crystal units. Overtones higher than five are rarely used because R_1 also increases rapidly with overtone number, and some tunability is desirable in order to allow setting the oscillator to the desired frequency.

Wide-tuning-range voltage-controlled crystal oscillators (VCXOs) use fundamental-mode crystal units of large C_1. Voltage control is used for the following purposes: to frequency or phaselock two oscillators; for frequency modulation; for compensation, as in a temperature-compensated crystal oscillator (TCXO, see below); and for calibration (i.e., for adjusting the frequency to compensate for aging). Whereas a high-stability, ovenized 10-megahertz (MHz) VCXO may have a frequency adjustment range of $\pm 5 \times 10^{-7}$ and an aging rate of 2×10^{-8} per year, a wide-tuning-range 10-MHz VCXO may have a tuning range of ± 50 parts per million (ppm) and an aging rate of 2 ppm per year.

In general, making an oscillator tunable over a wide frequency range degrades its stability because making an oscillator susceptible to intentional tuning also makes it susceptible to factors that result in unintentional tuning. For example, if an oven-controlled crystal oscillator (OCXO) is designed to have a stability of 1×10^{-12} for a particular averaging time and a tunability of 1×10^{-7}, then the crystal's load reactance must be stable to 1×10^{-5} for that averaging time. Achieving such load-reactance stability is difficult because the load reactance is affected by stray capacitances and inductances, by the stability of the varactor's capacitance versus voltage characteristic, and by the stability of the voltage on the varactor. Moreover, the 1×10^{-5} load-reactance stability must be maintained not only under benign conditions, but also under changing environmental conditions (temperature, vibration, radiation, etc.). Therefore, the wider the tuning range of an oscillator, the more difficult it is to maintain a high stability.

The Quartz Crystal Unit

A quartz crystal unit's high Q and high stiffness make it the primary frequency and frequency stability determining element in a crystal oscillator. The Q values of crystal units ($Q^{-1} = 2\pi f_s R_1 C_1$) are much higher than those attainable with other circuit elements. In general-purpose crystal units, Qs are generally in the range of 10^4 to 10^6. A high-stability 5-MHz crystal unit's Q is typically in the range of two to three million. The intrinsic Q, limited by internal losses in the crystal, has been determined experimentally to be inversely proportional to frequency (i.e., the Qf product is a constant). The maximum $Qf = 16$ million when f is in MHz.

Quartz (which is a single-crystal form of SiO_2) has been the material of choice for stable resonators since shortly after piezoelectric crystals were first used in oscillators — in 1918. Although many other materials have been explored, none has been found to be better than quartz. Quartz is the only material known that possesses the following combination of properties:

1. it is piezoelectric ("pressure electric" — *piezein* is the Greek word meaning "to press")
2. zero-temperature-coefficient resonators can be made from quartz plates when the plates are cut properly with respect to the crystallographic axes of quartz
3. of the zero-temperature-coefficient cuts, one, the SC cut (see below), is "stress compensated"
4. it has low intrinsic losses (i.e., quartz resonators can have very high Qs)
5. it is easy to process because it is hard but not brittle and, under normal conditions, it has low solubility in everything except the fluoride etchants
6. it is abundant in nature
7. it is easy to grow in large quantities, at low cost, and with relatively high purity and perfection

Of the man-grown single crystals, quartz, at more than 2000 tons per year (in 1990), is second only to silicon in quantity grown.

Quartz crystals are highly *anisotropic*, that is, the properties vary greatly with crystallographic direction. For example, when a quartz sphere is etched in hydrofluoric acid, the etching rate is more than 100 times faster along the fastest etching-rate direction, the Z direction, than along the slowest direction, the slow-X direction. The constants of quartz, such as the thermal-expansion coefficient and the temperature coefficients of the elastic constants, also vary with direction. That crystal units can have zero temperature coefficients of frequency is a consequence of the temperature coefficients of the elastic constants ranging from negative to positive values.

The locus of zero-temperature-coefficient cuts in quartz is shown in Fig. 4. The X, Y, and Z directions have been chosen to make the description of properties as simple as possible. The Z axis in Fig. 4 is an axis of threefold symmetry in quartz; in other words, the physical properties repeat every 120° as the crystal is rotated about the Z axis. The cuts usually have two-letter names, where the

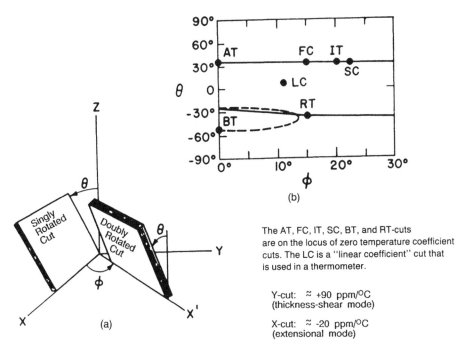

The AT, FC, IT, SC, BT, and RT-cuts are on the locus of zero temperature coefficient cuts. The LC is a "linear coefficient" cut that is used in a thermometer.

Y-cut: ≈ +90 ppm/°C
(thickness-shear mode)

X-cut: ≈ -20 ppm/°C
(extensional mode)

FIG. 4 Zero-temperature-coefficient cuts of quartz: *a*, definitions of the angles of cut θ and ϕ; *b*, locus of zero-temperature-coefficient cuts.

"T" in the name indicates a temperature-compensated cut; for instance, the AT cut was the first temperature-compensated cut discovered. (The SC cut is also temperature compensated; the name of this cut resulted from it being the only stress-compensated cut known.) The highest-stability crystal oscillators employ SC-cut or AT-cut crystal units. The FC, IT, BT, and RT cuts are other cuts along the zero-temperature-coefficient locus. These cuts were studied in the past for some special properties, but are rarely used today.

Because the properties of a quartz crystal unit depend strongly on the angles of cut of the crystal plate, in the manufacture of crystal units, the plates are cut from a quartz bar along precisely controlled directions with respect to the crystallographic axes. After shaping the quartz to required dimensions, metal electrodes are applied to the wafer, which is mounted in a holder structure. Figure 5 shows the two common holder structures used for resonators with frequencies greater than 1 MHz. (The 32-kilohertz [KHz] tuning fork resonators used in quartz watches are packaged typically in small tubular enclosures.)

Because quartz is piezoelectric, a voltage applied to the electrodes causes the quartz plate to deform slightly. The amount of deformation due to an alternating voltage depends on how close the frequency of the applied voltage is to a natural mechanical resonance of the crystal. To describe the behavior of a resonator, the differential equations for Newton's laws of motion for a continuum, and for Maxwell's equations, must be solved with the proper electrical and mechanical boundary conditions at the plate surfaces. Because quartz is

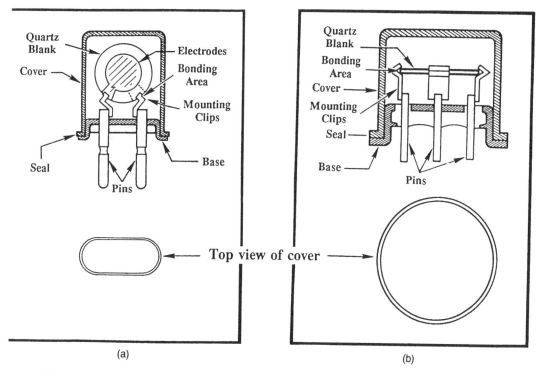

(a)   (b)

FIG. 5 Typical constructions of AT-cut and SC-cut crystal units: *a*, two-point mount package; *b*, three- and four-point mount package.

anisotropic and piezoelectric, with 10 independent linear constants and numerous higher-order constants, the equations are complex, and have never been solved in closed form for physically realizable three-dimensional resonators. Nearly all theoretical works have used approximations. The nonlinear elastic constants, although small, are the source of some of the important instabilities of crystal oscillators, such as the acceleration sensitivity, the thermal-transient effect, and the amplitude-frequency effect, each of which is discussed in this article.

As the drive level (the current through a crystal) increases, the crystal's amplitude of vibration also increases, and the effects due to the nonlinearities of quartz become more pronounced. Among the many properties that depend on the drive level are the resonance frequency, the motional resistance R_1, the phase noise, and frequency versus temperature anomalies (called *activity dips*), which are discussed in another section of this article. The drive-level dependence of the resonance frequency is called the *amplitude-frequency effect*. The frequency change with drive level is proportional to the square of the drive current. Because of the drive-level dependence of frequency, the highest-stability oscillators usually contain some form of automatic level control in order to minimize frequency changes due to oscillator circuitry changes. At high drive levels, the nonlinear effects also result in an increase in the resistance. Crystals also can exhibit anomalously high starting resistance when the crystal surfaces possess such imperfections as scratches and particulate contamination. Under such conditions, the resistance at low drive levels can be high enough for an oscillator to be unable to start when power is applied.

Bulk-acoustic-wave quartz resonators are available in the frequency range of about 1 KHz to 500 MHz. Surface-acoustic-wave (SAW) quartz resonators are available in the range of about 150 MHz to 1.5 gigahertz (GHz). To cover the wide range of frequencies, different cuts, vibrating in a variety of modes, are used. The bulk-wave modes of motion are shown in Fig. 6. The AT-cut and SC-cut crystals vibrate in a thickness-shear mode. Although the desired thickness-shear mode will exhibit the lowest resistance, the mode spectrum of even properly designed crystal units exhibits unwanted modes above the main mode. The unwanted modes, also called "spurious modes" or "spurs," are especially troublesome in filter crystals, in which "energy-trapping rules" are employed to maximize the suppression of unwanted modes. These rules specify certain electrode geometry to plate geometry relationships. In oscillator crystals, the unwanted modes may be suppressed sufficiently by providing a large enough plate diameter to electrode diameter ratio, or by contouring (i.e., generating a spherical curvature on one or both sides of the plate).

Above 1 MHz, the AT cut is commonly used. For high-precision applications, the SC cut has important advantages over the AT cut. The AT-cut and SC-cut crystals can be manufactured for fundamental-mode operation up to a frequency of about 200 MHz. (Higher than 1 GHz units have been produced on an experimental basis.) Above 100 MHz, overtone units that operate at a selected harmonic mode of vibration are generally used. Below 1 MHz, tuning forks, X-Y and NT bars (flexure mode), $+5°$ X cuts (extensional mode), or CT-cut and DT-cut units (face-shear mode) can be used. Tuning forks have become the dominant type of low-frequency units due to their small size and low cost. Hundreds of millions of quartz tuning forks are produced annually for quartz watches and other applications.

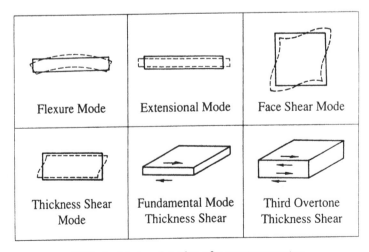

Flexure Mode	Extensional Mode	Face Shear Mode
Thickness Shear Mode	Fundamental Mode Thickness Shear	Third Overtone Thickness Shear

FIG. 6 Modes of motion of a quartz resonator.

Generalized Atomic Oscillator

Description

An *atomic frequency standard* is a device that determines frequency from some property of a simple atomic system. The term is not restricted to devices in which the frequency derives from neutral atoms, but is also applied to devices based on molecules and ions. The terms *atomic clock* and *atomic oscillator* are often used synonymously with the term atomic frequency standard. However, sometimes the term *clock* is used to mean a frequency standard connected to a counter—a device that provides time as well as frequency. There are several good reviews of atomic frequency standards (9-11).

Atomic frequency standards must be understood in terms of the concepts of quantum mechanics. The properties of simple atomic systems cannot assume arbitrary values. For example, the energies of the bound states of an atomic system are constrained to discrete values called *energy levels*. When an atomic system changes energy from an excited state to a state with lower energy, it emits a quantity of electromagnetic energy called a *photon*, the frequency of which is determined by the energy difference between the two states. If the energy of the upper state is E_2 and the energy of the lower state is E_1, the photon frequency is given by Planck's law

$$v = \frac{E_2 - E_1}{h}$$

where h is Planck's constant. The atomic standard produces an output signal the frequency of which is determined by this intrinsic atomic frequency, rather than some property of a bulk material.

The advantages of atomic oscillators all stem from this feature. Intrinsic atomic properties are more easily reproduced than collective properties, endowing atomic frequency standards with the property of accuracy. Atomic systems are easy to isolate from unwanted perturbations, which results in very small sensitivities to temperature, pressure, and other environmental conditions. The low level of interaction also results in extremely sharp resonance features, and reduces errors due to imperfections in the electronics. All atoms of an element are identical, and atomic properties are time invariant, which makes it possible to build very stable devices. Finally, it is surprisingly easy to measure atomic properties and build practical devices suitable for a wide variety of applications.

Atomic frequency standards are categorized in several ways; most often, they are referred to by the type of atom: hydrogen, rubidium, or cesium. Actually, these three categories are based on the same type of atomic interaction, but there are great practical differences in their implementation. Some atomic frequency standards, called oscillators, are active, in which case the output signal is derived from the radiation emitted by the atom. Others are passive; the atoms are then employed as a discriminator to measure and control the frequency of an electronic oscillator, such as a quartz oscillator. The third classification follows the method of interaction. In atomic beams, the atoms are observed "on the fly"; they pass through the interaction region and are not used again. In contrast, storage devices contain some type of cell that holds the atoms to be observed for a much longer time. In some cases, the atoms are recycled.

Atomic Spectroscopy

The energy levels of an atom are generally classified according to their physical origin. For example, the principal levels of an atom are associated with the radius of the "orbit" of an electron about the nucleus. These levels have the largest atomic energy separations. The principal energy levels are subdivided as a result of the quantization of the angular momentum of the atom. The angular momentum due to the motion of a particle, such as an electron, is called *orbital angular momentum*. Even when their motion is such that there is no orbital angular momentum, atomic particles may possess an intrinsic angular momentum or spin and a proportional intrinsic magnetic moment. This is another concept unique to quantum mechanics. The principal levels are first divided according to the shape of the electron "orbits." Still finer division occurs as a consequence of the particular orientation of the electron's spin and the spin of the nucleus.

The photons emitted when atoms change states among the principal energy levels are usually in the infrared and higher energy regions of the electromagnetic spectrum. The frequencies of these very energetic photons are too high for practical electronic devices. However, very narrow spectroscopic features associated with principal energy levels have been obtained in the laboratory and are useful for relative measurements. Atomic frequency standards are feasible because of the splitting of the ground state of the atom. Next lower, in terms of energy, is the fine structure of the atom, which results from the interaction of the spin of the electron with the magnetic field due to the motion of the electron through the nuclear electric field. This structure is thousands of times smaller than the separation of the principal energy levels. Laboratory atomic frequency standards based on fine structure in calcium and magnesium have been built, but the fundamental frequencies of the atomic transitions are higher than 600 GHz, which is very difficult to synthesize (12).

A finer energy splitting than the spin-orbit coupling is produced by the interaction of the electron and nuclear spins; this is called the *hyperfine structure*. The ground state of a hydrogen-like atom (e.g., H, Li, Na, K, Rb, Cs, and singly ionized Be) has a single unpaired electron in a symmetrical orbit. In this case, there is no orbital angular momentum and no fine structure. The energy splitting due to the intrinsic magnetic moments of the electron and the nucleus can be a million times smaller than the separation of the principal energy levels. The transition frequencies are quite convenient: 1.4 GHz for hydrogen, 6.8 GHz for rubidium, and 9.2 GHz for cesium. All commercial atomic frequency standards are based on the hyperfine spectroscopy of one of these three atoms.

Because the frequency of such a device is determined by the energy of interaction of a pair of magnetic moments, it is generally altered by any background magnetic field. The hyperfine states are also split into multiple energy levels, called *Zeeman sublevels*, depending upon the component of the total angular momentum in the direction of an applied magnetic field. Figure 7 shows the hyperfine structure of the hydrogen atom in an applied magnetic field. In the figure, the two states that have no angular momentum component along the direction of the applied field have quadratic dependence of energy on the mag-

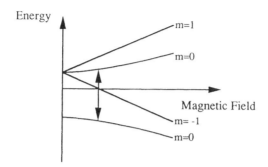

FIG. 7 Magnetic field dependence of the hyperfine doublet in the ground state of hydrogen (m is the quantum number corresponding to the component of the angular momentum along the direction of the applied magnetic field).

netic field for small field values. Thus, the frequency corresponding to a transition between these two levels has very small dependence on the magnetic field and is suitable for use in an atomic frequency standard. Atomic frequency standards utilize a small background magnetic field to allow the selection of the desired magnetic sublevels. The transitions between other levels have too high magnetic field dependence for frequency standard applications; variations in external magnetic fields, such as the earth's field, cannot be shielded sufficiently well to prevent them from disturbing the frequency of the device. The magnetic field dependent transitions are used to manufacture sensitive magnetometers rather than frequency standards.

The process of measuring the frequency of atomic transitions is called *atomic spectroscopy*. Atoms are prepared in one atomic state and then stimulated to change to the second state while some method of observing the transition is employed. The atoms must be stimulated to make the transition because the lifetime of the upper hyperfine state is very long; it is an appreciable fraction of the age of the universe in the case of hydrogen. By applying a field at the transition frequency, atoms can be stimulated to make a rapid transition between hyperfine levels. Typically, the atoms change state in a few milliseconds to one second, during which time they are observed.

Since the hyperfine energy separation is small compared to the thermal energy of atoms in a gas at room temperature, one expects to find a nearly equal number of atoms in the upper and lower hyperfine states, with slightly more atoms in the lower state. The number of atoms in the lower hyperfine level of room temperature hydrogen gas is 0.01% higher than the upper-state population. The number of atoms making transitions from the upper to the lower state and radiating photons is nearly equal to the number making transitions from the lower to the upper state and absorbing photons. Any effects of one process are nearly canceled by the other. In order to observe the process, it is necessary to produce a larger discrepancy in the populations of the two levels. After this state selection is performed, it is possible to observe the atomic transition in many ways. In cesium atomic frequency standards, the number of atoms making a transition is measured, whereas in hydrogen masers, the radiation emitted by the atoms is detected.

Practical Atomic Spectrometers

Conceptually, the simplest atomic frequency standard is the active oscillator, the functions of which are diagrammed in Fig. 8. Atoms in the upper hyperfine level of the ground state are stored in a microwave resonator. A fraction of the microwaves is coupled out and amplified for the output signal. In order for oscillation to take place, it is necessary for the population of upper-state atoms to be increased substantially compared to the lower-state atoms, a condition called *population inversion*. In the presence of microwave fields due to noise, the inversion causes the atoms to emit more power than they absorb. When the gain provided by the atoms exceeds the losses, the microwave fields build up until saturation limits the gain and a steady-state condition is reached. Each atom is exposed to a microwave field produced by previously emitting atoms and all the atoms are stimulated to emit approximately in phase with one another. The output signal is highly coherent; the dominant short-term noise is white phase noise. The line width of the atomic transition is determined by the observation time, which is limited by processes that destroy either the population inversion or the coherence of the atoms. The active hydrogen maser is an example of a practical atomic oscillator. It is available commercially and is the principal clock used for radio astronomy. In order to achieve high resolution, radio astronomers use long baseline interferometry, in which two or more radio telescopes that are widely separated make simultaneous observations, each one using a maser to provide the time base. The clocks must stay coherent to a small fraction of a radio-frequency (RF) cycle during the observation period. For X-band observations, this implies time errors less than 25 picoseconds (ps) for observation times from 15 minutes to several hours.

Passive atomic frequency standards are somewhat more complicated than active oscillators; they employ an atomic spectrometer the gain of which is insufficient to sustain oscillation. The spectrometer may even have a net loss. As diagrammed in Fig. 9, a microwave oscillator stimulates the atomic transi-

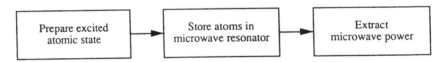

FIG. 8 Block diagram of a general active atomic oscillator.

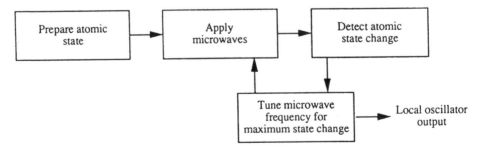

FIG. 9 Block diagram of a general passive atomic frequency standard.

tions and a control loop provides feedback to tune the oscillator to the frequency that maximizes the transition rate. The peak transition rate is detected by frequency modulating the microwave signal applied to the atoms. The population of one of the hyperfine states is synchronously detected using the modulating signal as the reference. A signal is obtained proportional to the difference between the microwave carrier frequency and the center of the atomic resonance. The passive approach is used when other constraints make it impossible to achieve self-oscillation. This happens, for example, when one tries to approach as closely as possible the ideal of an isolated atom at rest in free space, as in the case of the cesium beam frequency standard, and when it is desirable to minimize the size of the device, as in the rubidium frequency standard and the passive hydrogen maser.

Since all atomic frequency standards, both passive and active, derive their output signal from quartz oscillators, the performance of the atomic standards is significantly affected by the capabilities of those oscillators. In particular, the very-short-term frequency stability, the vibration sensitivity, the radiation sensitivity, and the sensitivity to thermal transients depend principally on the performance of the quartz local oscillator.

State Selection.   Magnetic state selection and optical pumping are both used to manipulate the hyperfine state populations. Magnetic state selection is based on the concepts that are discussed above. Examination of Fig. 7 shows that the energy of some of the Zeeman sublevels increases with an increasing magnetic field; the energy of the remainder decreases. When the atoms are passed through a region of strongly varying magnetic field, they experience a force proportional to the rate of change of the field with distance. The atoms that increase in energy with increasing magnetic field are deflected toward the region of strong field; the atoms that decrease in energy with increasing magnetic field are deflected in the opposite direction. The method of magnetic state selection is illustrated schematically in Fig. 10. Dipole, quadrupole, and hexapole optics all are used in atomic frequency standards to achieve the desired state population.

An alternative to magnetic state selection is optical pumping. This technique manipulates the populations in the hyperfine levels of the ground state by excit-

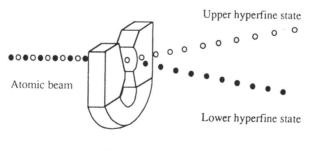

Dipole magnet

FIG. 10 Concept for the magnetic state selection of the upper and lower hyperfine states in an inhomogeneous magnetic field.

ing transitions to higher principal quantum states with infrared, or higher frequency, light. As shown by the diagram in Fig. 11, the atoms in one hyperfine level are excited optically to a higher state from which they decay spontaneously to both ground-state hyperfine levels. The population of the hyperfine state involved in the stimulated transition is rapidly depleted; the population of the second hyperfine level is enhanced. Optical pumping has both advantages and disadvantages compared to magnetic state selection. On the positive side, it can be accomplished in a more compact device and it can enhance the number of atoms in the desired state rather than just rejecting the atoms in the undesired state. On the negative side are increases in complexity and some additional performance-degrading mechanisms.

Detection of the Atomic Resonance.   Passive atomic frequency standards use a variety of methods to detect the atomic resonance. That is, the standards detect the atomic transition probability as a function of the frequency of the applied radiation. The earliest approach was the direct detection of the state populations. The Stern-Gerlach experiment, which first demonstrated the quantization of angular momentum, used this method. A small background magnetic field established a reference direction. According to classical physics, the angular momentum of the silver atoms along this direction could take on any value between plus and minus the total angular momentum. In the Stern-Gerlach experiment, a beam of silver atoms emitted from an oven passed through a dipole magnet and fell on a glass plate. The silver atoms were expected to deposit along a continuous line on the plate parallel to the magnetic field. However, only two spots were observed, demonstrating for the first time that the angular momentum along the field could have only two values. The amount of silver deposited indicated the populations of the two states.

An extension of this technique is still used. In a cesium-beam frequency standard, upper-state atoms initially may be selected using magnetic state selection. The application of microwaves near 9.2 GHz causes transitions to the lower state. A second magnet selects the lower-state atoms, which subsequently

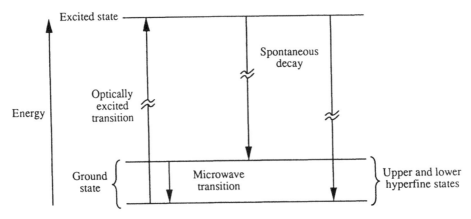

FIG. 11 Concept for optical pumping using a three-level system.

fall on a hot ribbon in the detection region. The cesium atom's ionization potential is significantly less than the work function for platinum or tungsten. Thus the cesium atoms give up an electron and are reevaporated from the wire as positive ions. Laboratory frequency standards directly collect the ions to indicate the number of atoms arriving at the detector. Commercial cesium-beam frequency standards filter the cesium ions using a mass spectrometer to exclude impurities emitted from the ionizer. The cesium-ion current is converted to an electron current by a specially treated surface on the first dynode of an electron multiplier. Since the conversion of cesium atoms to ions by the ionizer ribbon is nearly 100%, this method of detection provides nearly ideal performance.

The effect of the microwave frequency on the transition probability can also be detected by its effect on the absorption of the light used to pump the atoms optically, such as in a Rb gas cell atomic frequency standard. With no microwaves applied, all the atoms are removed from one of the hyperfine states and the stored atoms become nearly transparent at the optical frequency. The transmission is maximum (see Fig. 12-*a*). Tuning the microwaves near the transition frequency causes atoms to be transferred back to the depleted level; the absorption then increases. Maximum absorption occurs at the peak of the transition probability (see Fig. 12-*b*). The atomic resonance may also be detected by its effect on the microwaves themselves. The state-selected atoms behave very much like a filter. If upper-state atoms predominate, the microwave signal will experience amplification and phase shift. The microwave frequency that maximizes the transition probability may be determined by comparison of the signal passed through the atomic discriminator with a reference. This is the method used in a passive-hydrogen-maser frequency standard.

Systematic Limitations of Atomic Frequency Standards

There are fundamental and practical limitations on the ability of a device to reproduce an atomic frequency (13,14). Quantum mechanics and thermal noise limit the quality of the measurements by producing stochastic frequency variations. Imperfections in the electronics, stray electric and magnetic fields, and interactions with other atoms cause the frequency standard to exhibit aging and temperature and pressure sensitivities.

Atomic frequency standards must be designed to prevent performance limitation by the Doppler effect. The first-order Doppler effect is the change in the observed frequency when a source is in motion relative to an observer. It is often observed as the change in pitch of a train whistle when the train overtakes and passes a stationary observer. Since many atoms with different velocities must be observed in an atomic frequency standard, the first-order Doppler effect would produce broadening of atomic resonance as well as average frequency shifts. The observed Doppler frequency shift is equal to the ratio of the atomic velocity to the speed of light, approximately 1×10^{-6} for room temperature atoms, and would limit the maximum atomic Q to 1×10^6.

All atomic frequency standards use some form of first-order Doppler-shift cancellation in order to achieve Qs in the range of 10^7 to 10^9. One method that is always used in microwave standards is to excite the atomic resonance with a

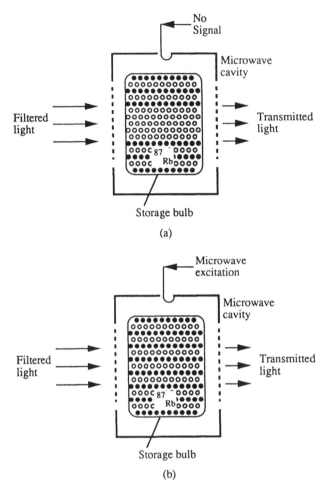

FIG. 12 Concept of microwave resonance detection using the optical density of the atomic medium: *a*, maximum transmission of light without microwave application; *b*, maximum absorption of light with microwave application.

standing wave produced by a microwave cavity. The atoms interact with microwaves of approximately equal intensity traveling in opposite directions and the Doppler shifts for these two interactions nearly cancel. Further cancellation is obtained when storage cells are used. The atoms in these devices change direction thousands of times while interacting with the microwave field. However, atom storage within a microwave resonator is insufficient to guarantee narrow, unshifted atomic lines. A more detailed analysis shows that the atoms must be confined to a region smaller than one-half wavelength. This restriction, known as the *Dicke regime*, prevents frequency modulation due to the Doppler shifts from absorbing all the power from the carrier (15). Atomic beam frequency standards are always designed so that the residual power flow in the microwave cavity is orthogonal to the atomic beam direction. This construction takes the place of the velocity averaging of storage-cell devices. Atomic beam standards must also operate in the Dicke regime.

There are numerous less fundamental perturbations to the observed atomic frequency. The atomic resonance is observed through the interaction of the atoms with the microwave field of a resonant circuit (cavity). If the cavity resonance frequency is not equal to the atomic frequency, the stronger fields at the cavity resonance frequency increase the probability of transitions at a frequency different from the true atomic resonance frequency, and the measured value is not equal to the true atomic frequency. In an active oscillator, the "pulling" is equal to the error in the cavity resonance reduced by the ratio of the cavity Q to the atomic Q. In a passive atomic frequency standard, the pulling may be reduced by the square of the Q ratio.

In optically pumped frequency standards, the atomic energy levels are affected by the light used for the optical pumping. This problem is a specific instance of a phenomenon called the *Stark effect*, which is the variation of atomic energy levels with applied electric fields. Other electric fields can usually be reduced to negligible levels, but the fields required for optical pumping are very intense and the resulting "light shifts" are significant.

Frequency shifts are also caused by interactions of the atoms with other matter. These interactions change the hyperfine interaction energy. *Pressure shifts* result from collisions with gas molecules. *Spin-exchange shifts* are caused by collisions of the subject atoms with each other. *Wall shifts* result from collisions of the atoms with the coating on a storage vessel. Additional changes in the measured maximum of the transition probability curve result from the presence of nearby atomic transitions. The tails of a neighboring transition result in a background slope, which distorts the measured shape of the desired transition.

Imperfections in the electronics cause additional errors in the measurement of the atomic transition frequency. Some of the problems are spurious microwave signals, voltage offsets and drifts, insufficient gain, insufficient dynamic range and frequency modulation of the VCXO by a time varying acceleration. In the end, performance is always limited by the ability of the electronics to find the center of the atomic line, and a major thrust of research and development is toward achieving higher atomic line Q.

Oscillator Categories

Quartz Oscillator Types

A crystal unit's resonance frequency varies with temperature. Typical frequency versus temperature (f vs. T) characteristics for crystals used in stable oscillators are shown in Fig. 13. The three categories of crystal oscillators, based on the method of dealing with the crystal unit's f vs. T characteristic, are XOs, TCXOs, and OCXOs (see Fig. 14). A simple XO does not contain means for reducing the crystal's f vs. T variation. A typical XO's f vs. T stability may be ± 25 ppm for a temperature range of $-55°C$ to $+85°C$.

In a TCXO, the output signal from a temperature sensor (a *thermistor*) is used to generate a correction voltage that is applied to a voltage-variable reactance (a *varactor*) in the crystal network (16). The reactance variations produce

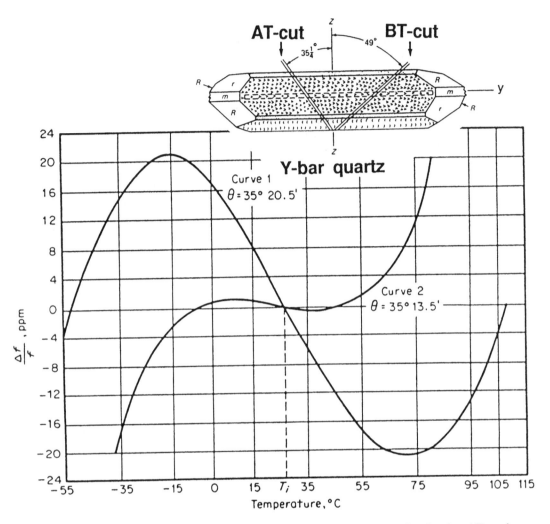

FIG. 13 Frequency versus temperature characteristics of AT-cut crystals, showing AT- and BT-cut plates in *Y*-bar quartz.

frequency changes that are equal and opposite to the frequency changes resulting from temperature changes; in other words, the reactance variations compensate for the crystal's *f* vs. *T* variations. Analog TCXOs can provide about a 20-fold improvement over the crystal's *f* vs. *T* variation. A typical TCXO may have an *f* vs. *T* stability of ±1 ppm for a temperature range of −55°C to +85°C.

In an OCXO, the crystal unit and other temperature-sensitive components of the oscillator circuit are maintained at a constant temperature in an oven (16). The crystal is manufactured to have an *f* vs. *T* characteristic that has zero slope at the oven temperature. To permit the maintenance of a stable oven temperature throughout the OCXO's temperature range (without an internal

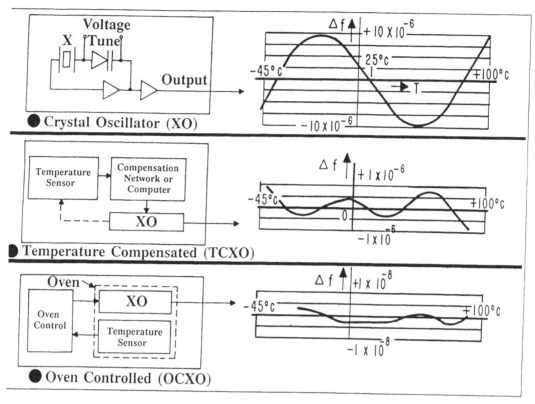

FIG. 14 Crystal oscillator categories based on the crystal unit's frequency versus temperature characteristic.

cooling means), the oven temperature is selected to be above the maximum operating temperature of the OCXO. OCXOs can provide more than a 1000-fold improvement over the crystal's f vs. T variation. A typical OCXO may have an f vs. T stability of $\pm 5 \times 10^{-9}$ for a temperature range of $-55°C$ to $+85°C$. OCXOs require more power, are larger, and cost more than TCXOs.

A special case of a compensated oscillator is the microcomputer-compensated crystal oscillator (MCXO) (17). The MCXO overcomes the two major factors that limit the stabilities achievable with TCXOs: thermometry and the stability of the crystal unit. Instead of a thermometer that is external to the crystal unit, such as a thermistor, the MCXO uses a much more accurate "self-temperature sensing" method. Two modes of the crystal are excited simultaneously in a dual-mode oscillator. The two modes are combined such that the resulting beat frequency is a monotonic (and nearly linear) function of temperature. The crystal thereby senses its own temperature. To reduce the f vs. T variations, the MCXO uses digital compensation techniques: pulse deletion in one implementation, and direct digital synthesis of a compensating frequency in another. The frequency of the crystal is not "pulled," which allows the use of high-stability (small C_1) SC-cut crystal units. A typical MCXO may have an f vs. T stability of $\pm 2 \times 10^{-8}$ for a temperature range of $-55°C$ to $+85°C$.

Atomic Frequency Standard Types

Atomic Standards in Production.   Three types of atomic standards are available to meet frequency stability requirements that exceed the performance capabilities of quartz crystal oscillators: rubidium, cesium, and hydrogen frequency standards. Rubidium oscillators offer better stability in the range from 100 seconds to a few hours, an improvement by a factor of ten in long-term aging, superior reproducibility, and lower sensitivity to the environment. Rubidium frequency standards are often employed in tactical applications or any time the clock must free run for a few hours to a day. Cesium standards offer further improvements. Compared to rubidium oscillators, they perform better for times longer than a few hours, generally do not suffer from frequency aging, are more reproducible, and have lower sensitivity to the environment. Cesium standards are used in strategic applications and wherever a clock is required to keep time autonomously for much longer than one day. The third type of standard, the active hydrogen maser, is used in applications requiring extreme coherence for periods of time from minutes to hours. Very long baseline interferometry and spacecraft tracking are the two principal applications.

Figure 15 is a schematic of a cesium beam tube. Spectroscopy is performed on cesium atoms in free flight through the microwave cavity in order to minimize environmental influences. A magnetic-state selector anterior to the cavity rejects atoms in the unwanted hyperfine level. A magnetic-state selector posterior to the cavity passes atoms having the state that was rejected in the first region. Thus, atoms must make a hyperfine transition in order to reach the hot-wire ionizer. A feature of this beam tube not discussed above is the U-shaped microwave cavity (*Ramsey cavity*). In this design, atoms are exposed to the microwave field, pass through a microwave-field-free region, and are then exposed to a second microwave field in phase with the first. As a result, the atomic line is an interference pattern that maximizes the resolution of the spectrometer. Even more important, the design of the Ramsey cavity reduces the sensitivity to magnetic-field inhomogeneities. Without it, the magnetic-shielding requirement would be impractical.

Figure 16 is a typical electronic schematic for the cesium frequency standard.

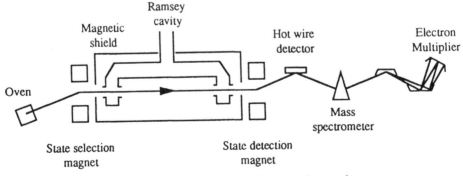

FIG. 15 Schematic drawing of a cesium beam tube.

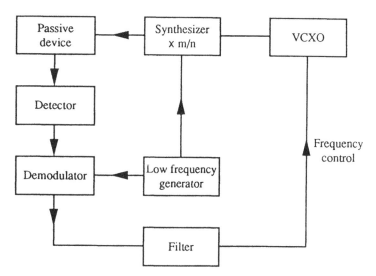

FIG. 16 Electronic schematic showing the frequency lock of a voltage-controlled quartz oscillator (VCXO) to an atomic (or other) resonance.

A 5-MHz VCXO provides the output signal. After audio frequency modulation, the signal from the VCXO is used to synthesize the microwave signal at 9192.631 MHz, which is then used to excite the hyperfine transition. The detected atomic beam current from the beam tube is applied to a phase-sensitive detector using the audio modulation as reference. After integration and additional loop filtering, the phase detector output is used to control the frequency of the 5-MHz VCXO. Figure 17 illustrates how phase-sensitive detection generates a control signal that passes through zero at the resonance frequency.

Cesium technology has been optimized for long-term timekeeping and frequency reproducibility at the expense of size, weight, power, and warm-up time. A healthy device shows no frequency aging, and the reproducibility is on the order of 2×10^{-12} for the life of the cesium beam tube. The dominant long-term noise is random-walk frequency, which limits the timekeeping to approximately 0.1 μs per month.

The atomic spectrometer of a typical rubidium standard is shown schematically in Fig. 18. Commercial rubidium standards, based on the hyperfine transition in the ground state of ^{87}Rb, employ optical pumping using the light emitted from a ^{87}Rb discharge lamp. The emission from this lamp contains light with wavelengths corresponding to the transitions from the two hyperfine ground states to the excited state. The light from the ^{87}Rb passes through a region containing ^{85}Rb. What makes rubidium standards practical is the fact that ^{85}Rb strongly absorbs the wavelength that would excite the upper hyperfine state and transmits most of the lower-state transition light. This filtered light is selectively absorbed by the lower-hyperfine-state atoms in the resonance cell. These atoms are excited to a third, optical, state, which decays to both the ground-state hyperfine levels. Atoms that return to the lower hyperfine level are excited again by the filtered light until they are converted into upper-hyperfine-level atoms. Thus, the filtered light optically pumps ^{87}Rb atoms contained in the microwave

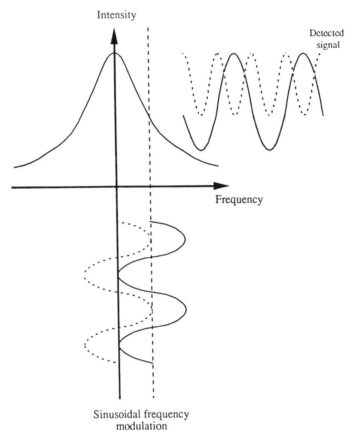

FIG. 17 Detection of a passive resonance using sinusoidal frequency modulation (FM). When the microwave center frequency equals the atomic frequency (dashed curves), the detected signal is at the second harmonic of the FM. When these two frequencies are unequal, there is a first harmonic detected signal. The phase of the detected first harmonic changes 180 degrees when the microwave frequency changes from one side of the atomic resonance to the other.

resonance cell into the upper hyperfine state. In order to make a compact device, the resonance cell completely fills a small microwave cavity. The filter can be implemented separately, but most often is integrated with the cell by adding ^{85}Rb. When microwaves are applied to the resonator cell, atoms are connected back from the upper to the lower hyperfine states. This increases the absorption of the filtered light that passes through the resonance cell. The microwave transition frequency is observed by detecting the filtered light that passes through the resonance cell. The Dicke criterion is satisfied within the resonance cell by using a buffer gas to localize the ^{87}Rb atoms during interaction with the microwave field. The buffer gas prevents the rubidium atoms from experiencing wall collisions that would introduce major perturbations. However, collisions of the Rb atoms with the buffer gas cause pressure shifts of the frequency. The resulting pressure and temperature coefficients can be set to

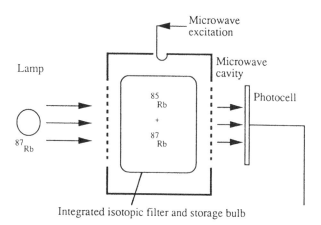

FIG. 18 Schematic drawing of the atomic spectrometer for a passive rubidium frequency standard.

zero at the nominal operating temperature by using a mixture of gases, such as nitrogen and argon, that have opposite-sign pressure shifts.

The rubidium-standard control electronics are very similar to that of the cesium standard. The transition frequency is 6.834 GHz and the line width is several hundred Hz. The detection of the resonance is performed by a photodetector that measures the transmission of pumping light through the cell. There is approximately a 1% decrease in transmission when the microwave frequency is at the peak of the transition probability.

Rubidium technology has been optimized for small size and fast warm-up. The smallest is 5 centimeters (cm) by 7.5 cm by 10 cm, has a warm-up time of less than 4 minutes from 25°C for 1×10^{-9} reproducibility, and the long-term frequency aging is 2×10^{-10} per year. Such a device does not have state-of-the-art frequency stability or phase noise. These performance parameters can be improved at the expense of increased size. In fact, a full-sized rubidium standard, the same size as a cesium standard, performs almost as well as a cesium standard.

Figure 19 is a schematic representation of an active hydrogen maser. Unlike cesium and rubidium, hydrogen occurs only in molecular form at room temperature. The necessary atomic hydrogen is produced using a radio-frequency discharge. Typically, the upper-state atoms are focused into a storage bulb using a hexapole state selector magnet. Eventually, spent atoms are pumped away. The bulb surface is Teflon coated to ensure nearly elastic collisions of the hydrogen atoms on the wall. This surface is so good that more than 10,000 bounces occur before the coherence of the hydrogen atoms is destroyed. Consequently, the observation time can exceed one second. As a result, the atomic line Q is greater than 1×10^9, the highest of the commercial atomic frequency standards. The density of hydrogen in the storage bulb is limited by hydrogen–hydrogen collisions that alter the hyperfine energy and thus produce frequency shifts. A high Q microwave cavity is required to achieve self-oscillation. The large size of most hydrogen masers (typical active hydrogen masers are ten times the volume of

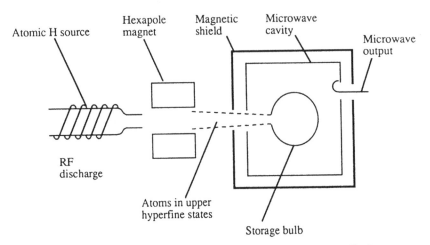

FIG. 19 Schematic drawing of an active hydrogen maser; RF = radio frequency.

cesium frequency standards) results from the 1.4 GHz microwave cavity and the surrounding layers of magnetic shielding and temperature-stabilizing ovens.

The control electronics for the active hydrogen maser is shown schematically in Fig. 20. The 1.42-GHz output from the atoms is amplified and mixed with a signal obtained by multiplying the output frequency to the microwave region. A final downconversion of this difference frequency using a reference synthesized from the output frequency completes the conversion of the atomic frequency to baseband. Feedback to the VCXO phaselocks it to the microwave signal originating from the atoms. In contrast to passive atomic frequency standards, which convert thermal noise at their preamplifier inputs to white frequency noise, the active maser converts this thermal noise to white phase noise.

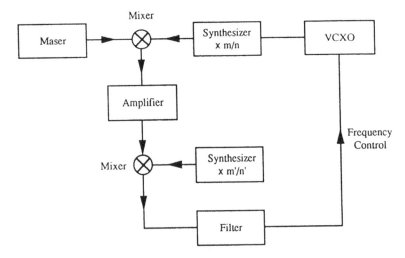

FIG. 20 Electronic schematic showing the phase lock of a voltage-controlled quartz oscillator (VCXO) to the atomic radiation emitted by an active maser.

As a result, the short-term stability of active masers varies inversely with the measurement interval, as opposed to the stability of passive standards, which varies inversely as the square root of the measurement interval.

Active masers have the best short-term frequency stability of all the atomic frequency standards. Typical performance is $5 \times 10^{-13}/\tau$, reaching a floor of better than 1×10^{-15}. However, the long-term performance is not as good as that of cesium frequency standards. Most masers have had substantial frequency aging, which has been attributed to cavity pulling. The frequency stability at one day is typically 1×10^{-14}.

Experimental Atomic Frequency Standards. At the time of this writing (1990), one of the most active areas of research in atomic frequency standards is atomic beam tube technology. The efforts are directed toward two different and contradictory objectives, but both employ optical-pumping technology. One goal is to develop a frequency standard with the low environmental sensitivity typical of the cesium standard, but having the size of a miniature rubidium standard (18). Optical pumping is used to replace the magnetic-state selection and detection. Elimination of the high magnetic fields makes it possible to design a more compact device. The linear beam of an optically pumped device makes more efficient use of the atoms emitted from the oven than the usual magnetic optics, and helps achieve good short-term stability. The second goal is to improve the accuracy, reproducibility, and long-term stability of the atomic beam frequency standard (19). The performance limitations of the existing technology are due to residual first-order Doppler shift, second-order Doppler shift, and pulling by neighboring lines. The linear beam of the optically pumped device reduces the complicated coupling between the velocity and the position of the atoms in the beam, making it possible to measure the Doppler effect better. Optical pumping can also be used to transfer almost all the atoms to the desired hyperfine state, thus reducing the size of and pulling by neighboring magnetic-field-sensitive transitions. This technique can be combined with advances made in alignment of the beam tube to reduce the pulling problem to negligible levels.

The most promising developments for improving long-term stability of atomic frequency standards are in the field of ion storage. The goal of this work is to achieve the low interactions of the beam-type frequency standard with the long observation times possible in a storage device. Singly ionized atoms, such as beryllium and mercury, have been stored using ion-trap technology (20,21). The ions experience no collisions due to the storage mechanism, but are confined by direct-current (DC) or RF electromagnetic fields. Mercury ion frequency standards have been built using the RF ion-trapping technique. The technology appears practical for commercial application in the near term. Other ion-frequency-standard research uses Penning traps that employ DC trapping fields. This approach is compatible with laser cooling of the ions—a technique that extracts energy and momentum from the ions by scattering a laser beam. Cooling the ions below 1°K reduces the magnitude of the Doppler effects and improves the accuracy and reproducibility of the frequency standard. The large magnets and complicated lasers used for laser cooling will relegate this approach to the laboratory for some time to come.

Oscillator Instabilities

Accuracy, Stability, and Precision

Oscillators exhibit a variety of instabilities. These include aging, noise, and frequency changes with temperature, acceleration, ionizing radiation, power supply voltage, and the like. The terms *accuracy, stability,* and *precision* are often used in describing an oscillator's quality with respect to its instabilities. Figure 21 illustrates the meanings of these terms for a marksman and for a frequency source. (For the marksman, each bullet hole's distance from the center of the target is the "measurement.") *Accuracy* is the extent to which a given measurement, or the average of a set of measurements for one sample, agrees with the definition of the quantity being measured. It is the degree of "correctness" of a quantity. Atomic frequency standards have varying degrees of accuracy. The International System of Units (SI) units for time and frequency (seconds and Hz, respectively) are obtained in laboratories using very accurate frequency standards called *primary* standards. A primary standard operates at a frequency calculable in terms of the SI definition of the second: "the duration of 9,192,631,770 periods of the radiation corresponding to the transition between the two hyperfine levels of the ground state of the cesium atom 133" (22). *Reproducibility* is the ability of a single frequency standard to produce the same frequency, without adjustment, each time it is put into operation. From the user's point of view, once a frequency standard is calibrated, reproducibility

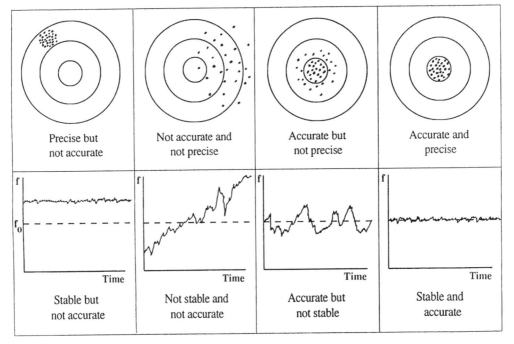

FIG. 21 Accuracy, stability, and precision examples for a marksman: *top,* and for a frequency source, *bottom.*

confers the same advantages as accuracy. *Stability* describes the amount something changes as a function of such parameters as time, temperature, shock, and the like. *Precision* is the extent to which a given set of measurements of one sample agrees with the mean of the set. (A related meaning of the term is used as a descriptor of the quality of an instrument, as in a "precision instrument." In that context, the meaning is usually defined as accurate and precise, although a precision instrument can also be inaccurate and precise, in which case the instrument needs to be calibrated.)

Aging

"Aging" and "drift" have occasionally been used interchangeably in the literature. However, in 1990, recognizing the "need for common terminology for the unambiguous specification and description of frequency and time standard systems," the International Radio Consultative Committee (CCIR) adopted a glossary of terms and definitions (23). According to this glossary, *aging* is "the systematic change in frequency with time due to internal changes in the oscillator," and *drift* is "the systematic change in frequency with time of an oscillator." Drift is due to aging plus changes in the environment and other factors external to the oscillator. Aging is what one denotes in a specification document and what one measures during oscillator evaluation. Drift is what one observes in an application. For example, the drift of an oscillator in a spacecraft might be due to (the algebraic sum of) aging and frequency changes due to radiation, temperature changes in the spacecraft, and power supply changes.

Quartz Oscillator Aging

Aging can be positive or negative (24). Occasionally, a reversal in aging direction is observed. Typical (computer-simulated) aging behaviors are illustrated in Fig. 22, where A(t) is a logarithmic function and B(t) is the same function but with different coefficients. The curve showing the reversal is the sum of the other two curves. A reversal indicates the presence of at least two aging mechanisms. The aging rate of an oscillator is highest when it is first turned on. At a constant temperature, aging usually has an approximately logarithmic dependence on time. When the temperature of a crystal unit is changed (e.g., when an OCXO is turned off and turned on at a later time), a new aging cycle starts. (See the section concerning hysteresis and retrace below for additional discussion of this on–off cycling.)

The primary causes of crystal oscillator aging are mass transfer to or from the resonator's surfaces due to adsorption or desorption of contamination, stress relief in the mounting structure of the crystal, and, possibly, changes in the quartz material. Because the frequency of a thickness-shear crystal unit, such as an AT cut or an SC cut, is inversely proportional to the thickness of the crystal plate, and because a typical 5-MHz plate is on the order of 1 million atomic layers thick, the adsorption or desorption of contamination equivalent to the mass of one atomic layer of quartz changes the frequency by about 1 ppm. Therefore, in order to achieve low aging, crystal units must be fabricated

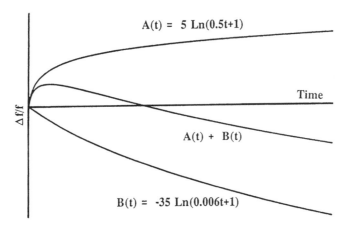

FIG. 22 Computer-simulated typical aging behaviors, where A(t) and B(t) are logarithmic functions with different coefficients.

and hermetically sealed in an ultraclean, ultra-high-vacuum environment. In 1990, the aging rates of typical commercially available XOs range from 5 ppm to 10 ppm per year for an inexpensive XO, from 0.5 ppm to 2 ppm per year for a TCXO, and from 0.05 ppm to 0.1 ppm per year for an OCXO. The highest precision OCXOs can age less than 0.01 ppm per year.

Atomic Standard Aging

One mechanism for frequency aging potentially affects all passive atomic frequency standards, that is, failure to control the aging of the internal VCXO. This source of aging is a consideration for passive frequency standards that use analog control electronics, in which case the integrators in the control loop have finite DC gain. The open-loop VCXO aging divided by the DC gain must be less than the noise-induced frequency changes; otherwise, frequency aging will be detectable. This requirement must not only be met in the new device, but also after the gain has been degraded by operation and exposure to the environment. For example, the gain of the electron multiplier used in a cesium standard degrades as a result of the unavoidable exposure to cesium, and the gain of a rubidium standard degrades due to the exposure of the photodetector to ionizing radiation. In the past, cesium standards exhibited this form of frequency aging. Today, "healthy" cesium standards exhibit no measurable frequency aging. Hydrogen masers exhibit frequency aging in the range from 1 to 10×10^{-12} per year. This aging is often attributed to pulling by the microwave cavity. In the past, it has been controlled through the use of very stable microwave cavities manufactured from ceramic materials. More recently, electronic control has been used to minimize the cavity frequency error in miniature active and passive hydrogen masers (25,26). Rubidium frequency standards exhibit the most aging, on the order of 1×10^{-10} per year. The largest effects result from changes in the light shift. The electrical discharge and elevated temperatures in the lamp cause diffusion of Rb into the glass and change its electrical properties. Over time, large changes in the lamp spectrum and intensity can take place, with associated changes in the output frequency. Additional aging can be produced by the diffusion of atmospheric helium into the resonance cell. The increasing helium partial pressure changes the pressure shift. As a result, all manufacturers use glass chosen for its relatively low helium-permeation rate.

Noise in Frequency Standards

Effects of Noise

Sometimes the suitability of oscillators for an application is limited by deterministic phenomena. In other instances, stochastic (random) processes establish the performance limitations. Except for vibration, the short-term instabilities almost always result from noise. Long-term performance of quartz and rubidium standards is limited by the temperature sensitivity and the aging, but the long-term performance of cesium and some hydrogen standards is limited by random processes.

Noise can have numerous adverse effects on system performance. Among these effects are the following: (1) it limits the ability to determine the current state and the predictability of precision oscillators (e.g., the noise of an oscillator produces time prediction errors of $\sim \tau \sigma_y(\tau)$ for prediction intervals of τ); (2) it limits synchronization and syntonization accuracies; (3) it can limit a receiver's useful dynamic range, channel spacing, and selectivity; (4) it can cause bit errors in digital communications systems; (5) it can cause loss of lock, and limit acquisition and reacquisition capability in phase-locked-loop systems; and (6) it can limit radar performance, especially Doppler radar.

Characterization of Noise

It is important to have appropriate statistical measures to characterize the random component of oscillator instability. The voltage from a precision oscillator can be written as

$$V(t) = [V_0 + \epsilon(t)] \sin[\omega_0 t + \phi(t)],$$

where
V_0 = nominal amplitude
$\epsilon(t)$ = amplitude fluctuations
ω_0 = nominal angular frequency
$\phi(t)$ = phase fluctuations.

The amplitude noise is usually small compared to the phase noise, and is not discussed further, but the validity of this assumption should always be verified. Two methods are described that are appropriate for characterizing the stochastic variations in the phase or frequency of an oscillator (27). If deterministic effects are present, other methods must be used to minimize or to remove them. Normally, short-term frequency stability measurements are made with the oscillator in a benign environment to reduce extraneous frequency perturbations. Building vibrations and temperature, pressure, and humidity fluctuations can affect the stability measurements. The effect of aging must usually be removed during stability analysis. Simple methods exist for estimating frequency aging in the presence of random-walk frequency noise, which is usually present.

One method of describing the phase noise of an oscillator is the spectrum of the noise. The spectral density $S_x(f')$ of a quantity x is its mean-square value per Hz Fourier frequency f'. The Fourier frequency is a fictitious frequency used in Fourier analysis of the signal. Zero Fourier frequency corresponds to the carrier, and a negative Fourier frequency refers to the region below the carrier. The integral of the spectral density over all Fourier frequencies from minus infinity to infinity is the mean-square value of the quantity. The spectral density of phase noise $S_\phi(f')$ is very important because it is directly related to the performance of oscillators in RF signal-processing applications. The single-sideband (SSB) noise power per Hz to total signal power ratio is often specified for oscillators instead of the phase spectral density. This ratio has been designated $\mathcal{L}(f)$. Recently, the definition of $\mathcal{L}(f)$ has been changed to one-half $S_\phi(f')$ (28). When defined this way, $\mathcal{L}(f)$ is equal to the SSB noise-to-signal ratio only as long as the integrated phase noise from f' to infinity is small compared to one rad^2. The phase spectral density depends on carrier frequency. When the signal from an oscillator is multiplied by n in a noiseless multiplier, the frequency modulation (FM) sidebands increase in power by n^2, as does the spectral density of the phase. Consequently, it is important to state the oscillator frequency together with any measurement of the spectral density of the phase. Sometimes oscillator noise is described in terms of frequency rather than phase. The instantaneous angular frequency $\omega(t)$ of an oscillator is the derivative of the total phase,

$$\omega(t) = \omega_0 + \frac{d\phi(t)}{dt}$$

The phase noise in precision oscillators is usually described in terms of a dimensionless instantaneous frequency fluctuation y, which is defined in terms of the angular frequency,

$$y \equiv \frac{\omega(t) - \omega_0}{\omega_0} = \frac{1}{\omega_0} \frac{d\phi(t)}{dt}$$

Since the frequency is the derivative of the phase, the spectral density of y is simply related to the phase spectral density.

$$S_y(f') = \left(\frac{2\pi f'}{\omega_0}\right)^2 S_\phi(f')$$

A relatively simple model is adequate to describe the noise in the precision oscillators discussed in this article. The model spectral density consists of a finite sum of terms proportional to the Fourier frequency raised to a positive or negative integer power,

$$S_y(f') = \sum_{\alpha} h_\alpha (f')^\alpha$$

Six processes are sufficient:

White phase modulation (PM)	$\alpha = +2$
Flicker PM	$\alpha = +1$
White FM	$\alpha = +0$
Flicker FM	$\alpha = -1$
Random-walk FM	$\alpha = -2$
Random-walk frequency aging	$\alpha = -4$

The even terms all can be produced by the integration or differentiation of white noise. Thus, models containing just these terms have simple Kalman filter or autoregressive integrated moving average (ARIMA) model equivalents. Under these circumstances, it is easy to design optimum filters for using the clocks in systems and control loops. If either of the flicker-noise processes appears to be present over a substantial Fourier-frequency interval, many lag-lead filters may be necessary to approximate the noise by filtering a white-noise source, making optimal system design very difficult.

The origin of white PM and white FM in a clock can usually be understood in terms of the physical and electronic design. Additive thermal noise in the buffer amplifiers is the usual source of white PM. Thermal noise in the control loop of a passive atomic frequency standard is a common source of white FM. The physical sources of the other noise processes are usually not known. The random-walk FM noise may only be apparent, that is, it may be the result of frequency fluctuations caused by changes in temperature, temperature gradient, pressure, humidity, magnetic field, or mechanical stress. If the spectrum were truly random walk in nature, the frequency would be unbounded, whereas the frequency of a precision oscillator usually does not change by more than a few line widths. Nevertheless, this term should be included in the oscillator model if the mean frequency of the oscillator appears to change. Similarly, random-walk frequency aging can be included in the model of an oscillator that has a variable frequency aging.

Although the spectrum is a powerful method of characterizing an oscillator, it is not very directly related to its timekeeping ability. The Allan variance $\sigma_y^2(\tau)$ is an alternate measure of frequency stability that is quite useful for this purpose. It is defined by

$$\sigma_y(\tau) = \frac{1}{2} E \left\{ \left[\frac{\phi(t + 2\tau) - 2\phi(t + \tau) + \phi(t)}{\omega_0 \tau} \right]^2 \right\}$$

where $E\{\ \}$ refers to the expectation value over the ensemble of possible observations. The traditional variance describes the deviation of a set of observations from the mean, but is not defined for noise processes more divergent than white frequency noise. On the other hand, the Allan variance describes the variation of the frequency from one measurement interval to the next with no dead time between intervals. As a result, it converges for both flicker frequency and random-walk frequency noise, but is not defined for random-walk frequency-aging noise. The root-mean-squared (rms) time error of a clock after a free-running interval τ is approximately $\tau\sigma_y(\tau)$. Thus, estimating the Allan variance provides a nonparametric method of characterizing the timekeeping ability of a clock.

The terms jitter and wander are used in characterizing timing instabilities in digital communications. *Jitter* refers to the high-frequency timing variations of a digital signal; *wander* refers to the low-frequency variations. The dividing line between the two is often taken to be 10 Hz. Wander and jitter can be characterized by the appropriate measurement of the rms time error of the clock. A 10-Hz low-pass filter should be used to remove the effects of jitter, if necessary. For very high Fourier frequencies or short integration times, it may be necessary to calculate the jitter from the spectrum rather than measure it directly. For example, the mean-squared phase jitter during a one-tenth second interval is the integral of the spectral density of phase over the Fourier frequency range from 10 Hz to infinity.

Noise in Crystal Oscillators

Although the causes of noise in crystal oscillators are not fully understood, several causes of short-term instabilities have been identified. Temperature fluctuations can cause short-term instabilities via thermal-transient effects (see the section below concerning dynamic f vs. T effects), and via activity dips at the oven set point in OCXOs. Other causes include Johnson noise in the crystal unit, random vibration (see the section below concerning accelerator effects in crystal oscillators), noise in the oscillator circuitry (both the active and passive components can be significant noise sources), and fluctuations at various interfaces on the resonator (e.g., in the number of molecules adsorbed on the resonator's surface).

In a properly designed oscillator, the resonator is the primary noise source close to the carrier and the oscillator circuitry is the primary source far from the carrier. The noise close to the carrier (i.e., within the bandwidth of the resonator) has a strong inverse relationship with resonator Q, such that $S_\phi(f')$ $\propto 1/Q^4$. In the time domain, $\sigma_y(\tau) \approx 2 \times 10^{-7}/Q$ at the noise floor. In the frequency domain, the noise floor is limited by Johnson noise, the noise power of which is $kT = -174$ dBm/Hz at 290°K. A higher signal (i.e., a higher resonator drive current) will improve the noise floor but not the close-in noise. In fact, for reasons that are not understood fully, above a certain point higher drive levels generally degrade the close-in noise. For example, the maximum "safe" drive level is about 100 μA for a 5-MHz fifth overtone AT-cut resonator with $Q \approx 2.5$ million. The safe drive current can be substantially higher for high-frequency SC-cut resonators. For example, $\mathcal{L}(f) = -180$ dBc/Hz has been achieved with 100-MHz fifth overtone SC-cut resonators at drive currents

≈ 10 milliamperes (mA). However, such a noise capability is useful only in a vibration-free environment, for if there is vibration at the offset frequencies of interest, the vibration-induced noise will dominate the quiescent noise of the oscillator (see the section below concerning acceleration effects in crystal oscillators).

When low noise is required in the microwave (or higher) frequency range, SAW oscillators and dielectric resonator oscillators (DROs) are sometimes used. When compared with multiplied-up (bulk-acoustic-wave) quartz oscillators, these oscillators can provide lower noise far from the carrier at the expense of poorer noise close to the carrier, poorer aging, and poorer temperature stability. SAW oscillators and DROs can provide lower noise far from the carrier because these devices can be operated at higher drive levels, thereby providing higher signal-to-noise ratios, and because the devices operate at higher frequencies, thereby minimizing the "20 log N" losses due to frequency multiplication by N. $\mathcal{L}(f) = -180$ dBc/Hz noise floors have been achieved with state-of-the-art SAW oscillators. Of course, as is the case for high-frequency bulk-wave oscillators, such noise floors are realizable only in environments that are free of vibrations at the offset frequencies of interest.

Noise in Atomic Frequency Standards

Most devices function in a way that can be described by classical physics. In this regime, the quality of measurements is limited only by thermal noise and the capability of the measurement equipment. Atomic frequency standards, on the other hand, depend on the quantum nature of the atom. This means that when a microwave field is applied to an atom, there is a probability that the atom will make a transition. Since all atomic frequency standards function by observing the effect of microwaves on the atoms, there is a variability in the outcome of each "observation" that limits the precision of the frequency standard. This noise enters into active and passive frequency standards somewhat differently, but results in white frequency noise in both cases. In an active oscillator, the noise outside the bandwidth of the atomic resonance integrates to produce phase shocks. Since the transition rate of the atoms is independent of the phase of the stimulating field, there is no correction of these phase errors, and the total phase displays a random walk, which is the same as white frequency noise. In a passive atomic frequency standard, the correction signal is proportional to the number of atoms and the frequency error of the microwave source. The fluctuations in detected signal due to transition probability have a white spectrum and are proportional to the square root of the number of atoms. Thus, even when there is no error in the microwave frequency, the control impresses frequency corrections on the VCXO. This results in white frequency noise inversely proportional to the square root of the number of atoms.

In addition, all atomic frequency standards have white phase noise due to additive thermal noise in the buffer amplifiers used to provide the output signal. They also display more divergent noises. All atomic frequency standards suffer from random-walk frequency noise. However, the source of this noise cannot be explained in the same physical manner as the white phase and frequency modulations. The random-walk frequency noise most likely results from the sum of many different frequency perturbations due to variations in temperature, pressure, humidity, and magnetic field. It is likely that a more complete understanding of each atomic frequency standard would show that the frequency variations described as random-walk frequency noise are actually deterministic. However, this is not possible today, and the stochastic description of the long-term frequency variations makes it possible to design appropriate filters for using the frequency standards optimally. Some atomic frequency standards suffer from (variable) frequency aging due to even closer coupling of the atoms to the environment.

Frequency versus Temperature Stability

Frequency versus Temperature Stability of Quartz Oscillators

Static Frequency versus Temperature Stability.   As an illustration of the effects that temperature can have on frequency stability, Fig. 23 shows the effects of temperature on the accuracy of a typical quartz wristwatch. Near the wrist temperature, the watch can be very accurate because the frequency of the crystal (i.e., the clock rate) changes very little with temperature. However, when the watch is cooled to $-55°C$ or heated to $+100°C$, it loses about 20 seconds per day because the typical temperature coefficient of the frequency of the tuning-fork crystals used in quartz watches is -0.035 ppm/$°C^2$.

The static f vs. T characteristics of crystal units are determined primarily by the angles of cut of the crystal plates with respect to the crystallographic axes of quartz. "Static" means that the rate of change of temperature is slow enough for the effects of temperature gradients (explained below) to be negligible. As Fig. 13 illustrates for the AT cut, a small change in the angle of cut (seven minutes in the illustration) can significantly change the f vs. T characteristics.

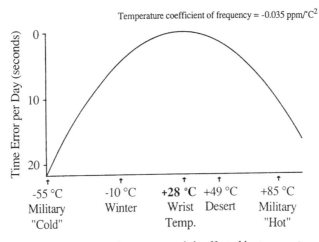

FIG. 23 Wristwatch accuracy as it is affected by temperature.

The points of zero temperature coefficient, the "turnover points," can be varied over a wide range by varying the angles of cut. The f vs. T characteristics of SC-cut crystals are similar to the curves shown in Fig. 13, with the inflection temperature T_i shifted to about 95°C.

Other factors that can affect the f vs. T characteristics of crystal units include the overtone; the geometry of the crystal plate; the size, shape, thickness, density, and stresses of the electrodes; the drive level; impurities and strains in the quartz material; stresses in the mounting structure; interfering modes; ionizing radiation; the rate of change of temperature (i.e., thermal gradients); and thermal history. The last two factors are important for understanding the behaviors of OCXOs and TCXOs, and are, therefore, discussed separately.

Interfering modes can cause "activity dips" (see Fig. 24). Activity dips can be strongly influenced by the crystal's drive level and load reactance. Near the activity-dip temperature, anomalies appear in both the f vs. T and resistance (R) vs. T characteristics. The activity-dip temperature is a function of C_L because the interfering mode usually has a large temperature coefficient and a C_1 that is different from that of the desired mode. Activity dips are troublesome in TCXOs, and also in OCXOs when the dip occurs at the oven temperature. When the resistance increases at the activity dip, and the oscillator's gain margin is insufficient, the oscillation stops. The incidence of activity dips in SC-cut crystals is far lower than in AT-cut crystals.

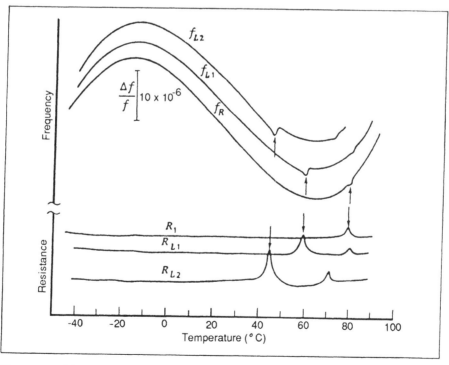

FIG. 24 Activity dips in the frequency versus temperature and resistance versus temperature characteristics, with and without C_L.

An important factor that affects the f vs. T characteristics of crystal oscillators is the load capacitor. When a capacitor is connected in series with the crystal, the f vs. T characteristic of the combination is rotated slightly from that of the crystal alone. The temperature coefficient of the load capacitor can greatly magnify the rotation.

The f vs. T of crystals can be described by a polynomial function. A cubic function is usually sufficient to describe the f vs. T of AT-cut and SC-cut crystals to an accuracy of ± 1 ppm. In the MCXO, in order to fit the f vs. T data to $\pm 1 \times 10^{-8}$, a polynomial of at least seventh order is usually necessary.

Dynamic Frequency versus Temperature Effects. Changing the temperature surrounding a crystal unit produces thermal gradients when, for example, heat flows to or from the active area of the resonator plate through the mounting clips. The static f vs. T characteristic is modified by the thermal-transient effect. When an OCXO is turned on, there can be a significant thermal-transient effect. Figure 25 shows what happens to the frequency output of two OCXOs, each containing an oven that reaches the equilibrium temperature in six minutes. One oven contains an AT-cut, the other, an SC-cut crystal. Thermal gradients in the AT cut produce a large frequency undershoot that anneals out several minutes after the oven reaches equilibrium. The SC-cut crystal, being insensitive to such thermal transients, reaches the equilibrium frequency as soon as the oven stabilizes.

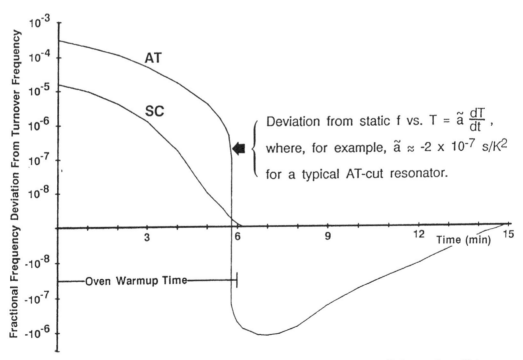

FIG. 25 Warm-up characteristics of AT-cut and SC-cut oven-controlled crystal oscillators (OCXOs).

In addition to extending the warmup time of OCXOs when crystals other than SC cuts are used, the thermal-transient effect makes it much more difficult to adjust the temperature of OCXO ovens to the desired turnover points, and the OCXO frequencies are much more sensitive to oven-temperature fluctuations.

The testing and compensation accuracies of TCXOs are also adversely affected by the thermal-transient effect. As the temperature is changed, the thermal-transient effect distorts the static f vs. T characteristic, which leads to apparent hysteresis. The faster the temperature is changed, the larger is the contribution of the thermal-transient effect to the f vs. T performance.

Thermal Hysteresis and Retrace. The f vs. T characteristics of crystal oscillators do not repeat exactly upon temperature cycling (29). The lack of repeatability in TCXOs, "thermal hysteresis," is illustrated in Fig. 26. The lack of repeatability in OCXOs, "retrace," is illustrated in Fig. 27. *Hysteresis* is defined as the difference between the up-cycle and the down-cycle f vs. T characteristics, and is quantified by the value of the difference at the temperature where the difference is maximum (30). Hysteresis is determined during a complete quasi-static temperature cycle between specified temperature limits. *Retrace* is defined as the nonrepeatability of the f vs. T characteristic at a fixed temperature (which is usually the oven temperature of an OCXO) upon on–off cycling an oscillator under specified conditions.

Hysteresis is the major factor limiting the stability achievable with TCXOs. It is especially so in the MCXO because, in principle, the digital compensation

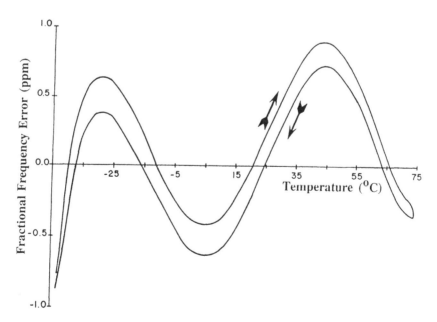

FIG. 26 Temperature-compensated crystal oscillator (TCXO) thermal hysteresis, showing that the f vs. T characteristic upon increasing temperature differs from the characteristic upon decreasing temperature.

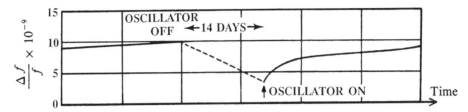

FIG. 27 Oven-controlled crystal oscillator (OCXO) retrace, showing that upon restarting the oscillator after a 14-day off-period, the frequency was about 7×10^{-9} lower than what it was just before turn-off, and that the aging rate had increased significantly upon the restart.

method used in the MCXO would be capable of compensating for the f vs. T variations to arbitrary accuracy if the f vs. T characteristics could be described by a single-valued function. Retrace limits the accuracies achievable with OCXOs in applications where the OCXO is on–off cycled. Typical values of hysteresis in TCXOs range from 1 ppm to 0.1 ppm when the temperature-cycling ranges are 0°C to 60°C, and −55°C to +85°C. Typical OCXO retrace specifications, after a 24-hour off period at about 25°C, range from 2×10^{-8} to 1×10^{-9}. Low-temperature storage during the off period, and extending the off period, usually make the retrace worse.

The causes of hysteresis and retrace are not well understood; the experimental evidence to date is inconclusive. The mechanisms that cause these effects include strain changes, changes in the quartz, oscillator circuitry changes, contamination redistribution in the crystal enclosure, and apparent hysteresis or retrace due to thermal gradients.

Frequency versus Temperature Stability of Atomic Oscillators

The frequency sensitivity of atomic oscillators to temperature changes results from actual atomic frequency changes and electronic effects. The temperature sensitivity of a typical miniature rubidium standard is 3×10^{-10} over the temperature range −55°C to +65°C. The pressure shift is probably the largest contributor to this temperature sensitivity. The small size of the frequency standard limits the quality of the rubidium-cell thermal control, and the temperature variations produce pressure changes of the buffer gas. In addition to the pressure shift, all the effects discussed below in the context of the other atomic frequency standards contribute to the temperature sensitivity of rubidium frequency standards.

The cesium standard has a much lower temperature coefficient than the rubidium standard. A typical value is 3×10^{-12} over the range 0°C to 50°C. The cesium atoms are better insulated from the environment than is the case with rubidium. Cavity pulling, residual Doppler shifts, and microwave-power changes all contribute to the residual variations of the atomic frequency. In addition, imperfections in the electronics result in an offset of the operating frequency from the atomic frequency that changes with temperature as the active-component offset voltages and currents change.

The temperature sensitivity of hydrogen masers is comparable to that of cesium standards. In active masers, the most significant contributor is cavity pulling. Stable ceramic cavities and four or five levels of thermal control are used to limit the thermal sensitivity. Active control of the frequency of the cavity is sometimes used, which reduces the requirement on thermal control. Electronic effects are not a problem for active masers, but do affect the frequency of passive masers.

Retrace in cesium standards varies between 5×10^{-13} and 3×10^{-12}. In miniature rubidium standards, the typical retrace is 5×10^{-11}.

Warm-up

When power is applied to a frequency standard, it takes a finite amount of time before the equilibrium frequency stability is reached. Figure 25, discussed above, illustrates the warmup of two OCXOs. The warm-up time of an oscillator is a function of the thermal properties of the oscillator, the input power, and the oscillator's temperature prior to turn-on. Typical warm-up time specifications of OCXOs and rubidium frequency standards (e.g., from a 0°C start) range from 3 minutes to 10 minutes. The warm-up times of cesium standards range from 30 minutes to 60 minutes. Hydrogen masers warm up in 4 hours to 1 day. Even TCXOs, MCXOs, and simple XOs take a few seconds to "warm up," although these are not ovenized. The reasons for the finite warm-up periods are that it takes a finite amount of time for the signal to build up in any high-Q circuit, and the few tens of milliwatts of power that are dissipated in these oscillators do change the thermal conditions within the oscillators.

Acceleration Effects

Acceleration Effects in Crystal Oscillators

Acceleration changes a crystal oscillator's frequency (31). The acceleration can be a steady-state acceleration, vibration, shock, attitude change (2-g tipover), or acoustic noise. The amount of frequency change depends on the magnitude and direction of the acceleration $\vec{A}$, and on the acceleration sensitivity of the oscillator $\vec{\Gamma}$. The acceleration sensitivity $\vec{\Gamma}$ is a vector quantity. The frequency change can be expressed as

$$\frac{\Delta f}{f} = \vec{\Gamma} \cdot \vec{A}.$$

Typical values of $|\vec{\Gamma}|$ are in the range of 10^{-9}/g to 10^{-10}/g. For example, when $\vec{\Gamma} = 2 \times 10^{-9}$/g and is normal to the earth's surface and the oscillator is turned upside down (a change of 2 g), the frequency changes by 4×10^{-9}. When this oscillator is vibrated in the up-and-down direction, the time-dependent acceleration modulates the oscillator's output frequency at the vibration frequency, with an amplitude of 2×10^{-9}/g. In the frequency domain, the modulation results in vibration-induced sidebands that appear at plus and minus integer multiples of the vibration frequency from the carrier frequency. Figure 28 shows the output of a spectrum analyzer for a 10-MHz, $1.4 \times 10^{-9}/g$ oscillator that was vibrated at 100 Hz and 10 g. When the frequency is multiplied, as it is in many applications, the sideband levels increase by 20 dB for each $10 \times$ multiplication. The increased sideband power is extracted from the carrier. Under certain conditions of multiplication, the carrier disappears (i.e., all the energy is then in the sidebands).

The effect of random vibration is to raise the phase-noise level of the oscillator. The degradation of phase noise can be substantial when the oscillator is on a vibrating platform, such as on an aircraft. Figure 29 shows a typical aircraft random-vibration specification (power spectral density [PSD] vs. vibration frequency) and the resulting vibration-induced phase-noise degradation. Acoustic noise is another source of acceleration that can affect the frequency of oscillators.

During shock, a crystal oscillator's frequency changes suddenly due to the sudden acceleration. The frequency change follows the expression above for acceleration-induced frequency change except, if during the shock some elastic limits in the crystal's support structure or electrodes are exceeded (as is almost always the case during typical shock tests), the shock will produce a permanent frequency change. If the shock level is sufficiently high, the crystal will break; however, in applications where high shock levels are a possibility, crystal units with chemically polished crystal plates can be used. Such crystals can survive

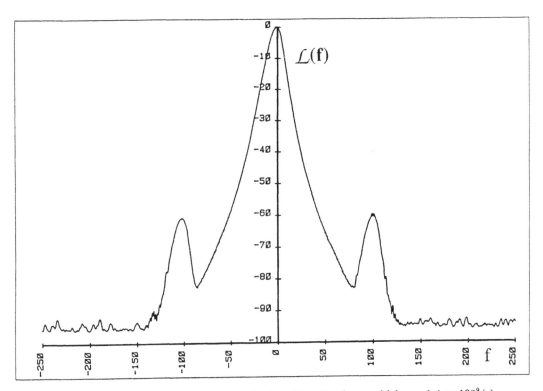

FIG. 28 Vibration-induced sidebands (g level $= 10g$, vibration sensitivity $= 1.4 \times 10^{-9}/g$).

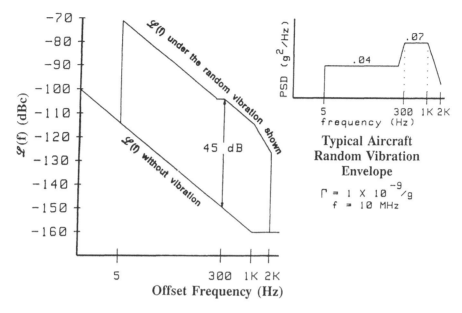

FIG. 29 Random-vibration-induced phase-noise degradation.

shocks in excess of 20,000 g (and have been fired successfully from howitzers [32]).

Acceleration Effects in Atomic Frequency Standards

Static acceleration is not much of a problem in atomic frequency standards (31). For example, miniature rubidium frequency standards exhibit acceleration sensitivity of approximately 2×10^{-12}/g. However, transient acceleration due to vibration can have significant effects. Control-loop time constants of a fraction of a second are typical, hence the sensitivity of the frequency to vibration throughout the audio range is primarily determined by the quartz crystal unit. Not only is the quality of the spectrum degraded, but the time errors due to random vibration accumulate. The time (or phase) errors do not average out because the white frequency noise is integrated to produce random walk of the phase and the time error increases proportionally to the square root of the elapsed time. An additional vibration-induced problem occurs because the microwave signal used to interrogate the atoms is produced by frequency multiplication, a process that transfers power from the carrier to the modulation sidebands. Under intense vibration, the carrier may essentially disappear, and loss of lock occurs.

Magnetic-Field Effects

Magnetic-Field Effects in Quartz Oscillators

Quartz is diamagnetic; however, magnetic fields can affect magnetic materials in the crystal unit's mounting structure, electrodes, and enclosure. Time-varying electric fields will induce eddy currents in the metallic parts. Magnetic fields can also affect components such as inductors in the oscillator circuitry. When a crystal oscillator is designed to minimize the effects of magnetic fields, the sensitivity can be much less than 10^{-10} per oersted. Magnetic-field sensitivities on the order of 10^{-12} per oersted have been measured in crystal units designed specifically for low magnetic-field sensitivity (33).

Magnetic-Field Effects in Atomic Frequency Standards

Atomic frequency standards are particularly sensitive to magnetic fields because the hyperfine frequency is proportional to a magnetic-interaction energy. All the commercial atomic frequency standards use atomic transitions with small quadratic magnetic-field dependence at the operating magnetic field of the unit—typically a few thousandths of an oersted. Several layers of magnetic shielding are used to provide a stable magnetic environment. Typical sensitivity for a miniature rubidium standard is 2×10^{-11} per oersted change in the external field. Typical cesium standards have magnetic-field sensitivity of 2×10^{-12} per oersted. Both rubidium and cesium are available in high-performance versions that have 10 times smaller magnetic-field sensitivity obtained by using additional shielding. The magnetic-field sensitivity of hydrogen masers is comparable to that of cesium standards.

Radiation Effects

Radiation Effects in Quartz Oscillators

Ionizing radiation changes a crystal oscillator's frequency primarily because of changes the radiation produces in the crystal unit (34). Under certain conditions, the radiation will also produce an increase in the crystal unit's equivalent series resistance. The resistance increase can be large enough to stop the oscillation when the oscillator is not radiation hardened.

Figure 30 shows a crystal oscillator's idealized frequency response to a pulse of ionizing radiation. The response consists of two parts. Initially, there is a transient frequency change that is due primarily to the thermal-transient effect caused by the sudden deposition of energy into the crystal unit. This effect is a manifestation of the dynamic f vs. T effect discussed above. The transient effect is absent in SC-cut resonators made of high-purity quartz.

In the second part of the response, after steady state is reached there is a permanent frequency offset that is a function of the radiation dose and the nature of the crystal unit. The frequency change versus dose is nonlinear, the change per rad being much larger at low doses than at large doses. At doses above 1 kilorad (Krad) (SiO_2), the rate of frequency change with dose is quartz impurity-defect dependent. For example, at a 1 megarad (Mrad) dose, the frequency change can be as large as 10 ppm when the crystal unit is made from natural quartz; it is typically 1 to a few ppm when the crystal is made from cultured quartz, and it can be as small as 0.02 ppm when the crystal is made from swept cultured quartz.

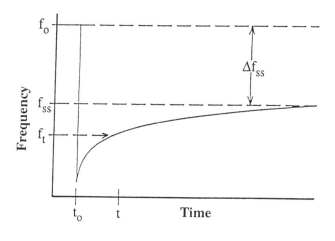

$$\Delta f'_{ss}/rad^* \simeq \begin{cases} 10^{-11} \text{ for natural quartz (R increase can stop oscillation)} \\ 10^{-12} \text{ for cultured quartz} \\ 10^{-13} \text{ for swept cultured quartz} \end{cases}$$

* for 1 Mrad dose

FIG. 30 Crystal oscillator's response to a pulse of ionizing radiation: f_0 = original, preirradiation frequency; Δf_{ss} = steady-state frequency offset (0.2 hours to 24 hours after exposure); f_t = instantaneous frequency at any time t.

The impurity defect of major concern in quartz is the substitutional Al^{3+} defect with its associated interstitial charge compensator, which can be an H^+, Li^+, or Na^+ ion, or a hole. This defect substitutes for an Si^{4+} in the quartz lattice. Radiation can result in a change in the position of weakly bound compensators, which changes the elastic constants of quartz and thereby leads to a frequency change. The movement of ions also results in a decrease in the crystal's Q (i.e., in an increase in the crystal's equivalent series resistance). If the oscillator's gain margin is insufficient, the increased resistance stops the oscillation.

Sweeping is a high-temperature, electric-field-driven, solid-state purification process in which the weakly bound alkali compensators are diffused out of the lattice and replaced by more tightly bound H^+ ions and holes. In the typical sweeping process, conductive electrodes are applied to the Z surfaces of a quartz bar, the bar is heated to about 500°C, and a voltage is applied so as to produce an electric field of about 1 kilovolt per centimeter (kV/cm) along the Z direction. After the current through the bar decays (due to the diffusion of impurities) to some constant value, the bar is cooled slowly, the voltage is removed, and then the electrodes are removed. Crystal units made from swept quartz exhibit neither the radiation-induced Q degradation nor the large radiation-induced frequency shifts. Swept quartz (or low-aluminum-content quartz) should be used in oscillators that are expected to be exposed to ionizing radiation.

At low doses (e.g., at a few rads), the frequency change per rad can be as high as 10^{-9} per rad (35). The low-dose effect is not well understood. It is not impurity dependent, and it saturates at about 300 rads. At very high doses (i.e., at $\geqslant 1$ Mrad), the impurity-dependent frequency shifts also saturate because, since the number of defects in the crystal are finite, the effects of the radiation interacting with the defects are also finite.

When a fast neutron hurtles into a crystal lattice and collides with an atom, it is scattered like a billiard ball. A single such neutron can produce numerous vacancies, interstitials, and broken interatomic bonds. The effect of this "displacement damage" on oscillator frequency is dependent primarily upon the neutron fluence. The frequency of oscillation increases nearly linearly with neutron fluence at rates of 8×10^{-21} neutrons per square centimeter (n/cm^2) at a fluence range of 10^{10}–10^{12} n/cm^2, $5 \times 10^{-21}/n/cm^2$ at 10^{12}–10^{13} n/cm^2, and $0.7 \times 10^{-21}/n/cm^2$ at 10^{17}–10^{18} n/cm^2.

Radiation Effects in Atomic Frequency Standards

To the extent that they occur fast compared to the loop time constant, the transient effects of radiation on the crystal oscillator are unaffected by the frequency control loop in the atomic oscillator. Longer-term effects are reduced by the loop gain up to a limit imposed by the effect of the radiation on the control electronics. For example, a military rubidium standard irradiated with a dose of 600 rad (Si) at a rate of 1×10^{10} rad (Si) per second suffered an offset of 6×10^{-8} after 1 second, but recovered to better than 5×10^{-11} in 20 seconds. Both rubidium and cesium standards have been radiation hardened to levels appropriate for use on the Global Positioning System (GPS) satellites. The physics packages are intrinsically insensitive to radiation dose, but the electronics may require latchup protection, additional gain margin, and other similar modifications for use in radiation environments.

Other Effects on Stability

Ambient pressure change (as during an altitude change) can change a crystal oscillator's frequency if the pressure change produces a deformation of the crystal unit's or the oscillator's enclosure (thus changing stray capacitances and stresses). The pressure change can also affect the frequency indirectly through a change in heat-transfer conditions inside the oscillator. Humidity changes can also affect the heat-transfer conditions. In addition, moisture in the atmosphere will condense on surfaces when the temperature falls below the dew point, and can permeate materials such as epoxies and polyimides, and thereby affect the properties (e.g., conductivities and dielectric constants) of the oscillator circuitry. The frequency of a properly designed crystal oscillator changes less than 5×10^{-9} when the environment changes from one atmosphere of air to a vacuum.

All atomic frequency standards are indirectly sensitive to pressure to the extent that the change in the thermal conductivity of the air modifies the thermal gradients within the unit. Hydrogen masers can be directly affected by pressure changes through the cavity pulling if the microwave cavity is not isolated from these changes or compensated. Rubidium frequency standards are directly affected by pressure changes as a result of the distortion of the rubidium cell and the consequent change in density of the buffer gas. The sensitivity of rubidium frequency standards to variations in ambient pressure is approximately 1×10^{-13} per torr. The typical cesium standard specification is "altitude: $< 2 \times 10^{-12}$ change up to 12.2 km (40,000 ft.)" (36).

Electric fields can change the frequency of a crystal unit. An ideal AT cut is not affected by a DC voltage on the crystal electrodes, but "doubly rotated cuts," such as the SC cut, are affected. For example, the frequency of a 5-MHz fundamental-mode SC-cut crystal changes 7×10^{-9} per volt (V). Direct-current voltages on the electrodes can also cause sweeping, which can affect the frequencies of all cuts.

Power-supply and load-impedance changes affect the oscillator circuitry and, indirectly, the crystal's drive level and load reactance. A change in load impedance changes the amplitude or phase of the signal reflected into the oscillator loop, which changes the phase (and frequency) of the oscillation. The effects can be minimized through voltage regulation and the use of buffer amplifiers. The frequency of a "good" crystal oscillator changes less than 5×10^{-10} for a 10% change in load impedance. The typical sensitivity of a high-quality crystal oscillator to power-supply voltage changes is 5×10^{-11}/V; that of a rubidium frequency standard is 5×10^{-12}/V.

Gas permeation under conditions where there is an abnormally high concentration of hydrogen or helium in the atmosphere can lead to anomalous aging rates. For example, hydrogen can permeate into "hermetically" sealed crystal units in metal enclosures, and helium can permeate through the walls of glass-enclosed crystal units and through the walls of the glass bulbs of rubidium standards.

Interactions among the Influences on Stability

The various influences on frequency stability can interact in ways that lead to erroneous test results if the interfering influence is not recognized during testing. For example, building vibrations can interfere with the measurement of short-term stability. Vibration levels of 10^{-3} g to 10^{-2} g are commonly present in buildings. Therefore, if an oscillator's acceleration sensitivity is 1×10^{-9}/g, then the building vibrations alone can contribute short-term instabilities at the 10^{-12} to 10^{-11} level.

The 2-g tipover test is often used to measure the acceleration sensitivity of crystal oscillators. Thermal effects can interfere with this test because, when an oscillator is turned upside down, the thermal gradients inside the oven can vary due to changes in convection currents. Other examples of interfering influences include temperature and drive-level changes interfering with aging tests; induced voltages due to magnetic fields interfering with vibration-sensitivity tests; and the thermal-transient effect, humidity changes, and load-reactance temperature coefficients interfering with the measurement of crystal units' static f vs. T characteristics.

An important effect in TCXOs is the interaction between the frequency adjustment during calibration and the f vs. T stability (37). This phenomenon is called the *trim effect*. In TCXOs, a temperature-dependent signal from a thermistor is used to generate a correction voltage that is applied to a varactor in the crystal network. The resulting reactance variations compensate for the crystal's f vs. T variations. During calibration, the crystal's load reactance is varied to compensate for the TCXO's aging. Since the frequency versus reactance relationship is nonlinear, the capacitance change during calibration moves the operating point on the frequency versus reactance curve to a point where the slope of the curve is different, which changes the compensation (i.e., compensating for aging degrades the f vs. T stability). Figure 31 shows how, for the same compensating C_L versus T, the compensating f vs. T changes when the operating point is moved to a different C_L. Figure 32 shows test results for a 0.5 ppm TCXO that had a ± 6 ppm frequency-adjustment range (to allow for aging compensation for the life of the device). When delivered, this TCXO met its 0.5 ppm f vs. T specification; however, when the frequency was adjusted ± 6 ppm during testing, the f vs. T performance degraded significantly.

Oscillator Comparison and Selection

The discussion that follows applies to wide-temperature-range frequency standards (i.e., to those that are designed to operate over a temperature range that spans at least 90°C). Laboratory devices that operate over a much narrower temperature range can have better stabilities than those in the comparison below.

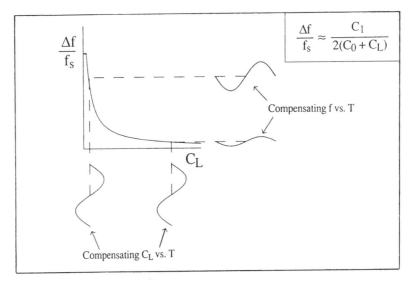

FIG. 31   Change in compensating frequency versus temperature due to C_L change.

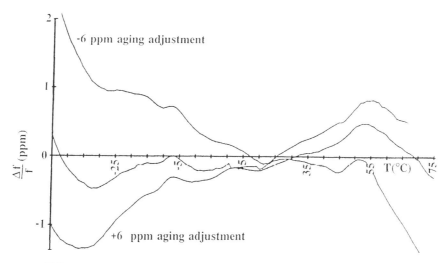

FIG. 32 Temperature-compensated crystal oscillator (TCXO) trim effect.

Commercially available frequency sources cover an accuracy range of several orders of magnitude — from the simple XO to the cesium-beam frequency standard. As the accuracy increases, so does the power requirement, size, and cost. Figure 33, for example, shows the relationship between accuracy and power requirement. Accuracy versus cost would be a similar relationship, ranging from about $1 for a simple XO to about $40,000 for a cesium standard (1990 prices). Table 1 shows a comparison of salient characteristics of frequency standards. Figure 34 shows the comparison of stability ranges as a function of averaging time. Figure 35 shows a comparison of phase-noise characteristics, and Table 2 shows a comparison of weaknesses and wear-out mechanisms.

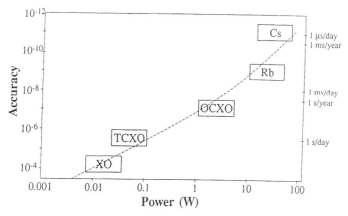

FIG. 33 Relationship between accuracy and power requirements (XO = crystal oscillator; TCXO = temperature-compensated crystal oscillator; OCXO = oven-controlled crystal oscillator; Rb = rubidium frequency standard; Cs = cesium beam frequency standard).

TABLE 1 Salient Characteristic Comparison for Frequency Standards

	Quartz Oscillators			Atomic Oscillators		
	TCXO	MCXO	OCXO	Rubidium	RbXO	Cesium
Accuracy* (per year)	2×10^{-6}	5×10^{-8}	1×10^{-8}	5×10^{-10}	7×10^{-10}	2×10^{-11}
Aging/year	5×10^{-7}	2×10^{-8}	6×10^{-9}	2×10^{-10}	2×10^{-10}	0
Temperature Stability (range, °C)	5×10^{-7} $(-55 \text{ to } +85)$	2×10^{-8} $(-55 \text{ to } +85)$	1×10^{-9} $(-55 \text{ to } +85)$	3×10^{-10} $(-55 \text{ to } +68)$	5×10^{-10} $(-55 \text{ to } +85)$	2×10^{-11} $(-28 \text{ to } +65)$
Stability, $\sigma_y(\tau)$ $(\tau = 1 \text{ s})$	1×10^{-9}	1×10^{-10}	1×10^{-12}	3×10^{-11}	5×10^{-12}	5×10^{-11}
Size (cm³)	10	50	20–200	800	1200	6000
Warmup time (min)	0.1 (to 1×10^{-6})	0.1 (to 2×10^{-8})	4 (to 1×10^{-8})	3 (to 5×10^{-10})	3 (to 5×10^{-10})	20 (to 2×10^{-11})
Power (W) (at lowest temperature)	0.05	0.04	0.25 – 4	20	0.35	30
Price (~$)	100	1000	2000	8000	10,000	40,000

*Including environmental effects (note that the temperature ranges for rubidium and cesium frequency standards are narrower than for quartz).

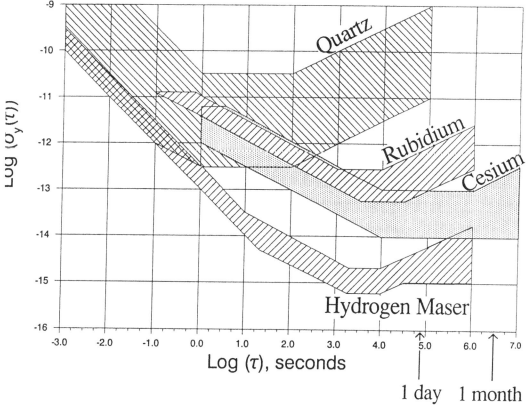

FIG. 34 Stability as a function of averaging time comparison for frequency standards.

Characteristics are provided in Table 1 for the rubidium crystal oscillator (RbXO), a device intended for applications where power availability is limited, but where atomic frequency standard accuracy is needed. It consists of a rubidium frequency standard, a low-power and high-stability crystal oscillator, and control circuitry that adjusts the crystal oscillator's frequency to that of the rubidium standard. The rubidium standard is turned on periodically (e.g., once a week) for the few minutes it takes for it to warm up and correct the frequency of the crystal oscillator. With the RbXO, one can approach the long-term stability of the rubidium standard with the low (average) power requirement of the crystal oscillator.

The major questions to be answered in choosing an oscillator include

1. What frequency accuracy or reproducibility is needed for the system to operate properly?
2. How long must this accuracy be maintained (i.e., will the oscillator be calibrated or replaced periodically, or must the oscillator maintain the required accuracy for the life of the system)?
3. Is ample power available, or must the oscillator operate from batteries?
4. What warmup time, if any, is permissible?

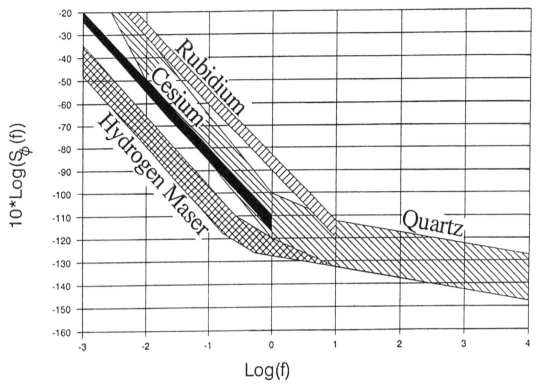

FIG. 35 Phase instability comparison for frequency standards.

TABLE 2 Weakness and Wear-out Comparison for Frequency Standards

	Weaknesses	Wear-out Mechanisms
Quartz	Aging	None
	Rad hardness	
Rubidium	Life	Rubidium depletion
	Power	Buffer-gas depletion
	Weight	Glass contaminants
Cesium	Life	Cesium-supply depletion
	Power	Spent-cesium gettering
	Weight	Ion-pump capacity
	Cost	Electron multiplier
	Temperature range	

5. What are the environmental extremes in which the oscillator must operate?
6. What is the short-term stability (phase-noise) requirement?
7. What is the size constraint?

In relation to the second question, what cost is to be minimized, the initial acquisition cost or the life-cycle cost? Often, the cost of recalibration is far higher than the added cost of an oscillator that can provide a calibration-free life. A better oscillator may also allow simplification of the system's design.

The frequency of the oscillator is another important consideration, because the choice can have a significant impact on both the cost and the performance. Everything else being equal, an oscillator of standard frequency, such as 5 MHz or 10 MHz, will cost less than one of an unusual frequency, such as 8.34289 MHz. Moreover, for thickness-shear crystals, such as the AT cut and SC cut, the lower the frequency, the lower the aging and the noise floor. Since at frequencies much below 5 MHz thickness-shear crystals become too large for economical manufacturing, and since all the highest-stability oscillators use thickness-shear crystals, the highest-stability commercially available oscillator's frequency is 5 MHz. Such oscillators will also have the lowest phase-noise capability close to the carrier. There are also some excellent 10 MHz oscillators on the market; however, oscillators of much higher frequency than 10 MHz have significantly higher aging rates and phase-noise levels close to the carrier than do 5 MHz oscillators. For lowest phase noise far from the carrier, where the signal-to-noise ratio determines the noise level, higher frequency crystals (e.g., 100 MHz) can provide lower noise because such crystals can tolerate higher drive levels, thereby allowing higher signal levels.

Time Transfer

Time-transfer techniques provide an additional method of maintaining synchronization among remote locations that complements the use of independent clocks. In fact, most systems derive time using both external time references and internal clocks. The former provide long-term accuracy and interoperability; the latter provide autonomous capability in the absence of the external references. A variety of time-transfer techniques are in frequent use today: telephone, LORAN-C, GPS, WWV/VB, Geostationary Operational Environmental Satellites (GOES), and two-way satellite (38). They vary in capability from a few milliseconds to a few nanoseconds.

Radio broadcast services, such as WWV and WWVB in the United States, disseminate time with modest accuracy and stability. The high-frequency broadcasts between 2.5 and 20 MHz are usually received after reflection from the ionosphere. As a result, variability in the path delay limits the accuracy to a few milliseconds for most users. The broadcasts contain time "ticks," time code, and voice announcements referenced to the National Institute of Standards and Technology (NIST) time scale. Commercial WWV receivers are available.

Improved performance is provided by the GOES system, the timing signals of which are transmitted from the satellites and may be received throughout most of the continental United States. Commercial time-code receivers are available that provide timing accuracies of approximately 100 μs limited by uncertainties in satellite position and receiver delays. The NIST time-scale is also the reference for the GOES timing signals.

The LORAN-C navigation system may be used to obtain time-transfer accuracies of approximately one microsecond referenced to the United States Naval Observatory (USNO) time-scale. These low-frequency broadcasts are propagated via ground wave, which is much more stable than the sky-wave propagation of the high-frequency (HF) broadcasts. Commercial timing receivers are available that simplify time recovery from LORAN.

The GPS also disseminates time referenced to UTC via the USNO time-scale. Because of the precise knowledge of the satellite positions, the GPS Standard Positioning Service's C/A code is capable of disseminating time that is accurate to approximately one hundred nanoseconds. Several commercial GPS timing receivers are available that provide completely automatic operation. GPS may also be used in a differential mode, often called common view, to provide improved synchronization capability (39). For sites located within several thousand kilometers (km) of one another, timing errors due to errors in the ephemeris and the propagation delay are approximately equal. Thus, when the absolute GPS times of arrival of simultaneously observed satellite signals are subtracted from one another, the differential accuracy improves to several tens of nanoseconds.

The highest accuracy time synchronization is obtained via two-way satellite techniques (40). Both the propagation errors and the delays through the receiver are calibrated by transmitting time in both directions between two sites. Each site measures the difference between the time of arrival of the pulse from the other site and the time of the local clock. The difference in the measurements made at the two ends provides the relative time of the two local clocks. The effects of the transmitter and receiver delays, the uplink and downlink propagation delays, and the delays through the satellite are substantially canceled. As a result, time-synchronization accuracy of a few nanoseconds has been obtained using commercial communication satellites and very small aperture terminals (VSAT). A custom spread-spectrum time-transfer modem is necessary.

Clocks and timing receivers can be combined in a timing system to provide a broader range of timing capabilities than either one can provide alone (41). Such a system uses the received timing signal to calibrate the local clock and learn its time, frequency, and frequency aging. When the timing signal is unavailable, the local clock acts as a "flywheel." Its free-running operation starts using the time and frequency provided by calibration versus the external source. The frequency may subsequently be updated periodically for the predicted frequency aging. This procedure produces the minimum possible free-running timing errors. Commercial "disciplined oscillators" now provide all these functions in an integrated package.

Relativistic effects become significant when nanosecond-level time-transfer accuracies are desired, and when clocks are widely separated or have high velocities (38). For example, at 40° latitude a clock will gain 9.4 nanoseconds per day (ns/day) when it is moved from sea level to a 1-km elevation, and the clocks in

GPS satellites (12-hour period circular orbits) gain 44 μs/day when compared to their rates on earth prior to launch.

Specifications, Standards, Terms, and Definitions

Numerous specifications and standards exist that relate to frequency standards. The major organizations responsible for these documents are the Institute of Electrical and Electronics Engineers (IEEE), the International Electrotechnical Commission (IEC), the CCIR, and the U.S. Department of Defense, which maintains the Military Specification (MIL-SPEC) system. A listing of "Specifications and Standards Relating to Frequency Control" can be found in the final pages of the *Proceedings of the Annual Symposium on Frequency Control*. In the 1990 *Proceedings*, for example, 79 such documents are listed (42). Many of the documents include terms and definitions, some of which are inconsistent. Unfortunately, no single authoritative document exists for terms and definitions relating to frequency standards. The terms and definitions in the CCIR glossary (23), in IEEE Standard 1139-1988 (28), and in MIL-O-55310's Section 6 (30) are the most recent; they address different aspects of the field, and together form a fairly good set of terms and definitions for users of frequency standards.

The most comprehensive document dealing with the specification of frequency standards is MIL-O-55310 (30). The evolution of this document over a period of many years has included periodic coordinations between the government agencies that purchase crystal oscillators and the suppliers of those oscillators. The document addresses the specifications of all the oscillator parameters discussed above, plus many others. This specification was written for crystal oscillators. Because the output frequencies of atomic frequency standards originate from crystal oscillators, and because no comparable document exists that addresses atomic standards specifically, MIL-O-55310 can also serve as a useful guide for specifying atomic standards.

MIL-STD-188-115, *Interoperability and Performance Standards for Communications Timing and Synchronization Subsystems,* specifies that the standard frequencies for nodal clocks shall be 1 MHz, 5 MHz, or 5×2^N MHz, where N is an integer. This standard also specifies a 1-pulse-per-second timing signal of amplitude 10 V, pulse width of 20 μs, rise time less than 20 ns, fall time less than 1 μs, and a 24-bit binary-coded decimal (BCD) time code that provides Coordinated Universal Time (UTC) time of day in hours, minutes, and seconds, with provisions for an additional 12 bits for day of the year, and an additional 4 bits for describing the figure of merit (FOM) of the time signal. The FOMs range from BCD Character 1 for better than 1 ns accuracy to BCD Character 9 for "greater than 10 ms of fault" (43).

For Further Reading

Reference 8 contains a thorough bibliography on the subject of frequency standards to 1983. The principal forum for reporting progress in the field has been the *Proceedings of the Annual Symposium on Frequency Control* (42). Other publications that deal with frequency standards include *IEEE Transactions on Ultrasonics, Ferroelectrics, and Frequency Control, IEEE Transactions on Instrumentation and Measurement, Proceedings of the Annual Precise Time and Time Interval (PTTI) Applications and Planning Meeting* (44), and *Proceedings of the European Frequency and Time Forum* (45). Review articles can be found in special issues and publications (46–49).

References

1. Kinsman, R., Gailus, P., and Dworsky, L., Communications System Frequency Control. In: *The Froehlich/Kent Encyclopedia of Telecommunications*, Vol. 4 (F. E. Froehlich and A. Kent, eds.), Marcel Dekker, 1991, pp. in press.

2. Abate, J. E., et al., AT&T's New Approach to the Synchronization of Telecommunication Networks, *IEEE Commun.*, 35–45 (April 1989).

3. Pan, J., Present and Future of the Synchronization in the U.S. Telephone Network, *IEEE Trans. Ultrasonics, Ferroelectrics, and Frequency Control*, UFFC-34: 629–638 (November 1987).

4. Dixon, R. C., *Spread Spectrum Systems*, Wiley, New York, 1976.

5. Smith, W. L., Precision Oscillators. In: *Precision Frequency Control*, Vol. 2 (E. A. Gerber and A. Ballato, eds.), Academic Press, New York, 1985, pp. 45–98.

6. Bottom, V. E., *Introduction to Quartz Crystal Unit Design*, Van Nostrand Reinhold, New York, 1982.

7. Gerber, E. A., and Ballato, A. (eds.), *Precision Frequency Control*, Academic Press, New York, 1985.

8. Parzen, B., *Design of Crystal and Other Harmonic Oscillators*, Wiley, New York, 1983.

9. Hellwig, H. H., Microwave Frequency and Time Standards. In: *Precision Frequency Control*, Vol. 2 (E. A. Gerber and A. Ballato, eds.), Academic Press, New York, 1985, pp. 113–176.

10. Audoin, C., and Vanier, J., Atomic Frequency Standards and Clocks, *Journal of Physics E: Scientific Instruments*, 9:697–720 (1976).

11. Ramsey, N., History of Atomic Frequency Standards, *Journal of Research, NBS*, 88:301–320 (1983).

12. Strumia, F., et al., Mg Frequency Standard: Optimization of the Metastable Atomic Beam, *Proc. 28th Ann. Symp. Frequency Control*, 350–354, NTIS accession no. AD-A011113 (1974).

13. Holloway, J. H., and Lacey, R. F., Factors which Limit the Accuracy of Cesium Atomic Beam Frequency Standards, *Proc. Int. Conf. Chronometry*, 317–331 (1964).

14. Wineland, D. J., et al., Results on Limitations in Primary Cesium Standard Operation, *IEEE Trans. Instrum. Meas.*, IM-25:453–458 (1976).

15. Dicke, R. M., The Effect of Collisions Upon the Doppler Width of Spectral Lines, *Phys. Rev.*, 89:472–473 (1953).

16. Frerking, M. E., Temperature Control and Compensation. In: *Precision Frequency Control*, Vol. 2 (E. A. Gerber and A. Ballato, eds.), Academic Press, New York, 1985, pp. 99–111.

17. Schodowski, S. S., et al., Microcomputer Compensated Crystal Oscillator for Low Power Clocks, *Proc. 21st Ann. Precise Time and Time Interval (PTTI) Applications and Planning Meeting*, 445–464 (1989). Available from the U.S. Naval Observatory, Time Services Department, 34th and Massachusetts Avenue, NW, Wash-

ington, DC 20392. Details of the MCXO are also described in a series of five papers in the *Proc. 43rd Ann. Symp. Frequency Control*, IEEE Catalog No. 89CH2690-6 (1989).

18. Lewis, L. L., Miniature Optically Pumped Cesium Standards, *Proc. 45th Ann. Symp. on Frequency Control*, IEEE Cat. No. 91CH2965-2, in press.

19. Giordano, V., et al., *Proc. 43rd Ann. Symp. Frequency Control*, 130–134, IEEE Catalog No. 89CH2690-6 (1989).

20. Wineland, D. J., et al., Progress at NIST Toward Absolute Frequency Standards Using Stored Ions, *Proc. 43rd Ann. Symp. Frequency Control*, 143–150, IEEE Catalog No. 89CH2690-6 (1989).

21. Cutler, L. S., et al., Initial Operational Experience with a Mercury Ion Storage Frequency Standard, *Proc. 41st Ann. Symp. Frequency Control*, 12–19, NTIS accession no. AD-A216858 (1987).

22. XIIIth General Conference of Weights and Measures, Geneva, Switzerland, October 1967.

23. International Radio Consultative Committee (CCIR), Recommendation No. 686, Glossary. In: *CCIR 17th Plenary Assembly*, Vol. 7, Standard Frequencies and Time Signals (Study Group 7), CCIR, Geneva, Switzerland, 1990. Copies available from International Telecommunication Union, General Secretariat — Sales Section, Place des Nations, CH1211 Geneva, Switzerland.

24. Meeker, T. R., and Vig, J. R., The Aging of Bulk Acoustic Wave Resonators, Oscillators and Filters, *Proc. 45th Ann. Symp. on Frequency Control*, IEEE Cat. No. 91CH2965-2, in press.

25. Walls, F. L., and Persson, K. B., A New Miniaturized Passive Hydrogen Maser, *Proc. 38th Ann. Symp. Frequency Control*, 416–419, NTIS accession no. AD-A217381 (1984).

26. Peters, H. E., Design and Performance of New Hydrogen Masers Using Cavity Frequency Switching Servos, *Proc. 38th Ann. Symp. Frequency Control*, 420–427, NTIS accession no. AD-A217381 (1984).

27. Barnes, J. A., et al., Characterization of Frequency Stability, *IEEE Trans. Instrum. Meas.*, IM-20:105–120 (1971).

28. IEEE, IEEE Standard Definitions of Physical Quantities for Fundamental Frequency and Time Metrology, *IEEE Std. 1139-1988*.

29. Kusters, J. A., and Vig, J. R., Thermal Hysteresis in Quartz Resonators — A Review, *Proc. 44th Ann. Symp. on Frequency Control*, 165–175, IEEE Catalog No. 90CH2818-3 (1990).

30. U.S. Department of Defense, Military Specification, Oscillators, Crystal, General Specification for, MIL-O-55310. The latest revision is available from Military Specifications and Standards, 700 Robbins Ave., Bldg. 4D, Philadelphia, PA 19111-5094.

31. Vig, J. R., et al., *The Effects of Acceleration on Precision Frequency Sources (Proposed for IEEE Standards Project P1193)*, U.S. Army Laboratory Command Research and Development Technical Report SLCET-TR-91-3, March 1991. Copies available from National Technical Information Service, 5285 Port Royal Road, Sills Building, Springfield, VA 22161; NTIS accession no. AD-A235470.

32. Filler, R. L., et al., Ceramic Flatpack Enclosed AT and SC-cut Resonators, *Proc. 1980 IEEE Ultrasonic Symp.*, 819–824 (1980).

33. Brendel, R., et al., Influence of Magnetic Field on Quartz Crystal Oscillators, *Proc. 43rd Ann. Symp. Frequency Control*, 268–274, IEEE Catalog No. 89CH2690-6 (1989).

34. King, J. C., and Koehler, D. R., Radiation Effects on Resonators. In: *Precision Frequency Control*, Vol. 2 (E. A. Gerber and A. Ballato, eds.), Academic Press, New York, 1985, pp. 147–159.

35. Flanagan, T. M., Leadon, R. E., and Shannon, D. L., Evaluation of Mechanisms for Low-Dose Frequency Shifts in Crystal Oscillators, *Proc. 40th Ann. Symp. Frequency Control*, 127–133, NTIS accession no. AD-A235435 (1986).

36. See, for example, "The HP 5061B Cesium Beam Frequency Standard: The World's Most Accurate Commercially Available Primary Standard with Proven Performance and Reliability—Technical Data," Hewlett-Packard Company, Palo Alto, CA (June 1987).

37. Filler, et al., Specification and Measurement of the Frequency versus Temperature Characteristics of Crystal Oscillators, *Proc. 43rd Ann. Symp. Frequency Control*, 253–256, IEEE Catalog No. 89CH2690-6 (1989).

38. Allan, D. W., Frequency and Time Coordination, Comparison, and Dissemination. In: *Precision Frequency Control*, Vol. 2 (E. A. Gerber and A. Ballato, eds.), Academic Press, New York, 1985, pp. 233–273.

39. Allan, D. W., and Weiss, M. A., Accurate Time and Frequency Transfer During Common View of a GPS Satellite, *Proc. 34th Ann. Symp. Frequency Control*, 334–346, NTIS accession no. AD-A213670 (1980).

40. Howe, D. A., Ku-Band Satellite Two-Way Timing Using a Very Small Aperture Terminal (VSAT), *Proc. 41st Ann. Symp. Frequency Control*, 149–160, NTIS accession no. AD-A216858 (1987).

41. MacIntyre, A., and Stein, S. R., A Disciplined Rubidium Oscillator, *Proc. 40th Ann. Symp. Frequency Control*, 465–469, NTIS accession no. AD-A235435 (1986).

42. The proceedings of the Annual Symposium on Frequency Control have been published since the tenth symposium in 1956. The earlier volumes are available from the National Technical Information Service, 5285 Port Royal Road, Sills Building, Springfield, VA 22161; the later volumes, from the IEEE, 445 Hoes Lane, Piscataway, NJ 08854. Ordering information for all the *Proceedings* can be found in the back of the latest volumes (e.g., the *Proceedings of the 44th Ann. Symp. on Frequency Control* (1990) is available from the IEEE, Cat. No. 90CH2818-3).

43. U.S. Department of Defense, Military Standard, MIL-STD-188-115, *Interoperability and Performance Standards for Communications Timing and Synchronization Subsystems*. The latest revision is available from Military Specifications and Standards, 700 Robbins Avenue, Building 4D, Philadelphia, PA 19111-5094.

44. The *Proceedings of the Annual Precise Time and Time Interval (PTTI) Applications and Planning Meeting* are available from the U.S. Naval Observatory, Time Services Department, 34th and Massachusetts Ave., N.W., Washington, DC 20392-5100. The latest volumes are also available from the National Technical Information Service, 5285 Port Royal Road, Sills Building, Springfield, VA 22161.

45. The *Proceedings of the European Frequency and Time Forum* are available from the Swiss Foundation for Research in Metrology (FSRM), Rue de l'Orangerie 8, CH-2000 Neuchatel, Switzerland.

46. *IEEE Trans. Ultrasonics, Ferroelectrics, and Frequency Control*, UFFC-34 (November 1987).

47. *IEEE Trans. Ultrasonics, Ferroelectrics, and Frequency Control*, UFFC-35 (May 1988).

48. Kroupa, V. F. (ed.), *Frequency Stability: Fundamentals and Measurement*, IEEE Press, New York, 1983.

49. Sullivan, D. B., et al. (eds.), *Characterization of Clocks and Oscillators*, National Institute of Standards and Technology Technical Note 1337, National Institute of Standards and Technology, Boulder, CO 80303-3328.

SAMUEL R. STEIN
JOHN R. VIG

Printed and bound by CPI Group (UK) Ltd, Croydon, CR0 4YY

17/10/2024

01775696-0019